The Neural Mind

The Neural Mind

How Brains Think

George Lakoff and Srini Narayanan

The University of Chicago Press

Chicago and London

The University of Chicago Press, Chicago 60637
The University of Chicago Press, Ltd., London
© 2025 by George Lakoff and Srini Narayanan
Published 2025
Printed in the United States of America

34 33 32 31 30 29 28 27 26 25   1 2 3 4 5

ISBN-13: 978-0-226-83588-4 (cloth)
ISBN-13: 978-0-226-83589-1 (e-book)
DOI: https://doi.org/10.7208/chicago/9780226835891.001.0001

Library of Congress Cataloging-in-Publication Data

Names: Lakoff, George, author. | Narayanan, Srini (Srinivas Sankara), author.
Title: The neural mind : how brains think / George Lakoff and Srini Narayanan.
Description: Chicago : The University of Chicago Press, 2025. | Includes bibliographical
    references and index.
Identifiers: LCCN 2024046695 | ISBN 9780226835884 (cloth) | ISBN 9780226835891
    (ebook)
Subjects: LCSH: Cognitive neuroscience. | Cognitive science. | Neural circuitry. |
    Cognitive grammar.
Classification: LCC QP360.5 .L35 2025 | DDC 153—dc23/eng/20241127
LC record available at https://lccn.loc.gov/2024046695

♾ This paper meets the requirements of ANSI/NISO Z39.48-1992 (Permanence of Paper).

Contents

Preface

The immune system, the hypothalamus, the ventro-medial frontal
cortex, and the Bill of Rights have the same root cause.

ANTONIO DAMASIO IN *DESCARTES' ERROR*

Antonio Damasio's insight quoted here is profound. As a neuroscientist, he
knows that ideas don't float in the air. Nor are they abstractions that somehow
inhabit our minds. Ideas are not the logical formulas and set-theoretical models of philosophical logicians. Nor are they digital sequences of ones and zeros
that can be downloaded from or uploaded into our brains. Instead, ideas arise
from and reside in neural circuitry interacting with the social and physical
world. Neural circuitry *constitutes* ideas, all ideas that human beings have ever
had or will ever have. Ideas are therefore socially and physically grounded, as
much part of our bodies as the immune system or the ventromedial prefrontal cortex.

As a citizen, Damasio is acutely aware that ideas have power and can be
created. New ideas have changed the world, bringing us self-government,
human rights, justice, environmentalism, and other true advances in human
life and civilization, such as the Bill of Rights. That great document is not just
a sequence of letters on a page or, if read aloud, just a bunch of sounds. The
Bill of Rights, when understood, awakens in us ideas by activating the neural
circuitry constituting those ideas in us, in real human beings.

The Bill of Rights has its effect through language, because language is more
than just writing or speech sounds. The language of the Bill of Rights, like language in general, is meaningful. The meaning is not in the letters or sounds
themselves but instead is in the ideas constituted by the neural circuitry that
the language activates when it is read, heard, or remembered.

Beyond that, the Bill of Rights is not just any document. It is a moral document of the highest order, communicating through its language the ideas implied by a moral vision of universal human rights. This means that that the neural circuitry constituting those ideas is not just any bunch of neurons located anywhere in the body. Because of the details of their physical nature and their location in our physical bodies, those circuits can constitute *values*: moral ideas, the values behind democracy.

All of this is what the Damasio quotation is expressing, and the scientific issues are profound in every detail. The Damasio quotation expresses the premise—the *raison d'être and the starting point*—of this book.

Toward a Theory of the Human Mind

Our ideas reside in and are constituted by neural circuitry. How is this possible? Neural circuitry is physical; ideas are not. The central question of neural mind research is how nonphysical ideas can be constituted by physical neural circuitry. The theory of this book, though far from complete, is complete enough to allow us to grasp not just the basics of how the neural mind works but also a vast range of details.

The Neural Creation of Human Reality

The neural mind is surprising if not downright shocking at times. We don't passively see what comes to our eyes, feel what touches our skin, or hear what is presented to our ears. Our neural circuitry takes what is presented to our sensory organs and *creates* what we see, feel, and hear, mostly unconsciously and not directly or in any simple or straightforward way. In short, much of perception is created, although we are not consciously aware of the creation.

Neurons are just cells. Alone, they are incapable of thought. But combined into circuits hooked up to our bodies in just the right ways, they have allowed human beings to survive, think, communicate, and create all of culture.

What This Book Is About

This book presents a way of thinking about how the neural mind works. We will present models, evidence, and results from various disciplines that allow us to think systematically about how phenomena in language and thought work.

Thought does not occur neuron by neuron. It occurs at a systems level, the level at which neurons come together to form circuits and where simple circuits combine to form complex circuits. This is what allows simple thoughts to come together in complex thoughts. Our goal in this book is to ask how this might work: how different types of neural circuits could account for the detailed structure of specific types of thoughts.

Exactly how? That is what this book is about. We start in chapter 1 with an introduction to the basic hypothesis of the neural mind, combining research from cognitive science, computational modeling, and linguistics. Chapter 2 goes into the last three decades of research in cognitive linguistics and shows how thought and language rely on the interaction of the brain and the body with the social and physical environment. Chapter 3 is a technical chapter describing how research in sensorimotor coordination and control lead to our theoretical proposal of neural circuit types that connect the brain and body to the mind. We focus on the active nature of these circuits, which can be used in real-time guidance, homeostasis, and control but are also internally triggered in imaginative simulation during observation, understanding, and generation of thought and language. Chapter 4 is the payoff, a rethinking of several hard and open phenomena in language and thought from the neural mind perspective, showing novel ways in which such a perspective sheds light on and sometimes help solve these long-standing problems.

Neural mind research takes us deep into the very nature of thought as well as the language we use to express and communicate thought. And beyond that, neural mind research can provide insights into what we experience every day, showing that experiences don't just magically come to us; instead, our neural minds create and maintain experiences without our even noticing this process. The neural mind is what enables us to function normally in the everyday world, and beyond that the neural mind creates what we experience as powerful, beautiful, and profound. In short, the neural mind defines our worldview.

The Neural Filtering Effect

Defining our worldview is the positive work of the neural mind. The neural mind does negative work as well. Since we can only understand ideas that can be made sensible to us via our neural minds, ideas that are not sensible to our neural minds cannot be admitted to our minds and so are filtered out. In short, the neural mind is also a neural filter.

Language and Coordination

Social living ultimately depends on coordination between group members, and communication is necessary to make this possible. Evolutionary psychologists such as Tomasello (2022) and David-Barrett and Dunbar (2016) suggest that the efficiency of individuals increases when the group communication results in better joint action. Tomasello (2022) presents evidence from an evolutionary perspective that the emergence of agency and shared intentionality are precursors to human communication and that these aspects are grounded in organizations of animal behavior control designed to achieve goals in an increasingly complex physical and social environments. In recent work, the sociolinguist Nick Enfield (2022) makes a persuasive argument that the primary function of language is coordination, triggering neural circuits in the hearer that lead to persuasion, social affiliations, planning, and joint action. Coming from a computational neuroscience perspective, Buzsáki (2019) argues that the brain is active and generative and that human perception and cognition are best modeled and understood through neural circuits and pathways for action and control.

Our work on the neural mind cashes out these observations using converging evidence from multiple fields. We show through detailed mechanisms, evidence, and results from cognitive linguistics and neural modeling that control circuits for emotion, social cognition, and sensorimotor control form the socially shared end embodied common ground for communication.

The Neural Creation of Experience: Color

To show the power of the neural system, we begin with a dramatic example. We normally assume that what we see is what is present before us in the world. But there is a major dimension of perception that is experienced but is not there in the world external to living beings: color!

1

What It Takes to Create Human Thought

The Creation of Perception: Color

In the Marx Brothers' movie *Duck Soup*, Chico Marx, imitating Groucho's outrageousness, says, "Who ya gonna believe, me or your own eyes?"

It is usually assumed that what you see is what there is, that seeing is believing.

But the science of the neural mind tells us otherwise, that much of what we see, hear, and feel is not out there in the external world but instead is created by the body and brain, even the most elementary properties of physical objects. The neuroscience evidence is clear: in many cases, your lying eyes are what you believe.

THERE IS NO COLOR IN THE WORLD!

For some people, it comes as a shock to learn that there is no color in the world: no green in grass, no red in blood, no blue in the sky or the sea, no yellow in the sun or moon. None! Color is not out there, independent of beings experiencing the world.

Yes, we see beautiful sunsets, roses, great paintings, and parrots, all in color. Yet, the color we see in them is not what is there. In themselves, they have no color.

Objects in the world reflect and absorb combinations of wavelengths of light, but wavelengths are not colors. And we don't see individual wavelengths. We see colors, whole ranges of colors in a continuity of shades, but they're not there, however beautiful, enjoyable, and awe-inspiring they are. Here's how it works.

OUR BODIES AND BRAINS CREATE THE BEAUTY OF COLOR

Color is created within you from the interaction of two factors in the world and three factors in our bodies. The external factors are (1) reflected wavelengths of light, each with a corresponding intensity, and (2) nearby lighting. The internal factors are (1) color cones in the retina, (2) retinal ganglion cells comparing the outputs of the color cones and extending from the cones to the brain, and (3) neural circuitry connected to the ganglion cells.

The color cone classes are called L, M, and S because their peak outputs lie in the long-, middle-, and short-wavelength regions of the visible spectrum. Retinal ganglion cells have what are called *opposed* responses to cone types. The term "opposition" is defined as increased firing to an increase in activation of one cone type (on-response) and to a decrease in activation of a different cone type (off-response).

There are four varieties of such opponent retinal ganglion cells: L-on/M-off, M-on/L-off, S-on/(L+M)-off, and (L+M)-on/S-off. Adding to the complexity are differences in the sizes and time courses of the receptive fields of "on" cells.[1]

Without retinal color cones, ganglion cells, and connected neural circuitry in the brain, there would be no color.

COLOR CATEGORIES, WORDS, AND INFERENCES

A color category that we "see" covers a range of shades of a color. The green category, for example, covers the range of shades of green fading off to yellow on one side in the spectrum and blue on the other side. Basic color terms such as red, green, and blue name whole categories of shades of color that we "see" and take as being out there in the world. Basic color categories have names such as red, blue, green, purple, and so on. We can reason with them. If an object is red, it is not green.

How do shades of color arise for us? How can we "see" what is not there?

THE NEUROPHYSIOLOGY OF COLOR

Each color cone in the retina varies in the degree to which it reacts to given wavelengths. As wavelengths increase, a given color cone reacts more strongly (sends out stronger electrical signals) up to a maximum point and then, as wavelengths keep increasing, decreases in the signals it sends out. Each color cone, because of its chemical makeup, reacts in this way to different ranges of wavelengths: short, medium, and long. Each object usually reflects not just one wavelength of light but multiple wavelengths. Each color cone reacts to a different degree to each reflected wavelength. The combination of color cones and neural circuitry creates different shades of color for different combinations of wavelengths. There are millions of shades of color that our embodied brains create. They allow us to distinguish objects from one another partly by their shades of color, that is, on the basis of differences in the combinations of their reflected wavelengths of light. This is an evolutionarily useful ability for animals living in the world to have.

The point: Our color cones and neural circuitry create color ranges and color categories *in us* from wavelengths of light out there in the world. But those colors are not out there in the world independent of us.

Do our eyes and brains create colors in all of us in the same way?

No. Color cones are not all the same. Their chemistry is linked in a complex way to X chromosomes. Generally, men have one X chromosome, and women have two X chromosomes. Some chromosomal variations lead to defects in color vision: various forms of color blindness, the inability to see the full range of colors that "normal" people see. Other chromosomal variations lead to forms of color vision in which the responses of one or more color cones will be in a different place in the spectrum, with peak responses (and other responses) at different wavelengths. What this means is that many people actually "see" different colors. And on the whole, women are more sensitive than men in their ability to see color differences.

When two people differ in their color cone composition, there is a good probability that they will actually "see" different colors for the same object in a given situation. Many of us have had disagreements with a member of the opposite sex on whether a color is, say, blue or green, red or orange, orange or brown, black or navy, and so on. This can result in arguments in which each assumes that the other is just mistaken. *But the fact is that both may be "right," that is, that they "see" the same object as having different colors because they have somewhat different color cones in their retinas!*

Some women even have *four* color cones. They are called **tetrachromates**. In principle, a fourth kind of color cone ought to allow one to "see"—that is, create—a range of color shades that the rest of us are blind to. Researchers studying tetrachromates vary in their findings. Most have found little or no measurable effect of the fourth kind of color cone. Others have found measurable effects, that is, a small percentage of people who can "see" a range of colors the rest of us cannot. Research in such areas of color vision is ongoing, and there is no general agreement on tetrachromates.

Of course, different organisms may perceive color in completely different ways as well. For us as humans, light intensity and brightness are required to activate the cones, so we find it hard to see color in the dark. Several species of fish have highly sensitive cones that allow for fine-grained color perception without any light. In addition, humans may have a much less developed color vision system than many bird species. For example, scientists at Princeton University have discovered that hummingbirds (which are tetrachromates) may perceive an astonishing range of colors including combinations of short and long wavelengths called **nonspectral colors**. For humans, purple is the only example of a nonspectral color combining red (short) and blue (long) but not green (medium) wavelengths.[2]

We apply words such as "blue" and "green" to color categories consisting of a range of shades. The meaning of each word for us is the color category that we happen to "see."

Thus, the meanings of color words are influenced by three factors:

1. The wavelengths that a given object reflects (what's in the world)
2. The composition of the color cones in your retina (what's in your body)
3. The embodied neural circuitry that connects your brain to your color cones

What we see here are the three basic kinds of embodiment:

1. Bodily interactions in the world (we interact with objects)
2. The nature of bodies (we have color cones)
3. Embodied circuitry (which links the neural system to the color cones)

These forms of embodiment result in color categories (categories that are part of thought), words for color categories (language expressing thought), and reasoning in terms of color category structure. A typical form of reasoning about color might come, for example, in discussing with another person how to paint a room in your house: "That's too light, too close to pink; we need a deeper red but not blood red." "But won't that red clash with orange?" If the two of you see the same object as having different colors, you may have different senses of what colors will clash.

Paul Kay and Willett Kempton (1984) asked whether having a name for a color category changes one's perception of the color. The Tarahumara language has no color term distinguishing green from blue, while English does. Kay and Kempton showed experimentally that it does matter. In the experiment, English speakers were able to make color distinctions that Tarahumara speakers could not make. Lera Boroditsky (2009) asked a similar question about English and Russian speakers. The situation is opposite from that comparing English and Tarahumara: Russian has separate color terms for light blue (*goluboy*) and dark blue (*siniy*), while English has only "blue." Boroditsky showed that Russian speakers could make color distinctions that English speakers could not. And Russian speakers have no general color category for what English speakers call "blue."

A more dramatic result came from Aubrey Gilbert, Terry Regier, Paul Kay, and Richard Ivry (2005). Inspired by the Kay-Kempton finding, they tested the making of color distinctions in the right side of a person's visual field versus the left side. The right visual field is connected to the left hemisphere of the brain, which makes language distinctions. English speakers can perform reaction-time color experiments faster in the right visual field (seen with the aid of language) than in the left visual field. It is not merely the case that language affects color perception. Nor is it merely the case that color is created by color cones and the brain. What is remarkable is that we *create and use the color categories that we "see" differently, depending on which hemisphere of the brain is being used!*[3]

COLOR PERCEPTION IS NOT WAVELENGTH DETECTION

Philosophers might be tempted to say that we can keep the idea that words just fit the world as it is, provided that we redefine what we mean by "color": that is, that a particular color is just a combination of reflected wavelengths of light waves in the world and that the meaning of color terms is just some combination of isolated wavelengths. Since we do not "see" the separate individual wavelengths, it would be hard to make a case for such a proposal.

Actually, any such case would fail. Consider the color wheel (Figure 1). Note that we perceive red (top right) next to violet (top left): moving clockwise, the sequence is red, orange, yellow, green, cyan, blue, and violet.

Now, consider the wavelength ranges of the perceived colors (Table 1). If what we call "colors" were just their wavelength-intensity ranges, colors

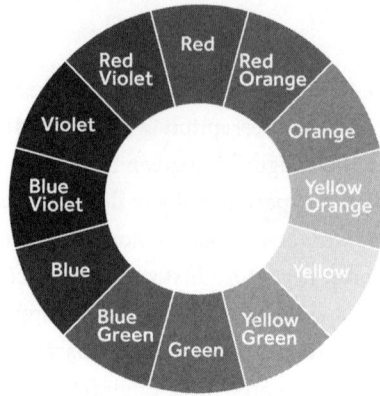

FIGURE 1. The color wheel shows the different colors and their closeness to each other in our perception. Note that the perceptual neighborhood of an individual color does not correspond in any direct way to the closeness in the wavelengths that characterize that color. (A high-resolution color figure can be seen at https://en.wikipedia.org/wiki/Color_wheel.)

TABLE 1. The colors of the visible light spectrum

COLOR	WAVELENGTH INTERVAL	FREQUENCY INTERVAL
red	~ 700–635 nm	~ 430–480 THz
orange	~ 635–590 nm	~ 480–510 THz
yellow	~ 590–560 nm	~ 510–540 THz
green	~ 560–520 nm	~ 540–580 THz
cyan	~ 520–490 nm	~ 580–610 THz
blue	~ 490–450 nm	~ 610–670 THz
violet	~ 450–400 nm	~ 670–750 THz

Note. nm = nanometer; THz = terahertz.

we perceive as adjacent on the color wheel should be made up of adjacent wavelength-intensity pairs. But they do not in the case of red and violet. Our brains appear to "bend" the wavelength sequence, making red (about 635 to 750 nanometers) adjacent to violet (about 400 to 450 nanometers). The colors we see as adjacent on the color wheel do not all correspond to adjacent wavelengths in the world!

A possible explanation of this color wheel comes from Paulus and Kröger-Paulus (1983). The authors argue that "the fundamental colors are estimated to be supersaturated violet, yellow-green and yellow-red" (p. 529). This would yield a color wheel as we see it with no bending.

The science of color shows that there is much more to embodiment than the oversimplified view that considers the brain as merely perceiving and acting upon what is out there. The beautiful colors that we "see" in the world are not out there. We create them, but not out of nothing. We create them out of combinations of reflected wavelengths, which *are* there. But we don't see wavelengths. We see color. And adjacent colors that we see may not correspond to adjacent wavelengths.

In addition, not all of us see the same colors, since color cones vary in how they work depending on our X chromosomes. Since women tend to have two X chromosomes and men tend to have one X chromosome, it is common for men and women to "see" different colors in the same scene. In gender-based disputes over whether an object is green or blue or is brown or orange, both can be right because a man and a woman may see different colors in the same scene. Color-cone differences can lead to differences of vision, to *seeing* different colors.

And words for color matter in making fine color distinctions but only in the left hemisphere (i.e., the right visual field).

In short, in the case of color, our bodies create what we "see," categorize, name, and understand, and we do so differently based on our physical differences.

What Are Ideas?

We see colors that are not there in the world external to our bodies that are created by our bodies and brains from things that *are* there but that we cannot see: the wavelengths and intensities of light reflected by objects. It takes color cones, ganglion cells, and neural circuitry to turn wavelengths with intensities into colors.

Like colors, ideas are not physical entities in the world. We have ideas of things in the world: ideas of trees and other plants, of animals, of mountains and rivers, of man-made artifacts such as houses and chairs, and of nonphysical entities such as institutions, numbers, emotions, and decisions. But although houses and rivers are physical, ideas of houses and rivers are not. You can live in a house but not in the idea of a house. You can swim in a river but not in the idea of a river. The idea of a river will not get you wet.

You can communicate ideas, but ideas don't move through the air. There are no physical entities called "memes" that float like viruses from person to person and country to country. When you "give" someone an idea, you still have it. Ideas are created by our bodies and brains and are constituted by our neural circuitry. Because we have no conscious access to our neural circuitry, most ideas are unconscious, with estimates ranging up to 98 percent of thought being unconscious and only about 2 percent of thought being conscious. Estimates differ, but in our experience this seems a reasonable ballpark estimate.

This book asks how ideas can be constituted by neural circuitry. In other words, how can we get ideas out of neurons, which are physical cells, living biological entities that function via electrochemistry? How can physical entities such as neurons create nonphysical ideas? And what is the role of the brain in communication? That is, how can the brain of one person reproduce the ideas in the brain of another person? These questions pose the central challenge of this book. Addressing such a challenge will require us to examine and integrate converging evidence from many of the cognitive and brain sciences.

IDEA BASICS

Let's start with a short first pass on what has been discovered in one branch of the cognitive and brain sciences, namely cognitive linguistics, which is about the nature of ideas. More details will be added in chapter 2.

Here are the elements of ideas:

- **Primary ideas**: Embodied schemas that arise naturally and are shared by people throughout the world, for example, schemas for up and down, near and far, and in and out. Such primary embodied schemas are used both in structuring perception and in reasoning.
- **Conceptual frames**: Mental structures for understanding circumscribed experiences and types of entities in terms of a whole structure, its parts, the relationships among the parts, and the relationships with other frames. Consider, for example, the shovel frame, which includes both the concept of a physical shovel and concepts for the uses of a shovel such as digging and shoveling.
- **Conceptual metaphors**: Mappings from frame to frame that allow one to conceptualize one frame in terms of another frame of a different kind. Conceptual metaphors tend to preserve the primary ideas of the source frame in the mapping to the target frame. For example, a politician in a political party is not physically in a physical container. Rather, a political party is metaphorically conceptualized as a container for its members.

- **Conceptual integration**: A mental mechanism for combining ideas, avoiding possible resulting contradictions by emphasizing some ideas over others.

An example of a primary embodied schema is the containment schema, which has an interior and an exterior, a boundary between them, and possibly a portal for moving between the exterior and the interior. Another example is the motion schema, with a mover, a source of motion, a path of motion, a direction of motion, and a goal of motion. Both of these schemas occur with the spatial use of the word "into" as in the sentence *Sarah went **into** a café*, in which Sarah is portrayed as the mover, with the café as the container and with the source of motion in the exterior of the café and the goal of motion in the interior of the café. What is inferred is that Sarah started outside the café and ended up inside the café. The container schema is not something you can see or get a fixed image of, since it can fit lots of very different kinds of containers: cups, bottles, houses, cities, valleys, the universe, chapters of books, institutions, etc.

A café is defined by a conceptual frame, the café frame. The café is not merely a container, that is, a bounded space, but also serves coffee drinks and certain foods and has a barista who makes the coffee, customers who are served coffee, and typically tables and chairs for the customers to use. Customers typically place their order, pay for what they ordered, get their order, and sit at the tables to drink and/or eat. It is also common for customers to bring laptops and work on them at the tables. The café frame is an instance of the part-whole schema, a whole with parts that are called "roles." Each frame is circumscribed; most kinds of things are *not* in any one individual frame. For example, tigers are not in the café frame, nor are surgeries. A café that happens to have a resident tiger as well as surgeons who are customers and might happen to provide certain simple surgeries is possible, but that would be a café that does not fit the typical café frame.

A sentence such as *Sara went into a depression* makes use of two conceptual metaphors. A depression is a state, and states are commonly understood metaphorically as bounded regions, that is, containers. A change of state can be metaphorically understood as a motion into a container, hence the words "went into." Conceptual metaphors are usually unconsciously used and are extremely common. We think using thousands of them. Here, the motion into a container frame is mapped onto the change-of-state frame. In addition, the term "depression" uses the conceptual metaphor that happy is up and sad is down. A physical depression is a bounded region in the ground that is lower than the region around it. The use of the word "depression" in psychology as a state in which one is sad is based on both the conceptual metaphor "sad is down" together with the conceptual metaphor that states are bounded regions in space.

Conceptual integrations are also so common that they go unnoticed. Take a sentence such as *Sarah wants to go to Heaven when she dies so she can play again with her long-dead grandmother*. We know that dead people can't play with one another. But Heaven is an imagined place where people can go after they die and function as they did when they were alive. Thus, our normal knowledge about being able to act after death is inhibited by our imagination (via simulation) of what Heaven is. Here our knowledge about playing is conceptually integrated with our imagined conception of Heaven.

These are simple cases of how (1) primary embodied schemas, (2) conceptual frames, (3) conceptual metaphors, and (4) conceptual integration are used in everyday thought, even of the simplest kind. Ideas are put together from these four general modes of thought.

Given that ideas are constituted by neural circuitry, a major challenge of this book is to show how this is possible and to show in detail what kinds of neural circuits characterize what kinds of ideas. To understand the challenge, we now turn to basic properties of neurons and neural circuits.

As you read the next section, think of the vast chasm between the biological structures and electrochemical processes occurring at the level of the neurons and neural circuits and what we have just seen about the nature of ideas. Our job in this book is to start to bridge that chasm, showing how the mechanisms of idea formation could arise from physical neural mechanisms. In short, what we experience as thought is a consequence of physical neural processes. But how does this work?

The Brain's Tool Kit

Research on how brains think takes place at many levels, from biochemical interactions at multiple levels in the anatomy of the cell to brain networks to whole systems that are embodied, that is, neural systems functioning both wholly within the brain and through connections to the living and acting body. Ideas and language may be embodied, but they are not reducible to neurons in themselves or to local brain regions (called **modules**). The nature of thought and language is at the higher systems level, a system of brain circuitry involving many parts of the embodied brain at once. Cognitive phenomena— whether individual ideas or systems of thought, individual words or larger grammatical structures—all occur at the systems level. Our computational models of how the brain functions are at the system level, although they are pieced together from smaller circuits that contribute to overall functioning.

There are now many ongoing efforts to explicate the connections between neurons in the brain (brain mapping), such as the BRAIN Initiative and its Brain Activity Map project, the European Commission's Human Brain Project, the National Institutes of Health's Human Connectome Project, and the Allen Brain Map project. Such research is showing in ever-greater detail the structural properties of overall neural networks computed from detailed recordings of multiple regions and their interconnections. The mapping goal has so far been largely medical to gain insight into treating brain diseases, strokes, and injuries in order to study traditional topics in cognitive psychology such as attention, memory, and consciousness.

Much less is known about what those networks do so far as meaningful ideas and language are concerned. As we have observed, the networks and the activity within them cannot tell us in detail how embodied cognition works, that is, which brain circuitry operations constitute which specific ideas, and which elements of language and how such circuitry affects behavior. To approach that problem, research such as that proposed in this book will be necessary.

UNDERSTANDING THE BRAIN VIA COMPUTER MODELS

Neuroscience depends on computer modeling of various kinds. In some cases, computers are needed to process masses of data and do statistics over the data. But in many cases, computer models are used to construct theoretical understandings of what is going on in the neural circuitry of the brain. For example, computer models of neural firing require **firing rules**, general principles used in modeling neural firing. Computational neuroscientists engaging in such modeling often talk metaphorically as if the neurons were "following" the firing rules. Neural modeling is also used to specify precisely and model the functions carried out by neural systems. Neuroscientists often talk as if the neural systems were "carrying out" the functions specified in their neural models. In short, neural modeling involves a mode of thought that goes back and forth between physical neural systems and computational models of neural systems, as if the models of the biological systems were *guiding* the biological systems.

We are acutely aware that this is so. However, there are many cases where the model is based on one range of data about real phenomena, while in another case the data is scarce and the preestablished model—or a hypothesized model—is used to understand phenomena where there is insufficient data. We too have found ourselves in such a situation especially in our use of the bridging model.

Our description of our computational neural models will necessarily be switching back and forth between modeling brain function and modeling specific phenomena in thought and language. To appreciate the computational modeling of relevant aspects of brain function, we have to take a good look at the level at which neural functioning is carried out. Biological neural structures function via complex chemistry and physics at various levels of complexity.

We begin with a simplified description of the details of neural activation and communication as a biological and electrochemical phenomenon.

We will start at the level of the cell.

THE CELL LEVEL

A neuron is a cell and, like other cells in the body, has a nucleus and a cell membrane, which allows substances to be ingested and excreted through cellular processes common to all cells. Extending from the cell body is an axon, a long tubelike structure ending in thousands of thin extensions with podlike terminals (called **synaptic terminals**) that contain **neurotransmitters**. From the membrane around the cell body branching outward are **dendrites**, extensive, extremely thin, branching treelike structures that serve as input terminals.

WHAT IS A "CONNECTION"?

We are born with between an estimated 85 billion and 100 billion neurons in our brains. Each neuron has between 1,000 and 10,000 connections to other neurons. That's close to a quadrillion—a thousand trillion—connections.

Neurons "connect" across synapses, short gaps filled with fluid between the axon terminals of one neuron (called the **presynaptic neuron**) and sites on the dendrites and cell body of the next neuron (called the **postsynaptic neuron**). The terms "pre" and "post" come from the fact that neural activation is conceptualized metaphorically as "traveling" from "pre" to "post."

Why is this a metaphorical concept? "Activation" is, via the metaphor, conceptualized as a substance that "flows" from neuron to neuron to neuron. This is an extremely useful scientific metaphor and is used throughout neuroscience and embodied neural computation, and we adopt it in this book. Our computational neural models, like all computational neural models, use this metaphor. There is a true story that is the basis of this metaphor: There are complex electrochemical processes that, taken overall, act is if there were a single "flow of activation."

Again, the connection metaphor is scientifically useful and is almost universally adopted within both neuroscience and computational neural modeling.

The connection metaphor leads naturally to the neural network metaphor. The network, like the internet, is not a physically connected network of wires or hollow tubes, as many neural mapping diagrams suggest. Such neural mapping diagrams use the conceptual metaphors of neural connections, neural networks, and neural flow that fit together to make an easy-to-understand picture. Again, this is an extremely useful collection of scientific metaphors, since it allows us to use computational models of neural systems. Moreover, the idea of the network makes more than a little sense because the axon terminals of a presynaptic neuron do get close—across a narrow liquid- and chemical-filled synapse—to the dendrites of postsynaptic neurons. The complex electrochemistry that supports the use of the scientific metaphors is real. But the details are tricky, to say the least.

At this point, let us look one level deeper at the details of the processes that comprise the firing of a neural cell. Many of these properties—such as the building up of activation, the threshold, and the timing and combination of inputs from multiple incoming synapses—will be represented abstractly as computational primitives in our model of the neural system.

SOME VERY BRIEF ELECTROCHEMISTRY

Ions are atoms that are either "missing" at least one electron and thus have a positive charge or have at least one more electron and so have a negative charge. For example, sodium ions (Na^+) and potassium ions (K^+) have one positive charge, while calcium ions (Ca^{++}) are missing two electrons and so have two positive charges. Chlorine ions (Cl^-) have one extra electron and so have a negative charge.

Each charged particle is surrounded by an electric force field. Positively charged ions have an electric field with lines of force pointing out from the center. Negatively charged ions have a force field with lines of force pointing in to the center. Thus, positive and negative charges are metaphorically conceptualized as "attracting" each other, with the positives pushing out and the negatives pulling in. Similarly charged ions are metaphorically understood as "repelling" each other. Again, the scientific metaphors are useful and commonly adopted.

A neuron has a cell membrane consisting of two lipid layers. The lipids are fatlike chemicals containing sugar called **cerebrosides**. The cell membranes have complex proteins embedded in them. Some are **transmembrane proteins**, large molecules going through the membrane that can either straighten or bend. When straightened, they form **molecular gates**, ion channels through which ions of a certain kind can pass. When bent, the transmembrane proteins block the passage of ions. In addition to ion channels, there are **ion pumps**, very complex proteins that, under certain voltage conditions, can transport ions; that is, they can be thought of metaphorically as pulling ions through rather than just allowing them to flow through.

Voltage is the ability to move charged particles. The cell membrane acts as an insulator, preventing the interaction between internal charged particles and external charged particles. The difference in charge between the total of charges inside the cell membrane and the charges outside of it is called a **potential difference**, measured in thousandths of volts (millivolts). At rest, a neuron has a resting potential of about −70 millivolts, that is, minus 70 thousandths of a volt. It is negative because, in addition to lots of positive ions inside the cell, there are a great many proteins with negative charge. The overall negative charge inside the cell body would tend to attract positively charged particles. Minus 70 millivolts may seem small, but 70 millivolts is one-fifteenth of a volt. For something as small as a neuron, that is a huge amount of voltage. Every neuron has that much.[4]

When the transmembrane proteins are straightened out, they form what are called "pores" or "channels" through which ions can move. Transmembrane proteins come in several types, each with a different structure that makes them specialized to allow only certain types of charged particles to pass. Ions involved in neural firing include sodium (Na^+), potassium (K^+), calcium (Ca^{++}), and chlorine (Cl^-).

Some transmembrane proteins change their shape quickly and easily. Others are slower to straighten and slow to bend back. Those differences are crucial for neural firing, as we shall see below.

HORMONES AND NEUROTRANSMITTERS

Many of the same chemicals in our bodies can have two different sources and functions, as either hormones or neurotransmitters. They are called

hormones when they are secreted by a gland into the bloodstream and work at a distance from their source. They are called **neurotransmitters** when they are emitted by a neuron and work across a synapse. Examples include dopamine, epinephrine, serotonin, acetylcholine, glutamate, GABA (gamma-aminobutyric acid), and so on.

The ion channels in neurons are central to the operation of the brain. There are two types of ion channels across cell membranes. The first, called **chemical channels,** contains receptors, molecules that chemically bind to other molecules called neurotransmitters. Neurotransmitters tend to be either excitatory or inhibitory; that is, they make receptor channels open to either positive excitatory ions, such as sodium (Na^+), potassium (K^+), and calcium (Ca^{++}), or to negative inhibiting ions, such as chlorine (Cl^-). Although there are dozens of neurotransmitters, the major ones are glutamate (excitatory), GABA (inhibitory), and acetylcholine, which can be either excitatory or inhibitory depending on the receptor molecules.

The second type of transmembrane ion channel works by electric voltage. Such **voltage channels** have transmembrane proteins that are normally bent but straighten out (and allow ions to pass through the membrane) when there is a voltage surge nearby.

The term "synaptic strength" refers to the capacity of a synapse to let ions into the cell body of a postsynaptic neuron. Synaptic strength depends on three factors:

1. The number of transmembrane proteins that allow positive ions to pass into the postsynaptic neuron
2. The number of neurotransmitters of the right kind released by the presynaptic neuron
3. The number of neurotransmitters of the right kind already in the synaptic fluid or introduced by other neurons

Synaptic strength is central to Hebbian learning, which was first pointed out by neuroscientist Donald Hebb (1972). Traditionally, Hebbian learning has been understood to work by the strengthening of synapses. Hebb hypothesized that when a presynaptic neuron fires shortly before a postsynaptic neuron fires, there is a causal connection between the presynaptic and postsynaptic firing. When such a causal connection occurs, Hebb hypothesized, the synaptic connection becomes stronger. The usual slogan for this phenomenon is "*Neurons that fire together wire together.*" And the more firing there is across a synapse, Hebb suggested, the more the synapse is strengthened, and thus the more the synapse between the neurons resembles a hypothetical wire connecting the neurons.

Correspondingly, when the two firing neurons do not fire together (close in time to each other and in the sequence presynaptic neuron before postsynaptic neuron), there is no causal relation between the two firings. When this regularly happens, Hebb claimed that the synaptic connection was weakened and might eventually be eliminated.

The change in synaptic strength is called **neural plasticity**. "Plasticity" is a general term used to highlight the fact that neural systems constantly change, and they change at the synapses, which can either get stronger or, if unused, weaker. There are multiple types of plasticity in nervous systems, and they may induce short-term or long-term changes in synaptic strength. Plasticity is important medically in the case of stroke and brain injuries, where whole systems of neural circuits need to gain new functions via synaptic changes in order to promote healing.

MULTIPLE MECHANISMS OF SYNAPTIC STRENGTHENING AND WEAKENING

There are several vital kinds of neural mechanisms in addition to the one we have discussed. The first concerns the role of genes. By processes metaphorically called **transcription** and **translation**, genes "direct" the formation of proteins. In every neuron, the transmembrane proteins are regularly absorbed into the cell body and are replaced by new transmembrane proteins formed by genetic transcription and translation processes. These genetic processes are controlled in complex ways that determine whether they create more or fewer transmembrane proteins. More transmembrane proteins means synaptic strengthening, since there will be more places for neurotransmitters to bind to and open channels for sodium ions to enter. Fewer transmembrane proteins means fewer such channels for sodium ions to enter.

Genetic factors also affect the growth of nerve fibers for generating new dendrites and axon extensions, both of which result in increased strengthening, since (1) more dendrites mean more sites for neurotransmitters to bind to and (2) more axon extensions mean more pods to open to release neurotransmitters. In both cases new synapses are formed, and new synapses for a neuron mean overall synaptic strengthening.

A second kind of neural mechanism for strengthening synapses is the chemical prolongation of the depolarization process. Any prolongation of depolarization means that more calcium^{++} channels remain open longer, and hence more vesicles get fused and more neurotransmitters are released.

Another kind of prolongation of depolarization can occur when a type of neurotransmitter (e.g., glutamate) that is released in the firing of a presynaptic neuron may flow to the presynaptic axon and bind there to ion channel receptors for that neurotransmitter. Those ion channels can open and admit more sodium or calcium ions, which would prolong the firing and hence strengthen the synapse.

A third mechanism involves neural back-signaling in a postsynaptic neuron, which results in postsynaptic dendrites releasing endocannabinoids that can bind to presynaptic cannabinoid G protein–coupled receptors. The result can be prolonging depolarization (strengthening the synapse via the release of more neurotransmitters into the synapse).

Another possibility is that this can influence, positively or negatively, the transcription and translation of genes that code for reuptake transporters. More reuptake transporters means fewer neurotransmitters left behind in the synapse and thus weakening. Fewer transporters means more neurotransmitters left behind in the synapse and thus strengthening.

In short, there are no single simple mechanisms for the strengthening and weakening of synapses. They are all chemically complex.

NEURAL RECRUITMENT

Even in the womb, babies move their arms and legs and suck their thumbs. That movement requires neural firing: neural firing in the brain and neural firing down the spinal cord through the limbs to the muscles. Neural connections are strengthened through use. A neural circuit is a network of connections that have been used enough for their synapses to have become permanently strengthened, effectively hard-wired. This is called **recruitment**. Neural recruitment is behind the acquisition of complex concepts. No thought of any complexity happens without it.

RECRUITMENT IS NEURAL NATURAL SELECTION

As your brain developed in the womb, the roughly 100 billion neurons and quadrillion neural connections were organized along various pathways, with a lot of redundancy. A lot!

In order for a fetus in the womb to suck its thumb or move its legs to kick, all the many neural circuits for thumb-sucking and leg-kicking had to be in the right places through random organization the first time that thumb-sucking and leg-kicking worked and for some time after that. Those connections that are in the right places to neurally allow thumb-sucking or leg-kicking could then be strengthened enough—naturally selected through use—to form permanent circuits (e.g., circuits for kicking one's legs and sucking one's thumb).

MAPS

Neuronal groups can occur as two-dimensional sheets called **topographic maps**, or just maps. They are thin, usually multilayered, and cover small areas of the cortex. Neurons in one sheet have connections to each other and to neurons in layered sheets, or sheets that can be adjacent, nearby, or at a greater distance.

What makes a map topographic? A topographic map preserves closeness: closeness of something, whether closeness in space, closeness of intensity, closeness of frequency, closeness of timing, and so on. Some maps are not topographic but instead have clumps of neurons, with each clump having different projections. The cortex is covered with maps.

Neuronal maps can also form structures that function as what we might call **map modules**, where a sequence or collection of maps combine, via the connections between them, to do a job that cannot otherwise be done by the brain.

The brain doesn't end with the head. The central nervous system of the body is an extension of the neural system of the brain. Neural maps are part of that extension. Neural maps occur throughout the brain and extend throughout the body into your arms, legs, and so on. As such, maps are crucial to the embodiment of the brain's neural system and the embodiment of concepts, that is, the way the concepts arise and continue to exist via bodily connections.

CLOSENESS IN SPACE

One of the best-studied maps is the retinotopic map. The primary visual cortex (V1) contains a map of the retina. Neurons next to one another coming out of the retina have axons that project through the thalamus to neurons in V1 (at the back of the brain) that are correspondingly next to one another. This is not magic; it is a matter of chemical gradients. In the womb, axons grow from the retina through the midbrain to the opposite side of V1. As they exit the retina, two chemicals secreted nearby attach to each axon terminal. The attaching molecules are secreted along two independent dimensions, with secretions starting thicker and getting thinner as they spread out. Like points in a Cartesian plane defined by two numbers, axons coming from retinal neurons each have fixed amounts of each secreted chemical. How much of each depends on their location on the retina. Two axons that are close in position on the retina are indeed close but are not the same in their amounts of the two chemicals.

The axons grow through the lateral geniculate nucleus (LGN) in the thalamic midbrain area to V1. When they reach V1, they encounter pairs of secreted chemicals that are receptors that bind to the original chemicals. Each axon lands and stays at the position where the receptors with the right concentrations of the two binding chemicals bind to the two chemicals on the axon terminal. Each axon stays exactly where the right pair of chemical bindings keep it in place. As a result, each two neurons that are next to one another coming out of the retina connect to neurons that are next to one another in V1. V1 is therefore a topographic map of the retina that preserves closeness in space.

Maps of this sort that preserve closeness occur throughout the brain and the body. It is worth stopping for a moment to consider how remarkable it is that such topographic maps in one part of the brain can correspond to topographic maps in another part of the brain.

OTHER TOPOGRAPHIC MAPS

The primary motor cortex is a *two*-dimensional topographic map of the *three*-dimensional distributions of muscles in the body. For example, neurons projecting to muscles in two adjacent fingers are adjacent to one another in the primary motor cortex of the brain. The somatosensory cortex of the brain is a topographic map of the skin and of places that we can feel in the body. The somatosensory cortex also preserves closeness of spatial location.

Think of how remarkable this is. For example, places on your arm that feel close to one another are connected to places in the brain that *are* close to one another.

The acoustic cortex is a tonotopic map of sound waves entering the ear. Hairs in the fluid of the inner air vibrate at different frequencies and intensities. Neurons connect the hairs to the acoustic cortex, making the acoustic cortex a map of frequency and intensity of sound waves.

There is actually more to it. The sound makes the stirrup bone of the ear (called the **stapes**) vibrate. The stirrup vibrates against a round membrane, inside of which is the liquid of the tympanic canal of the cochlea, which is a spiral structure, thin on one end and wide on the other. Along the spiral of the cochlea is the basilar membrane, along which 12,000 to 20,000 hair cells are distributed evenly. The hair cells go through the organ of corti, which is wrapped around the cochlea between the basilar membrane and the fluid of the tympanic canal. Along the tympanic canal is the tectoral membrane, a physical structure that vibrates with the sound waves in the canal.

The hair cells extend into the fluid. They are stiffer near the narrow end and wavier near the wide end. The stiffer ones vibrate to higher-frequency sound waves, and the wavier ones vibrate to lower-frequency sound waves. The vibrations trigger ion gates to open in the hair cells. Ions in the fluid rush through the gates, sending action potentials to the basilar membrane. The membrane has neural projections from the locations of the hair cells to locations in the acoustic cortex. These neural projections preserve spatial closeness to the locations on the membrane. The locations on the membrane correspond to the frequency of the sound waves, stimulating the hair cells. As a result, the acoustic cortex is a spatial map preserving the frequency of the sound waves stimulating the hair cells. And those sound waves are a direct reflection of the external sound waves vibrating the stirrup, which vibrates the round membrane. Thus, the acoustic cortex is a **tonotopic spatial map** of sound waves hitting the inner ear!

The insula is a viscerotopic map of the visceral organs: the lungs, heart, liver, stomach, sexual organs, colon, and kidneys. This is not a topographic map but rather a discrete map; neural projections from these organs go to specific spots on the insula and in the opposite direction from those spots on the insula to the organs. Connections from the prefrontal cortex to the insula allow for the monitoring and control of important aspects of these organs.

The middle temporal (MT) region in the parietal cortex detects motion. How? The MT region has topographic maps that receive projections from V1. There are MT maps that have neurons that are connected to adjacent neurons. As adjacent neurons in V1 are activated, they send activation to MT neurons. As activation spreads from neuron to neuron in MT, the connections among adjacent neurons create **a path of activated neurons**. Via this path, we perceive a corresponding path of motion in the external world.

What do topographic maps demonstrate? They are maps of how the neural system of the brain connects to the body. In other words, maps in the brain constitute an internal embodiment of the brain's external connections in the body. Such neural maps are one of the major forms of embodiment circuitry, that is, circuitry that allows us to conceptualize what our bodies are doing.

THE WHAT AND WHERE PATHWAYS

The V1 region is in the occipital lobe at the rear of the brain. V1 is farther to the rear than any other brain region. In perception, neural activation flows from V1 along various pathways to parts of the brain that are farther toward the front. Within V1, there is a pathway split between the **dorsal stream** (toward the top central areas of the brain) and the **ventral stream** (toward the bottom and side areas).

The dorsal stream goes from the V1 region to the V2 (secondary visual cortex) region just in front of it, then to the **dorsomedial** area (central toward the top) and to the V5 region (also called the MT region), which is active in motion perception. The dorsal stream, also called the **where pathway** and the **how pathway**, functions in location and motion (using topographic maps) and in control of the eyes and arms when they use visual information to guide reaching with the limbs or movements of the eyes (**saccades**).

The ventral stream goes from V1 to V2 and then goes forward and sideways to third visual cortex (V3) and the fourth visual cortex (V4), to the **inferior temporal cortex** ("inferior" for underneath and "temporal" for the side where the temples are). This is sometimes called the **what pathway**, since it is active in characterizing the structure of objects and other forms, such as letters and faces. Partly overlapping with the hippocampus, the ventral stream is also active in memory.

The what and where pathways appear to have a profound effect on our conceptual system and language. Spatial relations make use of what is called the **trajector-landmark** relation, with the trajector (what is located or moving) characterized via the what pathway and with the landmark (what the trajector is located or moving with respect to) characterized via the where pathway. Thus, in a sentence such as *Sara is in the café*, the café is called the "landmark," and Sara is the trajector. The trajector-landmark relation is central to all human conceptual systems and appears to arise because of the what/where pathway relationship.

WHY EARLY CHILDHOOD EDUCATION IS VITALLY IMPORTANT

By mid to late childhood, about half of the quadrillion connections that we are born with die off, the ones that are least used. That still leaves hundreds of trillions of connections available to form future circuits through use. Many of those early childhood connections have formed fixed circuits that will be there for life. That is why early childhood education is so important. Debates on the politics of early childhood education usually miss this crucial fact about the die-off of neural connections in mid to late childhood. The politics of early childhood education is rarely informed by neuroscience, though the neuroscience is crucial to the politics.

All of the circuitry that characterizes fixed concepts is acquired in this way. But we do not start with a blank slate. We exit the womb with embodied circuitry, which imposes a considerable amount of neural pathway structure on both possible thoughts and possible linguistic forms. We all also start life in the same world with basically the same kinds of bodies, with neural circuitry not just in the brain but also throughout the body and connected to the brain. Human beings share a great deal of common neural structure and common real-world experience (in a gravitational field, moving our bodies, perceiving, eating, and so on). Part of what we share is a mechanism for forming circuitry: neural recruitment via neural natural selection that operates via the strengthening and weakening of synapses.

The Body Is Neural

We normally think of the body as flesh and blood, bone and muscle, and heart, stomach, guts, and other internal organs. When we think of the body, we think of how the body looks and what normal and special physical things bodies can accomplish and what our particular bodies may or may not be able to do or be. Beauty and looks are seen as facets of the physical body. Being an athlete is seen as having a special body: more muscular, lithe, and controllable than an ordinary body. Old age is seen as the deterioration of those aspects of the body, in the extreme being unable to feed, dress, and bathe ourselves. Infanthood is seen as occurring before such bodily capacities develop. The accomplishments and character of the body are seen as separate from those of the mind and the brain. To be a paraplegic is seen as a loss of embodiment, as limited to a separable, separated, and separate mind.

All of this leaves out the fact that the body is neural and that the brain has been structured through evolution to run and be an implicit part of the body. The premotor and supplementary motor areas choreograph the motor actions that the body performs and send instructions in the form of neural signals to the motor cortex. The motor cortex sends neural signals to—and receives them from—the whole of the body via the spine and connections to the face.

The signals to the body go to motor synergies, finely structured complex circuits that connect to the motor neurons that are connected physically to the muscles, either activating and hence tensing muscles or inhibiting and hence relaxing muscles.

There are between 400 million and 600 million neurons in the gut (from the esophagus to the anus), which is up to three times as many neurons than in the brain of a rat. You're not conscious of your gut thinking, but the gut system produces about 95 percent of the serotonin and 50 percent of the dopamine found in your body. This system of neurons lining the walls of the alimentary canal is often called the second brain, or the enteric nervous system.

Each muscle in the body has an equal and opposite muscle, that is, each tensor muscle (whose activation makes the muscle tense) has a corresponding extensor muscle (whose activation makes that corresponding tensor muscle relax), and the motor neurons attached to those muscles are connected to each other by mutual inhibition. Thus, tensing a tensor muscle sends an inhibition signal to its extensor muscle, which releases the extensor muscle. Each extensor muscle, when activated, sends an inhibition signal to its tensor muscle, releasing the tensor muscle. Thus, the muscle system is thoroughly connected to the neural system.

In addition, the motor cortex is a two-dimensional topographic map of the body. For example, the muscles in each finger are connected to the brain's motor cortex in such a way that fingers next to each other are connected to cortical regions next to each other in the brain's motor cortex.

Circuitry at the base of the skull inhibits connections to the body during sleep so that the body doesn't move in response to dreams. There is an exception. Some people have a disease of this area, allowing signals during dreaming to get through to the body. Such people are called **thrashers**, since they may move their legs while dreaming of running, for example. Some otherwise normal people may react to powerful dreams by screaming out loud what they are screaming in their dreams. In such cases, it appears that especially strong brain activations may override inhibition at the base of the skull.

Furthermore, the somatosensory cortex of the brain is also a topographic map of the body, extending to one's skin and whatever parts of the body you can feel. Again, connections from bodily regions next to one another extend to the regions next to one another in the somatosensory cortex of the brain. And pain that is felt in the body is actually registered physically in the brain via neural connections from the body. And finally, internal organs are connected to the insula, a structure in the midbrain.

In short, the body is neural—made up of neural connections to brain regions every bit as much as it is made up of bone and muscle. And the brain is structured to connect to the body optimally, with connections to the body located more toward peripheral brain regions, while attentional and control circuitry is located more toward the front of the brain, the prefrontal cortex.

Dual Models

Neural mind research studies how neural circuits can constitute ideas. What got our research started was the observation by Jerome Feldman that there is an important shared property—the **gestalt property**—between frames and certain circuit types called **gestalt circuits**. The gestalt property is that the activation of one or more parts activates the whole. Bear in mind that a frame is an abstract theoretical device that we, as cognitive scientists, use as a tool to analyze human concepts. Such scientific tools are needed to study the nature of ideas; they are not the ideas themselves. But when we claim that a tool such as a particular frame is adequate for characterizing the structure of the idea we are studying, it is common scientific practice to attribute the conceptual tool for analysis to the idea itself, speaking of the idea itself as *having the inherent structure that the frame attributes to it.*

For example, consider the typical understanding of the concept of surgery. As cognitive scientists, we analyze that concept in terms of a surgery frame. The surgery frame is a whole that has a number of parts, or roles: a surgeon; a patient an operating room; an operating table, an anesthesiologist, a nurse, possibly other surgeons, and tools such as scalpels, needles and thread, and clamps. The roles also include various actions: the anesthesiologist anesthetizes the patient, the surgeon operates on the patient, the nurse assists the surgeon including passing tools to the surgeon, the surgeon uses a scalpel to cut into the patient, and so on.

Here we are attributing the analytic structure of the surgery frame that we use to analyze the concept of surgery to the inherent structure of the concept of surgery itself. The idea is that *if the analysis is adequate given the evidence, then the analysis reflects reality.* This is how routine science tends to operate, attributing to reality the technical analysis based on the evidence.

The entire surgery frame can be activated by hearing various parts. "He's been wheeled into the operating room" activates the surgery frame and tells you that "he" is the patient, that a surgeon will be operating on him, that he will be anesthetized, and so on. "He's about to be anesthetized" tells you that "he" is the patient and that the operation hasn't started. In short, *the activation of one or more parts of the frame can activate the whole frame* and with it all the typical parts. This is the gestalt property. All schemas and frames have the gestalt property.

Feldman observed that a neural circuit for a concept that is adequately analyzed in terms of a frame must have the same gestalt property. That is, a circuit with the gestalt property must have a part-whole structure, and the activation of one or more parts of the circuit can activate the whole circuit and with it the rest of the parts of the circuit. Such a circuit is called a gestalt circuit.[5]

A SIMPLE COMPUTATIONAL MODEL OF A GESTALT STRUCTURE

Feldman showed that mental structures with the gestalt property can be modeled computationally. The computational model operates on a graph with at least three nodes: one node is labeled the **gestalt node**, and the other nodes are labeled **role nodes**. Lines (or **edges**) of the graph connect each role node to the gestalt node. Each node has a number assigned to each of two parameters: **threshold level** and **activation level**. A node is activated if its activation level number is equal to or greater than the threshold level number. Activation is taken as "flowing" along the lines (edges) of the graph from node to node. Thus, the activation of each role node flows to the gestalt node. And the activation of the gestalt node flows to each role node. This computational model applies to both neural circuits with the gestalt property and to analytic tools used to study ideas such as schemas and frames that have the gestalt property.

Here is how computational modeling works. *The properties of the computational model are attributed to what is modeled.* The computational model applied to neural circuitry is seen as characterizing what "neural computation" means. Applied to cognitive analysis, the computational model is seen as characterizing cognitive computation, that is, computation done by thought mechanisms. Attributing the specific computational properties of a model to what is modeled is commonplace in the field of computational modeling.

Frames are general theoretical mechanisms for adequately analyzing the structure of thoughts. The gestalt property is a general property of frames. The computational model for frames, to adequately model frames, must therefore also have the gestalt property. The neural circuitry that constitutes each frame must therefore also have the gestalt property. And the computational model that models the neural circuitry for frames must also have the gestalt property.

A **dual model** is a single computational structure that can accurately model two different kinds of things at once: both the structure of an idea (as analyzed by cognitive scientists) and a plausible neural circuit hypothesis that constitutes the idea physically in the brain.

If the idea structure and the neural circuitry characterizing the idea share a property, then the dual model must also share that property.

The dual model we will be using is based on the following hypothesis:

- Thoughts are structured as activations of generalized neural circuits. These neural circuits are generalizations of neural functions for controlling the body and acting in the world.
- Thoughts are carried out by specific kinds of embodiment (sensory, motor, emotional, social) via the way that neural circuits carrying out those thoughts are embodied in those ways. For example, thoughts of movement are carried out by neural circuitry governing the motor system without any necessary bodily movement; that is, actual movement may be inhibited, while movement is mentally simulated.
- Thinking a specific thought corresponds to the activation of the circuits that constitute that thought.
- Mental connections between thoughts are neural connections between circuits.
- Generalized neural circuits constitute general cognitive mechanisms (such as frames, conceptual metaphors, action schemas, and image schemas).

The neural mind theory that we are presenting relies on dual models to form a "bridge" between neural circuitry and the ideas that the circuitry embodies.

The general strategy we followed in constructing the theory we are presenting was not to theorize idea by idea, one at a time, but rather to theorize at the level of the hypothesized general mechanisms for generating all ideas, mechanisms such as frames and conceptual metaphors. There has to be a match across (1) general mechanisms, (2) their neural circuitry, and (3) the dual computational models linking them, with the same general properties for one applying to all.

This is a match across things of different kinds. The general mechanisms come from cognitive science, which studies ideas. The neural circuitry is the real locus of ideas, as *constituting* the ideas. What we call thinking is what those neural circuits do. And the dual computational models reveal both the computational properties of the neural circuitry with the general mechanisms of computational analysis forming a bridge between the neural circuitry and the computational models of that circuitry.

The power of the theory arises from converging constraints from its multiple sources: the extensive and deep research into conceptual systems within the field of cognitive linguistics, the remarkable knowledge gained from neuroscience, and the advances made in artificial intelligence as applied to structured neural computation. By showing how they fit together, the theory draws on the scientific knowledge gained in all.

THE ACTIVE BRAIN AND SIMULATION

The brain is always active. Even when we appear to not be attending to an external stimulus, our default network is working away. The default network is a structure consisting of connected brain regions including the ventral medial prefrontal cortex, the posterior cingulate cortex, the inferior parietal lobule, the lateral temporal cortex, the dorsal medial prefrontal cortex, and the hippocampal formation (Buckner et al. 2008). A primary function of the default network is the imaginative simulation of hypothetical scenarios of various kinds: imagining what other people think, envisioning acting in future situations, remembering one's own life, and constructing what you should have said or done. We may not be thinking consciously about anything in particular, but the default network is working away.

In addition to the default network, there is also a task-related simulation that engages in ongoing active simulations, instant-by-instant simulations as you walk down a busy street trying not to bump into people or millisecond-by-millisecond simulations as you drive on a freeway looking forward, to the sides, in the mirrors, and over your shoulder deciding unconsciously and automatically via the default network whether to speed up or slow down, pass or drop back, or change lanes while reading signs or listening to the radio.[6]

Of course, we also perform and attend consciously to specific actions, see the actions of others, and plan future actions. And we conceptualize these actions and reason about them. Physical actions are carried out and understood via the neural circuitry in our motor systems. In an important recent account, Buzsáki (2019) argues that the inherent, active, and self-organizing dynamics of the brain guides both action and perception. The same neural circuitry used to perform actions can be used to perceive and understand the goals and intentions of others when they perform them or to imagine those actions. Conceptualizations of actions are thoughts, idea circuitry applied to action concepts. Carrying out a planned action is fitting a simulation of that planned action in real time with real movements. Practicing a musical instrument, an action in a sport, a dance routine requires carrying out simulated actions.

Even the ideas of static objects involve active simulation. Recognizing an object as a cup typically involves a simulation of picking it up with the particular grip (depending on whether there is a handle). Mimes can make you imagine physical objects that are not there by simulating the movements of using the object. Imagine a mime picking up and drinking a cup of tea.

Thinking is an action performed by the neural system. Understanding a passive experience—for example, sitting in the passenger seat of a car being driven by someone else—requires actively simulating it by the neural system even though it is barely active. Neural simulation goes on constantly in everyday life and in everyday thought.

LANGUAGE AND THE COORDINATION OF ACTIONS

There is a rich history of treating language and communication as action going back to the work of John L. Austin and Paul Grice and followed by work on speech acts by John Searle and others. Building and extending this hypothesis, there is an emerging view of language as a social act of joint action and coordination. This is exemplified in recent books by the evolutionary psychologist Michael Tomasello (2022) and the social linguist Nick Enfield (2022). Tomasello argues through evidence from evolutionary characteristics and behavior that early humans had to evolve a cooperatively rational agency and joint action primarily for survival. Enfield posits that language is irreducibly social and hypothesizes language as the interface between individuals that enables joint action and coordination. Language can be viewed as providing knobs in an interface between speakers and hearers. Words and constructions set these knobs to create and reinforce shared values and goals and to initiate and maintain joint action. Language thus acts on the hearer to persuade, move the hearer into action, and form affiliations and social bonds. In this sense, language is less about communicating facts and more about social coordination. The precision and details inherent in language are geared toward enabling joint action and not so much about communicating detailed facts about the world.

Actions of any complexity at all require coordination. Whether you are tying your shoes, playing a game, playing the piano, or cracking an egg, the coordination of arms, hands, fingers, and body orientation is required. There are a number of general types of coordination, and they each require specialized neural circuitry. There are computational models of that circuitry. The computational models with circuit diagrams will be discussed in detail in chapter 3. But for now, we at least need a sense of the most common and general types of coordinated circuits that are used in everyday life and what they mean for the theory we are proposing:

- **Sequencing circuits**. Playing the piano requires precise sequences of coordinated actions by your hands, fingers, and feet. Dance movements also require precise sequences, as does tai chi and turning a double play as a second baseman or shortstop.
- **Control circuits of various types**:
  - **Iteration**. In tai chi, you might repeat the cloud hands sequence four times. In drumming, a beat is created by iteration. In football, the offense lines up to start each play. In making an omelet you beat eggs, repeating fast rotations of a fork or whisk.
  - **Suspension/resumption/cessation**. In a complex process that requires iteration, you often have to test to see if you are done. So, you suspend what you are doing, and if not done you resume and repeat until you are done, and then you stop.
- **Concurrent circuits**. Many actions require doing two things at the same time. To pick up a glass of water to take a drink, you have to reach your arm toward the glass while opening your hand until you get close enough to the glass to close your hand around the glass.
- **Conditional circuits**. Many actions depend on other actions. In cooking wild rice, you bring the water to a boil, and when it starts boiling, you turn the heat down to a simmer.
- **Shifter circuits**. This is a combination of a conditional circuit with two different sequence circuits. While sequence 1 is active, a condition may be met that results in inhibiting sequence 1 and shifting to the activation of sequence 2. An example is sautéing a filet of salmon with the skin on. After oiling both sides of the filet and heating the pan, you lower the heat and put the salmon, skin side down, in the pan, and then you keep lifting the skin side, checking how brown the skin has become. If the skin is brown enough, you flip the salmon over and repeatedly check to see how well it is cooked on the inside by looking at the side. When done enough, you remove the salmon from the pan.

Here you are shifting from a sequence of checking skin brownness to a sequence of checking interior doneness.

- **Substitution circuits**. A regular routine is a default way to accomplish a goal. But if something gets in your way at some point, you may have to substitute a new action for a default action in order to accomplish the goal. For example, you might regularly plug a space heater into a wall outlet behind a standing lamp, putting the plug in from the side. But if a pile of boxes of books has been placed around the lamp, you may have to plug the heater in by putting your arm above and behind the boxes of books.

- **Priming circuits**. Priming involves activating a node in a circuit to a significant extent but keeping it below the threshold level so that very little more activation will result in the threshold being met.

WHY THESE COORDINATION TYPES MATTER

Each of these types of circuits characterizes a type of action and a type of action concept. Many actions are coordinated in some way or another. The coordination may be sequencing, iterating, suspending with resumption and cessation, concurrency, conditionality, shifting, substitution, or priming.

For each type of coordination, there must be a type of neural circuit that can perform that type of coordination. And there must be a form of neural computation that can provide a dual model of that type of coordination for both the neural circuitry carrying out the action and the action as we understand it by thinking about it. Circuit hypotheses for such neural circuitry and the computational models for that circuitry are provided in chapter 3.

Each circuit type is a property of both the dual model that uses it and the neural circuitry being modeled. In dual modeling, the same computational model must be able to model that property in both the neural circuit and the conceptualization of the action. In short, those computational circuits allow the dual models to carry out coordinated actions by modeling and simulating the hypothesized neural circuitry.

When we think about actions, we conceptualize them; that is, we form action concepts. Each type of coordinated action concept has the properties that distinguish one type of coordinated action from another. That is, we distinguish sequencing from concurrency, concurrency from substitution, and so on. Those circuits are the computational models that simulate those distinctions and allow us to reason accurately about each type of coordinated action.

Dual computational modeling in our theory provides the link between neural circuitry and the ideas that we have when that circuitry is activated. Functional neural generalizations such as coordination circuits for acting in the world and controlling the body (including social and emotional functions) constitute the neural embodiment of cognitive theoretical constructs structuring human ideas such as frames, metaphors, and schemas.

As such, coordination circuits provide a clear example of how dual modeling works in the theory we are proposing and why the theory takes the form it does.

Generalizations

The brain has evolved to control the body interacting with the physical and social development. The more peripheral parts of the brain have relatively direct connections to the body. For example, the visual cortex at the rear has connections to the eyes. The acoustic cortex, near the temples, has connections to the ears. The motor cortex across the top in the middle has connections to the muscles via subcortical (underneath the cortex) and brain stem structures to the spinal cord and from there to the muscles. The somatosensory cortex, just in front of the motor cortex, has connections to the skin and wherever on the body you can feel. The insula, a ringlike structure in the center below the cortex, has connections to the internal organs (e.g., heart, lungs, liver, kidneys). These brain-body connections go in both directions, with the body providing constant feedback to the brain. This is part of the neural structure we are all born with.

As we shall see, these all require control circuitry elsewhere in the brain. Where would such control circuitry be located? What would we expect? We would expect the control circuitry to be located in the prefrontal cortex, the most frontal area of the brain! We would expect this because the prefrontal cortex fits so economically to the connections to the body, which are farther back, to the sides, above, and below. What is remarkable is that superb research by neuroscientists Mark D'Esposito, David Badre, and their coworkers has actually shown this experimentally for certain kinds of behavioral rules.

Moreover, they went a step further. They showed via functional magnetic resonance imaging (fMRI) research that circuitry carrying out "rules" governing the operation of the body are structured hierarchically within the prefrontal cortex, with the most general toward the front (**rostral**), the more specific in the middle, and the most specific toward the back (**caudal**), a rostral-to-caudal (front-to-back) organization within the prefrontal cortex![7] They have shown that this works for four types of cases: general versus specific (1) behavioral rules, (2) conceptual relationships, (3) goals, and (4) policies that govern goals. In addition, more recent work by Badre et al. (2009) and Bhandari and Badre (2020) at Brown University show that abstract task structure involves circuits connecting midlateral prefrontal cortex with inferior parietal cortex.

THE INEVITABILITY OF GENERALIZATION

Human concepts come with a general case. That is, they are mostly about general ideas such as "cup" rather than specific cases such as the antique cup in your cupboard inherited from your great-grandmother. Indeed, to understand the specific cases, you need to understand general cases.

Concepts arise via neural recruitment. In order to be "learned" via synaptic strengthening, the networks of connections must be in place and functioning correctly *before* the strengthening takes place. Remember, you can only kick in the womb if the right networks of neurons required for kicking are in place. As you kick, they are gradually strengthened, forming circuits used in further kicking.

EMBODIED PRIMITIVES AND GENERALIZATION

Neural circuits for physical object concepts are recruited on the basis of experience with specific examples. Why should there be a general versus specific divide in recruitment?

Our hypothesis concerns embodied primitives that are universally learned via universal experiences very early in life. They include concepts such as containment, motion, control of bodily movement, purposeful action, force exertion, force resistance, undergoing force, contact, adjacency, front-back, up-down, near-far, center-periphery, and so on as well as combinations of these to form complexes: frames, conceptual metaphor, conceptual integrations, etc.

Consider houses. They are all containers, with interiors, exteriors, and portals to move between the exteriors and interiors. They all have solid walls and doors that exert a resistant force, protecting those inside from weather and other factors outside. They all have places to prepare and eat food, rooms to sleep in, and locations for commonplace, everyday embodied activities. The concept of a house is general.

The particular houses tend to have different specific details, different shapes, materials, furniture, decorations, etc., which are not part of the general house.

What a house is will differ from culture to culture. If there are houses and if not, there will be general concepts for tepees, igloos, yurts, apartments, and so on. But in each case the concept is general and learned via recruitment.

THE SPECIFICS AND THE GENERAL

There are many kinds of houses: ranch houses, Mediterranean houses, Victorian houses, Cape Cod houses, duplexes, mansions, cabins, and so on. The concept of a house fits them all. When you learn them, the general concept is activated plus other details that are added, subtracted, or modified. When you activate your concept of a Victorian house, you activate your general concept of a house and add details, different details than those for a ranch house.

Of course, there has to be circuitry linking such specifics to the general case. For example, there must be **neural bindings** between the general roof and the Victorian roof, between the general door and the Victorian door, and so on. To get an overall concept of a Victorian house, all those bindings must fit together at once. That is, there must be a Victorian integration circuit with activation links to all those Victorian bindings.

In certain forms of modern architecture, there is a large common area containing the kitchen, dining area, and living area. To conceptualize this common area as such, there would presumably be neural bindings linking the understanding of the common area of the specific house to the general roles kitchen, dining room, and living room of our understanding of the general house.

INTEGRATION AND GENERALIZATION

Badre, D'Esposito, and their coworkers have shown that degrees of generalization fit a hierarchy in the brain, with activation of the most general concepts being more frontal (rostral, which rhymes with "nostril") and the most specific concepts being least frontal (caudal). Moreover, general concepts are used in more specific concepts, with the specific concepts adding specific details to the more general concepts.

We take this to mean that there must be neural circuitry linking the added specific details to the neural circuitry for the general concepts. That circuitry linking the specific additions to the general circuitry may extend from more frontal (rostral) locations to less frontal (caudal) locations.

Our hypothesis suggests that the general-to-specific additive circuitry must do the following:

- activate the circuitry of the general schema with all of its roles, and
- bind the circuitry for each specific detail to the circuitry for the corresponding role of the general schema.

THE ANATOMY OF GENERALIZATION

The brain is always active, monitoring the environment and acting, projecting, and predicting the consequences of actions. Generalization helps this process in a fairly natural way. Indeed, if large sequences of actions produce desirable results reliably, the brain can package them into general rules and habits. The brain is always seeking to strengthen—and therefore make more permanent—meaningful, repeated, prototypical experiences and routines. Once strengthened enough to become permanent, these routines (such as speaking a language) become largely unconscious and automatic. Of course, situations that are exceptions to these rules would require additional processing, that is, additional control to inhibit the rules in exceptional situations. Conceptual frames—scripts—are habits and are examples of such packaged circuitry.

This makes anatomical sense. Since the most general rules are more rostral (located neurally toward the front of the brain), those rules that are less general will, of course, have more specific content and will therefore have connections to more embodied areas toward the rear and outside areas of the brain, that is, those areas that are less frontal (less rostral and more caudal).[8] This also has a functional explanation based on how the neural system works. In fact, using neurally motivated learning rules such as Hebbian learning, of which spike-timing-dependent plasticity is a temporally asymmetric form, it is possible to predict how the system learns to associate and package these stable and familiar patterns reliably and automatically. Chapter 3 goes into more detail on this process and how it also may play a role in language learning.

EMBODIMENT AND PATH-DEPENDENT LEARNING

The brain is a physical system. As such, it is subject to the law of conservation of energy and appears to work by the principle of least energy. When there is more than one possibility for recruitment, the circuit that uses the least energy is presumably the one recruited.

As a result, generalizations arise naturally. It takes less overall energy to use a single general circuit in understanding many specifics than to characterize each specific case neurally.

If this is true, as it appears to be, then embodiment—how thought is embodied neurally—plays a major role in recruitment and generalization. What are called **primary concepts**, concepts used throughout our conceptual systems, are embodied. Primary concepts such as motion, containment, and force are all present in the neural circuits all of us acquire early, perhaps as early as in the womb.

For example, a container schema has an interior, an exterior, a boundary, portals, and contents. Every instance of a house is also an instance of a container. But the container schema may be there before the conception of a house is learned. When you learn what a house is, you have known for some time what a container is. Containment is also an integral part of the concept of a bottle or a sippy cup. That is, when you are learning a concept that makes use of the general schema for a primary concept, you don't learn the primary concept anew. You use the one that is already there in your brain circuitry. This means that there are fewer synapses to change in acquiring such a concept and presumably less energy needed. In short, it is easier to learn concepts that make use of already learned concepts, a principle that psychologists have long observed.

In a human conceptual system, this would result in concepts defined via complex networks with more general to less general branchings, and less general to more general chains terminating in embodied schemas. Such networks and chains of circuits are what we have called **integration circuits**. Integration circuits optimize the overall flow of neural activation.

UNCONSCIOUS NEURAL INTEGRATION

What we consciously perceive—what we see, hear, or feel—is often not exactly what comes to us through our sense organs. The brain makes changes in sensory input, *unconsciously fitting preestablished neural patterns* before vision, hearing, and touch become conscious. These changes usually take about 80 to 100 milliseconds (a twelfth to a tenth of a second) although sometimes as much as 500 milliseconds (half a second).

Why should unconscious **neural integration** to fit existing circuitry occur? A possible answer is **neural optimization**, that is, least energy use. It may take less energy to make small changes to fit a preexisting circuit than to recruit an entirely new circuit.

MULTIMODALITY AND CONCEPTUAL INTEGRATION

Much of thought is **multimodal**, bringing together embodied circuitry from different modalities: vision, motor control, touch, hearing, communicating, and thinking. **Multimodal integration**—the formation of consistent, integrated circuits across different modalities in different brain regions—is central to consciousness.[9] Conscious thought—a small part of all thought—tends to be integrated into a consistent whole, while much of unconscious thought is fragmented.

Multimodality brings two or more embodied regions of the brain together, allowing fixed circuitry in one region to use fixed circuitry in another region. It is common for experiments that show what might be called **multimodal integration dominance**, in which circuitry in one modality adjusts to circuitry in another modality. Examples will be discussed in the section "Creating Perception: Seeing, Feeling, and Hearing" below, with examples coming from vision with motion, touch with motion, and sound with vision. But an everyday example is useful: When you are driving you adjust your steering, acceleration, and braking (control of movement) to what you see, and you adjust the force of your touch (on the accelerator and brakes) to what steering, acceleration, and braking require.

We should bear in mind that multimodality is anything but an accident of evolutionary history. Multimodality contributes to neural optimization, since it enables the brain to exploit both redundancy across modalities and complementarity in information from different modalities, that is, the ways in which information from different modalities fit together to provide an integrated understanding. The what and where pathways define this multimodality.

THE TAKEAWAY: THE PHYSICAL LOGIC OF NEURAL LEARNING

This section is a first pass on many issues discussed in this book. The neural basics just listed fit together according to what might be called a **physical logic**.

This section began with the connection metaphor: Neurons don't touch; they don't literally connect. The connection metaphor and the circuit metaphor allow neuroscientists to use knowledge about electric circuits to model with great accuracy the way networks of neurons function in the brain and the body. The use of metaphorical models to describe with mathematical precision the functioning of physical systems is at the heart of the physical sciences.

The metaphors "firing" and "connection strength" allow neuroscientists to use the circuit metaphor to model neural learning, the acquisition of new skill and knowledge via change in the strength of connections between neurons. Such models contain learning rules that specify the change in strength of connections between neurons when an incoming neuron to the connection fires. The idea of such a rule is of course metaphorical, using the familiar notion of a rule to characterize what a complex neural structure does. An important rule is Hebbian learning: "*neurons that fire together wire together.*" This creates circuits on the basis of neural firing close in time and space. Reward-based reinforcement and other Hebbian paradigms such as spike-timing-dependent plasticity will be discussed in later sections as rules that model the creation and maintenance of circuits where firing goes in one direction.

The idea of recruitment allows neuroscientists to use the concept of connection strengthening to explain how hundreds of trillions of partly random connections can lead, beginning in the womb, to the learning of functional neural circuits via strengthening. The recruitment metaphor leads to the concept of neural natural selection, a version of Darwin's central insight applied to neural learning. Recruitment has the consequence that you cannot learn something from nothing: the neurons have to be in just the right place in advance before strengthening—and hence "learning"—can take place. Just as in evolution, species have to be in the right ecological niches before selection can take place, with selection being the optimization of the fit of species to ecological niches. The fittest are those that best fit the niches they happen to be in. Survival of the fittest means that those species that best fit niches where they are will be the most likely to survive.

The insight that new learning depends on prior learning is crucial. We have circuits that have already been "learned," that is, that have already become fixed via recruitment and strengthening through use. Newly acquired circuits can be thought of as changing the neural ecology of the neural system. New learning is easier if already existing circuits are used with minimal additions, that is, with minimal new recruitment via strengthening. This is the idea of best fit or, in other words, neural optimization. The theory that neural optimization happens via least energy use allows for an explanation, not just a description. The principle of conservation of energy in physics automatically chooses least energy use to explain neural optimization and cases of best fit.

By the time a child is five years old, a vast amount of neural circuitry has become fixed in both the brain and the body. What a child learns before the age of five is crucial. And neural learning—the strengthening of synapses—happens all the time, everywhere. Neural learning, which is more than just learning subject matter and is mostly unconscious, includes learning cultural attitudes, interpersonal relationships, an understanding of the self, food preferences, and negatives such as trauma, social awkwardness, and resistances of all sorts. That's why early childhood is so important. Early learning sets the stage for later learning by providing constraints on what neural circuitry and how much energy are required for later learning.

Neural learning naturally leads to generalization. There are three factors:

1. From early childhood we have learned embodied primitives including motion, containment, force, and many others. A house is a container; its walls are solid and resist force, and the motion is from outside to inside and back. We get a general understanding of a house using those primitives.

2. Different houses have different specifics. We learn those specifics by repeated experience. When the details are different and not repeated, those specifics are not learned.

3. Our neural anatomy links the periphery of the brain to the body. This leaves the prefrontal cortex as a potential control center, and it has been discovered that our more general circuits occur closer to the front of the prefrontal cortex.

Thus, *our anatomy makes generalization inevitable*. Cognitively, every detail of a specific structure must fit a semantic role in the general structure. For example, a highly detailed Victorian roof must fit the semantic role "roof" in the general conceptual structure for a house.

What we have just discussed is largely unconscious, simply because we have no direct conscious access to our neural circuitry. Most ideas are not just unconscious; they are complex as well and make use of circuitry extending across multiple brain regions. Unconscious circuitry operates on a millisecond timescale. It takes 80–100 milliseconds (a twelfth to a tenth of a second) for unconscious percepts (sight, sound, touch, etc.) to become conscious. This is so fast that we cannot be consciously aware of what is happening.

Consciousness is integrated. Those unconscious circuits have to fit together in a way that makes sense if they are to become conscious. In chapter 3, we show examples of some simple kinds of integration circuits. Integration circuitry fits together other circuitry so that it makes sense and so prepares the integrated circuitry for conscious understanding. The scientific study of integration circuits and their neural realization is at very early stages, and making progress to uncover these circuits is a central open task for the neural mind. This book represents only the first step of what we hope will be a productive direction for the field.

In short, we don't just understand what we are conscious of. Understanding is always complex. There is circuitry linking and integrating various modalities: for example, vision, motor control, and touch. We regularly use such multiply connected brain regions. These are brought together by integration circuitry that allows them to function as integrated wholes. Our normal understanding of the world is thus unconscious, fragmented, and distributed across various brain regions, while conscious thought is unified and arises via the integration of unconscious fragments across multiply connected brain regions.

The what and where pathways extend across multiply connected, multimodal regions. Different conceptual work is done in these regions, with the what pathway characterizing trajectors (entities located or moving) and the where pathway characterizing landmarks (where the trajectors are located or what they are moving with respect to). The conceptual systems and grammars of all languages make use of the what and where pathways, and human anatomy shapes how we think and communicate via trajectors and landmarks.

In short, the basic neural traits that we began by just listing—recruitment, strengthening, integration, circuit formation, and so on—fit together via a physical logic that characterizes the most basic integrated mechanisms of thought, with the least-energy principle naturally giving rise to generalizations: neural circuitry shared by circuitry for specific cases.

Functional Generalization

This book presents the first steps toward a theory of the neural mind, that is, a theory of how ideas can arise from and be constituted by neural circuitry. In other words, it is a theory of how it is possible for physical brains to think, how thought can be physical in nature, and what configurations of neural circuitry can physically constitute corresponding thoughts.

Ideas and the language that expresses them are relatively stable. That is why communication—imperfect in certain cases—is possible in most everyday interactions. At the neuron-by-neuron level, neural systems vary considerably from person to person and from time to time. Yet, neural systems function to produce the relative stability of ideas and language. A functional neural theory claims that there are real **functional generalizations** over the neuron-by-neuron details; that is, the neuron-by-neuron circuits actually work according to those physical generalizations, and such physical generalizations accurately characterize what is sufficiently stable in ideas and language to permit ordinary, everyday thought and communication.

Why should this be true? Why should there be physical generalizations over neural structures that characterize stable details of thought and language?

Human beings have to communicate. Communication requires memory: stable, remembered structures of thought and language. The great stability of ordinary concepts and language in everyday communication requires our complex neural systems to accommodate to the real-world stability of language and thought. The fact that stable ideas and language are constituted by our neural systems means that our neural systems really do function at a level of generalization to produce that stability. In short, functional generalization at the neural level produces stability in thought and language.

WHAT ARE NEURAL FUNCTIONS?

To control the body, the brain needs neural circuitry that integrates sensory information from many sources (vision, sound, touch, smell) in order to move the body so as to function physically in the world. Functional control of the body requires at least the following:

- keeping bodily functions within a reasonable range—temperature, heartbeat, breathing, eating, and so on—which is known as homeostasis;
- interacting with the world, including other beings, physical objects, physical locations, and so on;
- sensing one's physical and social environment;
- monitoring internal well-being and ill-being and one's emotions;
- predicting the effects of one's actions and those of others;
- choosing actions that enhance well-being, both yours and those of others; and
- coordinating movements and interactions on the basis of all of the above.

To achieve such things, the brain has to integrate many different sources of information (e.g., vision, hearing, smell, touch, appetite, and internal state sensors such as temperature) to maintain homeostasis, to make predictions about the environment (social as well as physical), to estimate their impact on well-being (your own and others') and make decisions, that is, to select actions and carry them out to achieve purposes.

Doing this requires fine-tuned connections between different parts of the brain, connections among parts of the brain that

- sense the world,
- sense the internal bodily state,
- make predictions about future states of the world and estimate their impact on well-being,
- make decisions,
- carry out actions, and
- correlate the consequences of those actions with those predictions.

Such connections form systems that we call **functional circuits**.

FUNCTIONAL FIXED NEURAL CIRCUITS

A functional circuit comes into existence when it is recruited to carry out some function that is required of all human beings in everyday life. That function may be an action that one performs regularly, such as washing your hands or frying an egg.

The circuit becomes fixed when, through regular use, its synaptic strengths get so strong as to become permanent.

Because neurons function in groups, circuits tend to be made up of a network of nodes—each node is a neural ensemble—with pathways from node to node, that is, from ensemble to ensemble. An ensemble can be (1) just a group of neurons or (2) a group of neurons with local connections performing some localized function. Thus, a node can be anything from a single neuron (rare), a group of unconnected neurons, or a locally functioning ensemble with a complex structure. Such an ensemble does more than just pass input from neural firings along to the next node; the ensemble systematically changes input firings to different output firings.

Networks have connections from node to node. Circuits are networks in which the synapses linking the connected neurons in the network have connection strengths strong enough to make them effectively permanent. The network becomes a circuit when the network is activated often enough to permanently strengthen its connections.

Additionally, trauma may create such strong firings in the network as to strengthen the connections enough to turn the network into a permanent circuit, one in which the neural effects of the trauma can be repeated when the trauma circuitry is activated.

The theory of ideas that we are presenting in this book presupposes functional generalization. The idea of functional generalization is based on the basic knowledge about (1) ideas and the language expressing those ideas and (2) the neural circuitry that constitutes those ideas and the language expressing them.

An overwhelming part of the vocabulary and basic grammar of a language and the ideas they express are the same from speaker to speaker in everyday experiences. In general, English speakers will have no trouble understanding sentences such as these: *He knocked on the door. She has long hair. I'm eating an apple. Sara walked into the café.*

The details of low-level neural circuitry are constantly changing from time to time and vary widely across speakers of the same language. There are hundreds of millions of English speakers and millions of such common everyday sentences. Although the low-level neural details in the brain circuits of those hundreds of millions of speakers are wildly different, they must function in pretty much the same way to understand those millions of sentences. In other words, there must be neural generalizations over the hundreds of millions of wildly different low-level neural details that will function in the brains of millions of speakers to understand those millions of sentences.

There is nothing mysterious about this. We saw above that there are circuits carrying out functional generalizations about the way we act in the world. *Our hypothesis is that ideas, thought, and language use the same generalizations about everyday experience.*

The point is that although there is no way to study either the hundreds of millions of speakers or millions of sentences, we can nevertheless study the shared conceptual system and language of those speakers.

We and many others have studied the semantics of English in general and the cognitive mechanisms in general that generate the ideas expressed in those sentences; that is, embodied schemas, conceptual frames, conceptual metaphors, and conceptual integration. Those general and widespread mechanisms for generating ideas characterize the millions of ideas understood by speakers of the language.

Although we cannot study the functional neural circuitry required by the millions of ideas one by one, we can study the functional neural circuitry required by the small number of cognitive mechanisms that generate those ideas.

That is what we will be doing in chapters 2 and 3. In chapter 4, we take up a way to extend this theory from ideas to language that expresses those ideas, looking at the case of English.

Cascades

The computational bridge is designed to model flows of neural activation across the brain, flows that we call **cascades**.

We will be using the word "cascade" in two senses:

- **Active cascade**: A flow of activation in the functional circuitry in the brain that goes across brain areas. In typical cases, there is a bidirectional flow of activation extending from one embodied brain region to another. In an active cascade, many fixed circuits across diverse brain regions will be activated. In addition, spreading activation from those fixed circuits will usually activate further circuits required by those fixed circuits, adding to the cascade flow.
- **Cascade circuitry**: A network that goes across areas in the brain and is made up of fixed circuits that regularly take part in active cascades and define where the flow of activation goes in an active cascade.

A SIMPLE CASCADE

Let us consider a simple cascade from the domain of perception, described by the neuroscientist Stanislas Dehaene (2009) in his book *Reading in the Brain*. Dehaene took up an important question: We did not evolve as reading creatures. Animals don't read. Our brains did not evolve to read. How, then, is reading possible, and how does reading work? Dehaene's account is called "repurposing," or "recycling." We did evolve circuitry to recognize the shape and form of objects. We have adapted that circuitry to the relatively new purpose of reading. Here is Dehaene's account:

> The primary visual area (V1) appears to be largely dedicated to the detection of thin lines and object contours. By the secondary visual area (V2), neurons are already sensitive to combinations of lines with well-defined inclinations or curves. Farther up, in the posterior part of the inferior tem-

poral cortex (an area called TEO), they are tuned to simple combinations of curves. Neurons' selective response to, say, the shape of an F can be reduced to the detection of a simple conjunction of elementary curves, each placed at a relatively fixed location: the top bar, the upper left angle, the middle bar, and so on. In brief, the same scheme seems to be repeated at each hierarchical level: neuronal selectivity results from a conjunction of the more elementary features coded by neurons at lower levels. (Dehaene 2009, p. 136)

The phenomenon described by Dehaene is deep and important. The next questions to be answered are: how is this done and what kind of circuitry does it and on what timescale.

A READING CASCADE

Here is our more detailed hypothetical reconstruction (an educated guess) of how the brain carries out letter recognition. The recognition mechanism for letters and other shapes makes use of layers of topographic maps of the visual field, connected in sequence. Each map is a layer of neurons with closest connections to neurons in the next layer.

Consider the letter F. We know a lot about the shape of the letter. It has three lines, one vertical and two horizontal. The horizontals are shorter than the vertical. Each line is made up of a sequence of points. The vertical is to the left of the horizontals. The point at the left end of one horizontal connects to the point at the top of the vertical. The point at the left end of the other horizontal connects to about the midpoint of the vertical. That is a lot of information, but we perceive it as a whole all at once.

Here is how an F is recognized as an F, a whole letter and not just a bunch of activated nodes. The proposed connections are **center-surround cells**, **receptive fields**, and **orientation-sensitive cells**.

CENTER-SURROUND CELLS

The ganglion cells of the retina are connected to neurons at the retina in just the right way to form center-surround receptive fields. Here's how: Ganglion cell centers activate the closest neurons emerging from the retina. Ganglion cells surrounding those centers inhibit those central retinal neurons activated by the ganglion centers. The retinal centers are maximally activated when they are not inhibited by any surrounding ganglion cells. That is, when there is, say, a black center with a white surround, a white center with a black surround, a red center with a green surround, or a green center with a red surround, and so on. Those retinal neurons connect one to one, via the LGN in the brain's central region, to neurons in the primary visual cortex, V1. Connections from each eye go to the opposite region in V1: right eye to left half of V1, left eye to right half of V1. From this point on we will speak of nodes, or neural ensembles.

In short, each node in the first layer of V1 is connected to a center-surround receptive field in the ganglion cells of the retina. That is how they are embodied via body parts, in this case the eyes.

The connections are not one-way. There are far more connections going in the opposite direction, providing continual feedback between the eyes and the brain.

RECEPTIVE FIELDS

The concept of a receptive field is somewhat complex: it is a relationship between a neuron and some area in which the presence of a given stimulus will result in the firing of the neuron. The idea is that the neuron (or neuronal group) receives input enough to make it fire on the basis of the presence of something in a given area, or field. The term "receptive" comes from the idea of the neuron receiving input from an external stimulus.

Technically, as we have seen, that input consists of ions flowing in as a result of neurotransmitters or voltage changing the shape of transmembrane proteins.

The idea was introduced by experimentalists who were concerned with finding out what kind of stimulus in what places would cause a given neuron to fire or inhibit it. The field was the set of places where the presence of the stimulus caused neural firing.

The idea is general. First, it can apply to a region of bodily input: an area on the skin, a region in the retina (groups of rods and cones), or a range of hairs in the cochlea (or equivalently a range of sound frequencies that would activate those hairs). But a receptive field can also be a place (or places) in the brain such that neural activation there results in the activation of a neuron (or neural ensemble) elsewhere.

The idea is important because it helps us understand functioning in the brain. For example, Hubel and Wiesel (1962) discovered that in the retina there are center-surround receptive fields: that is, regions of rods and cones where activation of those in the center provides firing input to ganglion cells, while activation of those in the surround tend to inhibit the firing of those ganglion cells.[10] The result is to make a spot that you see tend to stand out against a different background.

Hubel and Wiesel also found that the centers of adjacent ganglion cells overlapped significantly. Moreover, when a short line (or bar) was presented to a region of the retina, the line covered the centers of adjacent ganglion cells, activating a linear sequence of ganglion cells. But the line was perceived as a whole! How?

1. The ganglion cells connect to neurons in layer 1 of the primary visual cortex, V1. The result is that each V1 neuron has the same receptive field in the retina as the corresponding group of ganglion cells.
2. Adjacent neurons in layer 1 of V1 have overlapping receptive fields in the retina, with overlapping centers.
3. A short linear sequence of neurons next to one another in layer 1 of V1 has a short linear sequence of overlapping centers in the retina.
4. A short straight line (or bar) presented to those overlapping centers in the retina will therefore result in the activation in layer 1 of V1 of a short sequence of neurons (or neural ensembles) next to one other.
5. The neurons in layer 2 of V1 form a topographic map of the neurons in layer 1 of V1 in such a way that an adjacent sequence of activated layer 1 neurons will activate a single neuron (or neural ensemble) in layer 2.

Thus, in the case of an F, there are three straight lines and hence three activated neurons (or nodes) in layer 2.

Now assume that there are connections from layer 1 to layer 3 in V1 such that layer 3 neurons only fire if they take input from two or more layer 2 neurons. This means that the points in the F where there are line overlaps (namely the two points on the left line at the top and middle) will result in the firing of the corresponding neurons in layer 3 maps.

Now assume that there are connections from layers 2 and 3 to layer 4 and that unique neurons (or neural ensembles) fire in layer 4 depending on the patterns of firings in layers 2 and 3. What constitutes an F is firing of that pattern of neurons across layers 1, 2, and 3.

This is oversimplified, of course. It leaves out the variations in shape and size on the letter F. But the point is that we comprehend, that is, we consciously recognize the F by virtue of a *cascade consisting of all of those layers at once*. Each layer and its inputs do something simple. But the combined simplicities yield the integrated complex pattern. This is how a simple cascade is formed.

There are many further complexities that arise out of simple cross-map connections. For example, short lines in a given region in layer 1 can be perceived in layer 2 as being in one of sixteen orientations: horizontal, vertical, at 45 degrees, and so on. Those oriented-line nodes form columns of adjacent orientations in layer 2.

a. In layer 2, there are three active nodes for F: one connected to the vertically oriented overlapping receptive field centers, which we see as the vertical line plus two active nodes connected to each of the two horizontally oriented overlaps of receptive field centers, which we see as the two horizontal lines.

b. In layer 3, there are two active nodes for F: those indicating the two points of overlaps between the vertical and horizontal lines.

c. In layer 4, there will be a single node indicating F. That node will be active when the nodes in (a) and (b) are active. The activity of that node will indicate the visual perception of F.

d. The connections between layers 4 and layers 2 and 3 go in both directions. Activating the node in layer 4 corresponds to imagining an F. That node will activate the corresponding nodes in layers 2 and 3, which will in turn activate the overlapping center-surround receptive fields in layer 1 that form the lines.

When all these connections across all of the layers are firing together, we recognize all at once the prominent center-surround nodes, the short lines, the longer lines, the orientations, the line intersections, and the patterns of line intersection. When they all fire together, the result is a cascade of neural firing.

- Recognizing all of these at once via a **neural cascade** constitutes recognizing a whole shape, that is, the visual gestalt structure, whether it is a letter, an object, or a face.

- Imagining the F happens when we begin with the layer 4 node and activate the nodes in layers 3, 2, and 1.

All of these connections anatomically go in both directions, linking the retina (which is embodied in the eye) to and from brain-internal layers. The F cascade can thus be triggered in two ways: in perception by an embodied region (the retina or V1) or in imagination by an internal region. Whether we are recognizing a perceived letter, object or face or imagining it, the activations flow back and forth between the interior of the brain and the embodied region all at once, putting together the entire visual structure.

For all of the neurons in these layers to be active at once, the two-way connections between layers must be active along the cascade long enough for the whole gestalt to be perceived or imagined. The layer-to-layer activations happen much faster than we can consciously perceive. It takes only a small number of milliseconds to establish a whole layer-to-layer-to-layer cascade going in both directions at once. Once established, all the parts are active at once.

In such a cascade, the connections from one layer to the next do only one tiny thing. But as more layers are added, the tiny contributions add up, forming new structures. Then those structures gradually add up. When they are activated all at once, the structural totality is created. Alone a single neuron can be in one location (e.g., the V1 map), can have connections to other neurons, and can either fire or not. That's all that an isolated neuron can do. It is the cascades of neurons that allow the formation of complex structures that we understand. That is the fundamental idea behind the theory of cascades.

GENERAL FEATURES OF CASCADES

From this simple cascade, we can see several features that we use throughout our formulation of cascade theory:

- Cascades are grounded in embodied circuitry.
  1. For thought, the embodiment is part of the emotional, conceptual, social cognition, and sensorimotor systems.
  2. For the form of a language, it is the sensorimotor system for phonology, that is, for speech and sound, for signing and perceiving signs (in signed languages), for reading and writing, and for ordering phonological elements and perceiving the order and the ordering hierarchies.
  3. The lexicon and grammar link phonological and conceptual cascades to create meaningful language.
- Connections in cascades can be bidirectional. You can perceive an image or language or create it.

- The component parts of a cascade can be perceived together all at once as parts of the whole. Or with different timing, the parts can be perceived sequentially.
- Individual circuits in the cascade can be activating, inhibiting, disinhibiting, or modulating.
- A node in a cascade can trigger an integration circuit that integrates information elsewhere in the brain.
- Cascades are learned via normal neural learning mechanisms: recruitment, synaptic strengthening, and natural neural change mechanisms.
- Cascades can be externally triggered in the case of perception and internally triggered in the case of imagination and understanding.
- Cascades can be combined, coordinated, and sequentially activated over time to form more complex cascades.

The theory of cascades is an important part of our approach to understanding how thought and language work systemically and globally in the brain. This is the opposite of the view that all brain processes are localized in isolated modules.

Timing is an important constraint. Each neural firing takes on the order of a few milliseconds, where a millisecond is one thousandth of a second. The time elapsed between one firing and the next is a small number of milliseconds. The chaining in cascades cannot be all that long in milliseconds. The maps and connections hypothesized are constrained by embodiment, by anatomical connections, by the types of connections, by the timing of neural processes, and by the time constraints on consciousness.

Visual perception takes between 80 and 100 milliseconds to become conscious and sometimes as much as 500 milliseconds. That has a possible implication for a theory of cascades. When we point out that in the above cascade all of the map-to-map activations have to be perceived all at once, we mean that *any real-world time differences have to be so small that we cannot consciously perceive the difference*. In other words, we consciously perceive the timing of neural events as simultaneous if we cannot consciously perceive any time difference between them. This puts a very real time constraint on hypothesized cascades.

TWO TYPES OF SIMULTANEITY

There is a big difference between what we consciously perceive as being all at once and what a neural system reacts to as all at once. Consider two nodes,

N1 and N2. When the rest of the brain cannot tell the difference in timing between the firing patterns of N1 and N2, those firing patterns are considered simultaneous even though there might be real time differences between them.

As we shall see, this notion of brain-based simultaneity plays an important role in the theory of neural language and concepts. The phenomenon of neural binding between nodes A and B requires that the rest of the brain not be able to tell the difference between the firing of node A and the firing of node B. In other words, *two nodes are understood as the same node if the rest of the brain cannot tell the difference between the firing or lack of firing of one and the firing or lack of firing of the other.*

Convergence-Divergence Zones

One of the early inspirations for the ideas in this book came from the remarkable neuroscience research on convergence-divergence zones (CDZs) by Antonio and Hanna Damasio in the late 1980s and early 1990s. Their work has greatly matured since then.[11] One of the most interesting things about their CDZ hypothesis is that it fits fundamental results about conceptual structure so well that it could be an important element in a common framework for bridging cognitive semantics with neural systems.

Before explaining why, however, we must first say a few things about what CDZs are.

A CDZ is a hypothesized neural control structure, a neural ensemble that integrates operations performed elsewhere in a number of different locations. The integration is done via temporal neural binding: control over the time-locked activation of neural patterns in those locations. Such temporal binding permits the integration of features computed in various other places into a single gestalt, a collection of circuits activated at once.

CDZs with the same neural architecture can control different functions depending on where in the brain they are located, that is, what they are connected to.

CDZs can be hierarchically and heterarchically linked. That is, hierarchical linkage means that neural ensembles in a CDZ can characterize general cases linked to more specific cases controlled elsewhere and can also characterize specific cases linked to more general cases controlled elsewhere. That is, CDZs are bidirectional. Heterarchical linkage is nonhierarchical linkage.

The early work on CDZs hypothesized that certain cognitive deficits associated with brain damage arise because lesions damage CDZs or connections between them, and early research established that this is correct. We will begin with that early research.

To get a feel for CDZs, consider what it takes to name a familiar face.

1. The face must be perceived as a whole, with various features (eyes, nose, mouth, chin, hair) integrated in just the right way (i.e., brought together) to form a single gestalt. There are patients who have difficulty integrating parts of a figure into a whole. For such patients, the CDZ hypothesis says that the CDZs integrating (i.e., bringing together) low-level facial features detected elsewhere are disrupted.

2. Let us call the face, perceived as a whole, a face-gestalt. The face-gestalt must then activate knowledge about a particular person. There are patients who can perceive a face as a whole visually but cannot connect it with knowledge about the person, although they can identify the face when either voice or gait is added. In this case, the CDZ provides only priming activation, which is at too low a level in itself to activate recognition, but the addition of voice or gait adds sufficient activation for recognition. There are other patients for whom voice or gait perception does not help face recognition. Here the CDZ hypothesis says that the CDZs connecting the face-gestalt with knowledge are disrupted.

3. Having processed the face as a gestalt and activated knowledge to know whose face it is, one must then be able to assign a name to the face as recognized. There are patients who can recognize a familiar face and give lots of information about the person (who may be a close relative or a celebrity) but cannot supply the name. Here the CDZ hypothesis says that the CDZ integrating proper names with recognized faces is disrupted or, in ordinary naming problems, requires additional activation from other knowledge.

In each case, there are CDZs integrating information computed in or controlled from other places. The CDZs form hierarchies. The face-gestalt CDZs are linked to knowledge about a particular person, voice perception, and gait perception via a person-recognition CDZ. And the person-recognition CDZs (CDZ-person) are linked to knowledge of proper names via person-naming CDZs (naming-CDZs). These CDZs thus form a hierarchical chain: face to person to name.

There are also some heterarchical (nonhierarchical) connections in this case. CDZ-person is linked with a voice-recognition CDZ (voice-CDZ) and a gait-recognition CDZ (gait-CDZ). Face-gestalt CDZs, voice-CDZs, and gait-CDZs are heterarchically, not hierarchically, linked. That is, they are on the same level; there is no hierarchical activation flow or chain.

The architecture of each CDZ is roughly the same. Connections to lower-level features or other CDZs go in both directions, forming feedback loops.

Consider, for example, face-gestalt CDZs:

- In perception, activation is initiated by the lower-level features and passed (up the hierarchy) to face-gestalt CDZs.
- In memory, the reverse is the case: activation initiates at a higher level (person-CDZ or higher) and is passed to the face-gestalt CDZ and then to lower-level features.

Antonio and Hanna Damasio refer to connections that go to CDZs higher in a hierarchy or sideways as feed-forward connections.

The CDZs themselves do not contain any representations. There are no representations of whole faces in a face-gestalt CDZ, no representations of recognized persons in a person-CDZ, and no representations of names or persons in a naming-CDZ. The CDZs do not represent; they only control the activations of neural ensembles elsewhere.

According to the CDZ hypothesis, this control is exercised through neural binding. For example, consider the face-gestalt CDZ, which controls face-gestalts. The face-gestalt CDZ *integrates* particular facial features (eyes, hair, nose, mouth, wrinkles) with what might be thought of as "slots" in a general schema of the human face. The association of the particular eyes with the eye slots, the particular mouth with the mouth slot, and so on is an instance of neural integration controlled by the face-gestalt CDZ.

In the Damasios' CDZ hypothesis, the proposed mechanism through which neural binding is hypothesized to take place is via time-locked activation of the various lower-level features that are bound together to form a particular face: a neural integration. The CDZ controls the firing of those lower-level activation patterns so that the patterns to be integrated fire in sync, producing an integrated whole.

In the case of a hierarchy of CDZs, one CDZ would control the firing in other CDZs, which control the firing in other CDZs and so on until the lower-level parts of the brain where the component features are computed is reached. These circuits are the cascades.

In the late 1980s when the Damasios formulated their hypothesis, they were using positron emission tomography (PET) scan data. The CDZ hypothesis made sense of commonplace PET scan data. When a person is performing a single relatively simple task, it was common for PET scans to reveal activation not just in one portion of the brain but also in many portions. On the CDZ hypothesis, the parts of the brain activated are linked via CDZs that are organized hierarchically and heterarchically. The idea is that a task of any complexity at all is not performed in just one place but instead involves a cascade of activation from the higher-association cortex to low-level regions in the visual, somatosensory, auditory, or other cortices.

CONVERGENCE-DIVERGENCE ZONES AND BASIC-LEVEL CATEGORIES

One of the most striking discoveries by the Damasios and their coworkers was that there are patients with face agnosia (the inability to recognize faces) who also cannot recognize certain basic-level categories.[12]

These are patients with lesions in the inferior temporal lobes who cannot integrate a percept of a particular face with the information that goes with that face, information that would permit recognition. Some of the same patients also cannot distinguish among prototypical basic-level animal concepts. That is, they cannot tell a raccoon from a wolf or a fox and refer to them as just animals, though they can recognize nonprototypical animals such as elephants and giraffes. Such patients can, however, recognize basic-level categories for artifacts and implements: knives, forks, chairs, tables, and so on.

Why should an inability to recognize particular faces accompany an inability to recognize categories of animals with prototypical shapes? The answer the Damasios give is that both involve (a) the ability to make fine distinctions among images that involve the complex integration of visual features and (b) the ability to pair such distinctions with specific knowledge. In the CDZ hypothesis, both disabilities involve disruptions of CDZs that integrate visual images with other knowledge and that consequently would be located in the same region of the brain.

And what about the fact that such patients can recognize other basic-level categories, those for implements and artifacts? According to the Damasios, those categories involve the information of not only visual images and specific information but also somatosensory (tactile) and motor information. Thus, knowledge of those categories would have to be accessed via CDZs controlling somatosensory and/or motor information, which would have to be located elsewhere in the brain in a place where inferior temporal lesions would not affect them. Examples of such locations include the posterior and lateral temporal cortex, some parietal cortices, and frontal cortices.

But if this is so, there should be patients with lesions in those areas who cannot recognize implements and artifacts but can recognize faces and prototypical animals. Indeed, there are such patients, as the Damasios report. That is, there is a double dissociation:

- Inferior temporal lesions but not more dorsal temporal or parietal lesions correspond to a loss of face recognition and prototypical basic-level animal categories.
- Dorsal temporal or parietal lesions can do the reverse. In all cases, superordinate categories (animal, object) are retained.

MOTOR-PROGRAM ANIMALS

There is one more remarkable finding in this data. Consider the case of cats and dogs, prototypical animals that most of us have motor programs for (e.g., petting) and horses, that many people, especially in rural areas, have motor programs for (e.g., petting, feeding, and riding). With the patients tested, these motor-program animals patterned just like implements, not like other animals. Patients with inferior temporal lesions did not lose the ability to recognize these animals. In brief,

- If categories are defined by just the integration of visual images and other knowledge, they will be accessed via CDZs in one sector of cortex.
- If categories are defined by the integration of visual images, other knowledge, and motor programs, they will be accessed via CDZs in another sector of cortex.

NAMING

There is a common belief among linguists that human beings have a lexical component in their brains, a single location where all words and their meanings are "stored." Using a word means just retrieving it from "storage," a kind of warehouse for meaningful words.

One can imagine an alternative: the forms of words (sound, speech, writing) show connections in the brain to different brain areas depending on how word meaning is embodied. Then, there would be no storehouse, no lexical component in the brain, no single place where all words and their meanings are stored.

The Damasios and their colleagues found a way to compare these opposite hypotheses using data on the naming of objects from patients who had brain lesions. Here is what they found.

Category recognition is distinct from category naming (Damasio and Tranel 1993). Two of the patients, AN-1033 and KJ-1360, have no problem with category recognition but only have naming deficits. The third patient, Boswell, has both recognition and naming deficits. Recall also that concepts involving tactile and motor activities (implement concepts, physical actions) are accessed via regions related to somatosensory and motor processing. On the other hand, concepts involving only visual images and other knowledge but not motor activity are accessed via the inferior temporal lobes. Recall as well that the language areas, which provide access to phonological forms, are located in the left hemisphere. One would therefore expect that CDZs for naming objects where concepts depend on visual images but not motor involvement would be in the left inferior temporal cortex.

On the other hand, one would expect CDZs for naming manipulable objects to be in the posterior temporal and inferior parietal cortices. The reason is that in those regions CDZs can access not only visual knowledge but also somatosensory knowledge and, indirectly, motor patterns of the hand.

What about the locations of CDZs for naming fruits? We would expect them to vary with the particular fruit, depending on how much visual, tactile, or motor pattern knowledge characterizes our interactions with that fruit. For example, consider a banana. It is a fruit that we learn a special motor program for, a program involving the hands. The CDZ hypothesis would lead us to expect that the banana concept would be stored in a CDZ with access to not only visual knowledge but also somatosensory and hand-based but not full-body motor knowledge. Finally, we would expect concepts for physical actions such as running and swimming to be accessed through still other CDZ sites, since the knowledge characterizing such actions involves both visuomotor and somatosensory patterns of whole-body actions, not just hand actions.

The hypothesis also generates a further expectation. Lesions are not strictly demarcated. Serious damage in one zone typically entails at least some lesser damage in nearby zones. Thus, where there is major damage to the inferior temporal lobes, it would not be at all strange to find lesser damage to such a nearby area as the posterior temporal region. Such differing amounts of damage would lead us to expect corresponding differences in deficits.

On this line of reasoning, we might expect KJ-1360, who has left premotor damage, to have a naming deficit for physical actions involving motor activity over the whole body (swimming, running). But we would not expect KJ-1360 to have a naming deficit for objects and other concrete entities such as animals, since they involve visual and tactile information but not whole-body motor information.

On the other hand, Boswell and AN-1033, who have left inferior temporal lesions, would be expected to have deficits in naming animals but not in naming physical actions. It would not be surprising for them to have lesser damage in the nearby posterior temporal region and hence a lesser deficit in naming tools and implements. This is essentially what Damasio and Tranel (1993) report.

Boswell successfully named 92 percent of the physical actions but only 24 percent of animals and 25 percent of fruits and vegetables, which is consistent with his expected deficit. He named 76 percent of tools and utensils, a minor deficit that is consistent with a lesser degree of damage in the nearby posterior temporal region. Incidentally, despite his problems with other fruits, he had no problem naming bananas.

Now consider KJ-1360. He got only 53 percent of the physical action names, an expected deficit given his premotor damage. His uniformly high scores otherwise are consistent with his lack of damage in other areas.

AN-1033 also fits the theory. He has the expected deficit in the non-motor-involved animal and fruit/vegetable categories: 51 percent and 54 percent, respectively, which indicates less serious damage than Boswell but significant nonetheless. AN-1033 has the expected lack of a deficit in the motor-involved physical action category: 96 percent. And his 70 percent performance on tools and utensils is, like Boswell's 76 percent, consistent with lesser damage in nearby areas. And like Boswell, though AN-1033 had severe problems naming other fruits, he had no problem with bananas.

GOODBYE TO THE WAREHOUSE THEORY

What do these results mean? They support the CDZ hypothesis and show that the warehouse theory is false. There is no lexical component in the brain, no single location where all words and their meanings are "stored." Words in the brain have connections to different brain areas depending on how word meaning is embodied. Moreover, the connections go through CDZs and branch out to diverse regions in the brain hierarchically and sideways at various points in the hierarchy.

Indeed, the very idea of storage is disconfirmed by CDZ research. Memories are not stored in any particular place. They can be triggered or controlled from a particular brain location, but the conceptual, visual, or motor content of the memories requires the activation of circuitry in multiple brain regions via different CDZs.

Moreover, CDZs unite memory and perception into a single theory. Overall perception occurs via the piecing together of separate perceptual features via the CDZ system. Memories occurs via the activation of one or more CDZs and are pieced together via the CDZ system, activating perceptual and cognitive neural circuitry, often fragmentary involving mechanisms of constructive simulation, which will be discussed in chapter 3. Memories are typically unconscious and only occasionally come to consciousness.

A COMMON FRAMEWORK

The Damasios' CDZ hypothesis was first proposed over a quarter of a century ago. Yet, its main features seem to have held up as more research has been done and seem prescient, almost predicting many of the conclusions we have come to.

- There is embodiment via connections from the brain to various locations in the body.
- Circuitry that is not directly embodied connects across the brain to embodied circuitry.
- CDZs have both hierarchical and lateral structure in connectivity.
- There are two-way connections entering and leaving CDZs.
- CDZs are part of cascades of connections that branch out (diverge) in one direction, with connections converging at the same points in the opposite direction.
- Neural binding works via two-way connections with synchronous firing of the bound nodes.
- The CDZ hypothesis creates neural integration of information from multiple modalities across multiple regions in the brain. CDZ architecture could form the underpinning of grammar and language. We will see more in later sections.

In addition, important discoveries of CDZs are continuing, especially in research on frontal-striatal circuitry.[13] These results suggest the possibility of a cascade with convergent projections from multiple cortical areas (prefrontal and parietal regions) to the striatum. This suggests an integration circuit of corticostriatal projections that combine reward, attention, and control during reward-based learning.

The theory of CDZs explains a great deal of neuroscience data correlating brain lesions and behavior. It also fits everything that we have found about neurally embodied cognition, which gives it even greater explanatory power.

The theory of CDZs is also a transformative theory in that it eliminates commonplace warehouse and storage theories that claim that memories and complex knowledge such as knowledge of language are "stored" at certain brain locations. There is a big difference between local storage and local triggering or control, which can explain high fMRI readings correlating with triggered memories and controlled behavior.

There is no lack of cognitive phenomena requiring a neural explanation. We now turn to several examples where deep explanations can be offered in the form of hypotheses but where rich experimental data is not available, at least not at present.

Explaining Basic-Level Concepts

ACTIVE EMBODIMENT

Embodiment is active in every sense: dynamically changing, situated in context, connected to the physical and social world, purposeful, and aesthetically attuned to what is awesome and pleasurable—or not—and with some personal sense of right and wrong. As John L. Austin (1962) and John Searle (1979, chap. 1) have observed, communicating is acting: informing, convincing, requesting, making commitments, establishing norms, and creating new situations in the world (as the US Supreme Court and Congress do). Embodied action, whether in thinking alone or in communication, involves at least the following causal capacities, which commonly occur together: enabling possibilities, creating probabilities, causing effects, and bringing about consequences. And embodied action in part defines the logics of those possibilities, probabilities, causes, effects, and consequences.

Active neural circuits also structure purposes. Purposes are constituted structurally by desires, resources, courses of action, difficulties, and goals, with successes and failures characterized via so-called reward circuitry, the neurally controlled release of hormones for pleasure, satisfaction, and dissatisfaction as well as circuitry controlling emotions. Active circuits coordinate actions, ordering them in sequence or performing them in parallel, structured *hierarchically* by the internal complexities of desire, types of resources (e.g., energy, strength, body structure, imagination), kinds of actions, and what it means to have purposes and satisfy them or not.

BASIC-LEVEL EMBODIED CONCEPTS

Even in the womb, we perform complex actions: kicking, turning over, and sucking our thumbs. Think of what a fetus's neural system, via integrated control over the body, has to perform for the fetus to suck its thumb. In one motion, the fetus has to make a fist with thumb up, lift the shoulder, bend the elbow, open the mouth, put the thumb into the mouth, suck on the thumb repeatedly, and then stop sucking, loosening the mouth's grip on the thumb and moving the thumb out of the mouth by opening the elbow and lowering the shoulder. Bit by bit it is complex, but overall the movement sequence forms a single **naturally integrated** schema, as natural as they come: what we will call a **basic-level embodied schema**.

After birth during childhood, we form categories of naturally embodied schemas: eating small pieces of food, sipping out of sippy cups, crawling, putting objects into containers, taking objects out of containers, throwing graspable objects, and grasping, holding, and carrying around a favorite object (a toy, a doll, a blanket). These develop into what are called **basic-level categories** of actions and objects, defined by motor programs, mental images, and gestalt perceptions: chairs, balls, cars, dolls, running, walking, jumping, throwing, dancing, and so on. These also have **cognitively special properties**: they have shorter names and are named earlier, recognized faster, and remembered more easily than higher-level superordinate categories (vehicles, furniture, exercises, and movements) as well as lower-level subordinate categories (sports cars, barber chairs, bobble-head dolls, bowling, pitching, waltzing, and so on).[14]

It is particularly interesting that natural embodied schemas, like basic-level categories, are multimodal; that is, they link together different kinds of bodily movements and perceptions, controlled via multiple brain areas. Given our bodies and how they function in the world, it is simpler, or more natural, or takes less energy to perform certain combinations of neural activity than their separate parts. Even with monkeys, mirror neuron researchers found that certain neuronal groups fire when the monkey picks up a real object but do not fire when the monkey performs the corresponding hand action without the object.[15] Certain neural circuits are tuned to more complex cases but not less complex cases. Why?

EXPLAINING BASIC-LEVEL PHENOMENA

There are two neural phenomena that explain why basic-level phenomena exist, that is, why certain complexes are more natural than their simpler parts:

- the unity of imagination and experience, and
- the neural linkage of action and perception.

IMAGINATION AND EXPERIENCE

Martha Farah (1989) showed that much of the same neural circuitry used in **seeing** an object is also used in **imagining seeing** an object, all the way down from higher cortex to the primary visual cortex. This result has had enormous implications, especially since connections are bidirectional. Perception is not just getting sense impressions from sensory organs. An image is not just a bunch of pixels, isolated neurons activated one by one. A visual perception has the same kind of structuring as an imagined mental image, which is never fully specified pixel by pixel but has an internal structure. Correspondingly, mental images are not disembodied, formed by some wholly separate capacity or independent module, as in the old theory of faculty psychology in which perception and imagination were assumed to be simply separate faculties.

Since then it has been shown that in the brain, much of the same is true of real and imagined actions. *Imagined kicking activates much of the same brain circuitry as real kicking.*

More recently, studies have shown that mental imagery and perception may use the same cascades with reverse information flows in the brain. Researchers at the University of Wisconsin, Madison,[16] used electrical activity to identify directions of information flow when subjects were asked to imagine a scene or watch a video of the scene. They found that during imagination there was an increase in information flow from more upstream (parietal cortex) areas to visual areas in the occipital lobe. The converse is true for the visual signal, which travels from the retina through the occipital lobe to parietal and other frontal regions. *In other words, imagination results in an internal triggering of visual regions in the brain in a top-down manner, from imagining (top) to seeing (down).* Such internal triggering of neural cascades is an important part of *connecting abstract reasoning and understanding neurally to embodied sensorimotor and emotional circuits.*

ACTION AND PERCEPTION

Mirror neuron phenomena, first discovered at the University of Parma in Italy, showed that when you see someone perform an action, some of the same neurons are firing as when you actually perform the same action.[17] Those neurons are partly in brain regions controlling action sequences and partly in brain regions integrating perception. For this to happen, there have to be two-way multimodal connections linking those distinct regions. There are various hypotheses as to what the exact circuitry is, but the phenomenon is real: there are multimodal action-perception neural links in the brain.

THEY OCCUR TOGETHER IN THE BRAIN

The **neural unity of imagination and experience** and the **neural linkage of action and perception** are not separate phenomena. Gestalt perception circuitry is involved in both action-perception and imagination-perception circuits. *When the appropriate circuitry overlaps we get basic-level phenomena, defined by motor programs, mental imagery and gestalt perception occurring together.*

Under what circumstances would they not occur together, where, for example, one would perform a grasping action with nothing there to grasp or a kicking action while not kicking anything? For natural grasping or kicking to occur in such cases, one would need circuitry for detailed imagination beyond the naturally used circuitry. Such circuitry can be and is learned — consciously by mimes, dancers, and athletes practicing but also unconsciously and naturally in cases governed by **metonymy**, or symbolic action, which occurs naturally in spontaneous gesture (such as thrusting a fist upward to indicate a victory in sports) and the meaningful gestures built into signed languages.

The basic-level cases are normal and natural, because in the brain action-perception and imagination-perception circuits do overlap naturally.

The fact that mental imagery and physical experience use the same neural circuitry suggests that the same circuitry is used in conceptualizing experi-

TABLE 2. Superordinate–basic subordinate concepts

SUPERORDINATE	BASIC LEVEL	SUBORDINATE
furniture	chair	rocking chair
mammal	dog	poodle

ence and, moreover, that active neural circuitry is central to any account of conceptual thought.

A POSSIBLE EXPLANATION FOR LATERAL PREFRONTAL DATA

Basic-level concepts are not just defined by the convergence of mental imagery, gestalt perception, and motor programs. They are also defined as being in the middle between more general superordinate concepts and more specific subordinate concepts (Table 2).

The lateral prefrontal cortex is organized front (rostral) to back (caudal). Early results by Badre et al. (2009) demonstrated that rostral areas showed more general control of behavior, while caudal areas showed more specific control of behavior. Badre et al. posited a front-to-back (rostral-to-caudal) hierarchy (general to specific) of control.

But research by Nee and D'Esposito (2016) showed a more subtle result. Here is how Nee and D'Esposito describe their findings:

> Higher-level cognition depends on the lateral prefrontal cortex (LPFC), but its functional organization has remained elusive. An influential proposal is that the LPFC is organized hierarchically whereby progressively rostral areas of the LPFC process/represent increasingly abstract information facilitating efficient and flexible cognition. However, support for this theory has been limited. Here, human fMRI data revealed rostral/caudal gradients of abstraction in the LPFC. Dynamic causal modeling revealed asymmetrical LPFC interactions indicative of hierarchical processing. Contrary to dominant assumptions, the relative strength of efferent versus afferent connections positioned mid LPFC as the apex of the hierarchy. Furthermore, cognitive demands induced connectivity modulations towards mid LPFC consistent with a role in integrating information for control operations. Moreover, the strengths of these dynamics were related to trait-measured higher-level cognitive ability. Collectively, these results suggest that the LPFC is hierarchically organized with the mid LPFC positioned to syn-

thesize abstract and concrete information to control behavior. (Nee and D'Esposito 2016)

In short, Nee and D'Esposito found that the mid-LPFC

- has a special status *between the more general front* (rostral) and *more specific rear* (caudal),
- *integrates* information and *controls* operations,
- has *stronger efferent (outward-going) connections* than afferent (inward-coming) connections,
- is correlated with higher-level cognitive ability, and
- synthesizes abstract (more general) and concrete (more specific) information to *control behavior.*

One possible consequence of this finding is that *the midlateral prefrontal cortex is the locus of control for basic-level concepts* that integrate mental imagery, gestalt perception, and motor control and lie between superordinate and subordinate concepts, not just conceptually but also anatomically in the brain.

At this writing, this hypothesis, linking basic-level concepts with the mid-LPFC apex of the LPFC anatomical hierarchy, has not been researched and remains only a suggestive hypothesis.

Sound Symbolism

Primary embodied schemas are real in experience: we consciously perceive **generalized containment** in a wide variety of "containers"—drawers, windows, rooms, buildings, doorways, cups, pots, bottles, bushes, caves, valleys, and city squares—allowing the use of the words "in" and "into." How the brain accomplishes this is a deep problem for neuroscience.

Unconscious neural integration, as we have just seen, is also real. Some remarkable experiments link the study of unconscious integration in language and conceptual systems to a further question for neuroscience, an extension of the question of *how primary embodied schemas work.* These phenomena seem to work together to produce a linguistic phenomenon: sound symbolism.

WHAT IS HIGH ABOUT HIGH TONES?

As early as the 1930s, C. C. Pratt (1930) and O. C. Timble (1934) studied the "spatial character" of "high" versus "low" tones. Literally, such tones arise from higher versus lower frequencies of sound, that is, *more versus fewer sound waves per unit of time.* There is no literal "height" involved.

Yet, actual height correlated with high tones does show up in a set of remarkable experiments done in recent years, such as one experiment by Maeda, Kanai, and Shimojo (2004). Their experiment used what vision scientists call **gratings**.

Imagine a computer-generated sine wave moving left to right on a computer screen. Now imagine such a sine wave moving right to left. Next, imagine the computer superimposing these two waves moving in opposite directions at the same speed. There is no preferred direction. That is a grating.

Take this as a visual stimulus. Now turn it 90 degrees so that it is oriented vertically, not horizontally. Again, there is no preferred direction. If subjects are asked to make a forced choice—is there motion up or down?—their responses are random, as one would expect.

What would happen if a rising tone (greater frequency, more sound waves per second) or a falling tone (lesser frequency, fewer sound waves per second) were presented with the motion-neutral grating either at same time (a zero millisecond delay) or with a delay of 100 milliseconds, a time within which unconscious neural integration takes place?

When the grating was presented at those times with an ascending (progressively greater frequency) tone, subjects perceived, that is, they consciously saw the neutral grating moving upward. The maximal effect was slightly after the grating was shown but within the 100-millisecond range that allows for conscious neural integration. In such cases, their brains encountered the motion-neutral grating first and then heard the sound a bit later within 100 milliseconds. The *slightly later* encounter with the sound affected the perception of the *slightly earlier* visual onset of the grating, adding motion that was not there in the external world. Similarly, when subjects were presented with a descending tone, they "saw"—that is, their brains *created a perception* of—the motion-neutral grating moving downward. This also worked with rising and falling intonations of various types of sound.

To test whether linguistic semantics was involved, they used (with Japanese subjects) the Japanese words *ue* meaning "up" and *shita* meaning "down" in place of the rising and falling tones. This time there was no significant difference between the two groups within 100 milliseconds; the meanings of the words had no effect within the 100-millisecond unconscious integration range. But when the words were presented after *a* 400 millisecond gap—known to allow a semantic understanding of the word, not just a perception of their sound—then the different meanings of the words had a slight positive effect. That is, when the words were perceived as just sounds without meaning, there was no perceived tone difference, and no motion was perceived.

Incidentally, when the grating was gradually rotated from vertical to horizontal position, the effect decreased, disappearing at the horizontal position.

The dramatic conclusion: Rising and falling tones *in sound* within the time range of unconscious neural integration causally affects visual perception *in space* in the corresponding rising and falling directions even when presented the sounds are presented slightly later than the visual presentation.

ATTENTION SHIFT

Rocco and Rich (2012) showed that "high and low tones induce attention shifts to upper or lower locations" (p. 339), depending on the degree of pitch height. As the pitch was raised, there was greater attention shift. There was no effect on horizontal attention shift.

Not surprisingly, cognition (consciously deciding to look at a certain location) can override such automatic attentional effects of rising versus falling tone.

PRELINGUISTIC INFANTS SHOW THE EFFECT

Walker et al. (2010) and Dolscheid et al. (2014) have shown that this effect occurs in prelinguistic infants. The Dolscheid et al. studies were done with Dutch four-month-olds. The infants, with no prior looking preference in the experiment, look upward significantly longer when presented with high tones and downward significantly longer when presented with low tones. In short, the cross-modal correlation between *high tone and attention upward* versus *low tone and attention downward* is there at four months, well before language.

THE MOTOR CORRELATION

Berkeley phonetician Keith Johnson made an interesting suggestion.[18] *When one is speaking in a high or rising tone, the glottis moves up. When speaking in a low tone, the glottis moves down.* Prelinguistic infants make rising and falling sounds with their glottises moving up and down. Of course, neither the prelinguistic infants nor most adults are consciously aware of the glottal motion, just as they are unaware of the attentional effects. But Johnson has suggested that the upward versus downward motion of the glottis may account for the difference between the perception of the tones as being high or low.

IS THERE A SPATIALIZATION OF THE MOUTH?

For glottal motion to be the explanation, there would have to be a *perceived spatialization of the mouth*. Our observations about the primary embodied schemas would have to apply to not just external objects (cups, windows, buildings, etc.) but also one's own body, to one's own mouth! The McGurk effect suggested that this might be true. *The perceived "da" is **pronounced in the mouth** along a continuum between the seen "ga" and the sounded "ba."* Are there other phenomena that suggest that we unconsciously spatialize our mouths?

The answer is yes.[19]

BLIPS AND BLOOPERS, CLIPS AND SIPS AND STRIPS

Richard Rhodes and John Lawler (1981) in their classic paper "Athematic Metaphor" cite a comment in the *Oxford English Dictionary* by James Murray, who originated that monumental work. Murray was working on words beginning with *bl-* such as "blow" in an effort to find Indo-European roots for them. In frustration he gave up, writing that none of them could be found earlier than the development of Germanic from Indo-European. Murray had come up against sound-symbolic words, words that arise at a given point in time because of sound-symbolic effects.

Syllables are divided naturally into two parts: onsets and rimes. The onset is the initial consonant or consonant cluster, such as the *bl-* in "blow," "blip," "bleep," "blab," "blurb," "blur," "blush," "blob," "blot," "blotch," "blister," "blast," "bladder," "blather," "bluster," and "blubber." The rhyme is *-ip*, as in "clip," "nip," "drip," "trip," "rip," "dip," "flip," "whip," "slip," and "sip." The *bl-* words *have an expanding outward motion*, as when *pronounced in the mouth*; the *-ip* words have a short path to a sudden stop, which is *what happens in the mouth*.

"Blip" is a word that came into existence with radar in the 1940s. A blip was the track of an airplane on a radar screen: a barely visible small burst of light that lit up, getting wider and lighter for a second and then dissipating. *Bl+ip* was a natural word for it: a widening burst of light (*bl-*) extending for a short time (*-ip*).

Then there are *-ap* words, pronounced flatter in the mouth. The *-ap* words have the meaning of *something coming in contact with a flat surface*: "rap," "tap," "clap," "snap," "flap," "slap," and so on. *Cl-* words are pronounced with *an opening and then coming together of the top and bottom of the mouth*. *Cl-* words fit that mouth shape: "clap," "clip," "clasp," "clump," "cleats," "cluster," "cloister," "close," "cleave," "cleavage," "claw," "clench," "clot," "clutch," and so on. *Cl+ap* is a perfect sound-symbolic word: *the flat surfaces of the hands coming together*.

The *str-* words have been cited by grammarians since the seventeenth century and have the meaning of *long and thin*: "string," "strip," "strap," "stripe," "streak," "stroke," "straight," "strait," "stretch," "strew," "stride," "striation," "stray," "strafe," "strop," "strum," and so on. *Str+ap* refers to an object that is *long and thin and is flat (pressed or hitting) against a flat surface*. "Strap" is a perfect sound-symbolic word.

We could go on, as Rhodes and Lawler do, with example after example of cases of either onsets or rimes whose meaning fits an image schema that structures the mouth. Then there are the **phonaesthemes**, sounds that fit a certain content associated with their pronunciation. Bergen (2001) includes the following lists, based on the Brown University corpus and Webster's *Seventh New Collegiate Dictionary*:

- *gl-* "light, vision": "glimmer," "glisten," "glitter," "gleam," "glow," "glint," "glance," "glare," "glass," "glaze," "glower," and so on; and
- *sn-* "nose, mouth": "snore," "snack," "snout," "snot," "snarl," "snort," "sniff," "sneeze," "sneer," "snicker," "snippy," "sniveling," "snorkel," "snuff," and so on.

As any linguist can tell you, these *gl-* and *sn-* sounds are not compositional, as with the "over" of "overdo" and the "re" of "redo," which combine with the meaning of "do." Bergen showed experimentally that these *gl-* and *sn-* sounds are indeed real meaning components. He demonstrated that "phonaesthemes, despite being non-compositional in nature, displayed priming effects much like those that have been reported for compositional morphemes. These effects could not be explained as the result of semantic or phonological priming, either alone or in combination" (Bergen 2001, 2004).

Phonaesthemes, like sound-symbolic onsets and rimes, have meaning: partial meaning within the meaning of the word. They do not characterize the whole meaning of the word. *Bl+ip* says nothing about radar or the metaphorical meanings that have evolved since radar. They do not predict the meaning of the word, but they have general meanings that fit the meaning. The meanings are *motivated, not predictable*. A motivated meaning for a word means that there is a reason why the meaning fits the word. Sound-symbolic words fit a pattern. The pattern is learned, and the parts are well known and embodied, making new sound-symbolic words not only easy to learn but also natural. Rhodes and Lawler point to the combination of the phonaestheme *dr-* "water": "drink," "drown," "drain," "drizzle," "drought," "dry," "drop," "dregs," "dredge," and so on. *Dr+ip*: a watery short movement to a sudden stop. Another excellent sound-symbolic word!

Given that embodied image schemas can structure visual perception, imagination, and motor control, it is not at all surprising that such image schemas could also structure the mouth.

THE POINT

Sound symbolism is embodied, cross-modal, and unconscious and uses embodied schemas that occur in visual perception, imagination, and motor control. Sound symbolism is part of the structure of language. As we shall see throughout this book, those are normal properties of conceptual thought and language.[20]

Creating Perception: Seeing, Feeling, and Hearing

We began this book with a fact that is not known to the general public, though it is well known in the brain and cognitive sciences. Most thought is unconscious. The ballpark estimate extends to around 98 percent, according to neuroscientist Michael Gazzaniga (as cited in Rock 2005). The neuroscientist Stanislas Dehaene (2014) in *Consciousness in the Brain* reviews experimental evidence for the overwhelming role of unconscious thought versus conscious thought. Nobel Prize winner Daniel Kahneman (2013) in *Thinking, Fast and Slow* documents massive experimental data showing how people make decisions unconsciously on the basis of fast, automatic, effortless, unconscious thinking. Kahneman's book, like this book, seeks to make us conscious of the extent and power of unconscious thought so that we can at least notice it and possibly not be misled by it. And Ezequiel Morsella and colleagues (2016) in "Homing in on Consciousness in the Nervous System: An Action-Based Synthesis" argue that "the primary function of consciousness is well-circumscribed, serving the *somatic nervous system*" (p. 1). In short, most of what we do is directed unconsciously, with consciousness playing a crucial but limited function. This assessment is very much in line with the results of cognitive linguistics. Most conceptual understanding is unconscious, as is most of the structure and use of meaningful language.

Conscious thought makes use of unconscious thought but not in a direct way. The brain integrates unconscious thought on its way to consciousness, often changing it significantly—even backward in time—often to fit existing patterns, and that the process of integration and even the fact of integration are not consciously noticed. Only the result becomes conscious.

To illustrate this, we begin with three experiments from perceptual research, brought to our attention by vision scientist Professor Shinsuke Shimojo (2014) of Cal Tech. Our point is to show that such processes are normal. Moreover, they apply not just to vision and not just to perception but also to everything you can be conscious of.

EXPERIMENT 1: BEEPS AND FLASHES

Subjects are shown a very short flash of light at the same time as they hear a beep. Later they are shown the same single flash of light accompanied by two beeps, 50 milliseconds (one-twentieth of a second) apart. Subjects *see two flashes accompanying the two beeps* even if they are told there is going to be only one flash. This only works for very short periods of time, such as 50 milliseconds.

The brain has unconsciously made one flash of light into the two that are consciously perceived. This is a case of unconscious neural integration, with the single flash integrated with each of the two beeps even though the flash occurred before the second beep. It doesn't work in the opposite direction. Fifty milliseconds is not enough time for the visual system to distinguish one flash from two flashes.

Moreover, the beep is presented to the acoustic cortex, which is in a very different brain region than the visual cortex, where the brain reacts to the light. In order for the flash to be "bound" to both beeps in the acoustic cortex, there must be an activated neural connection linking the visual cortex to the acoustic cortex (a neural binding). The neural binding must last at least 50 milliseconds for the single flash to be bound to both beeps.

What do we learn from this? *There is multimodality: neural connections from one sensory modality (vision) to another (sound).* Thus, there must be a visual-to-acoustic cortex binding circuit. Vision is neurally linked to sound. The binding must last at least 50 milliseconds. And the unconscious integration must take more than 50 milliseconds.

Subjects were briefly shown an oval black-and-white finely checked pattern. Using transcranial magnetic stimulation (TMS), a focused magnetic field is then applied very briefly to the subject's primary visual cortex. This results in a small oval mask where the perception is inhibited within the large patterned oval in the input. The subjects consciously perceived the large checked oval containing a small, darkened oval region (the masked region is called a **scotoma**) where the TMS is applied. The scotoma appeared like a small dark hole in the larger oval. This is repeated several times. Then *for five seconds* a large red oval is shown. Subjects *consciously perceived the small oval scotoma (the hole) as being red* even though the large red oval is shown *after* the scotoma TMS was shut off. The *large* red oval, shown for five seconds, was not consciously perceived at all. The red region was *entirely integrated backward in time* to fill in the *small* oval "hole" made by the TMS and otherwise was not present in consciousness.

In this case of *unconscious neural integration*, the brain unconsciously used the large red oval to fill in the scotoma hole backward in time! The 80 milliseconds was enough time for the subjects to perceive the large striped oval containing the scotoma, but *after* the 80 milliseconds, *the brain neurally bound the red of the red oval to the scotoma hole.* The red of the hole appeared consciously to subjects as if it were filling in the scotoma hole at the same time the hole appeared.

What do we learn? First, since the *color* red is activated in a somewhat different brain region than the *shape* of the scotoma hole, there must be a neural binding between the color region and the shape region. Such color-to-shape bindings have been known for some time. Second, the unconscious neural integration leading to consciousness must take more than the 80 milliseconds that the striped oval was shown, more like 100 milliseconds.

EXPERIMENT 3: THE CUTANEOUS JUMPING RABBIT

An apparatus (1) touches a spot on the forearm of the subject. Then, after a short time (40–200 milliseconds), the apparatus (2) touches *the same spot*. After the same short time, the apparatus then (3) touches the forearm a short distance away. The subject consciously experiences *the second touch as being in space (not time!) between the first and the third touch and not at the same spot as the first touch.* It is as if the second touched spot had jumped to a location on a straight line between the first and third spots.

We learn from this that (1) the unconscious neural integration can take up to 200 milliseconds; (2) the real second tap is inhibited from being registered in the brain in its real location; (3) the straight line pattern determines that the second tap will be on that line; (4) the time difference determines the location of the tap, as if the tap location *were moving at a constant speed*; and (5) the third tap determined the final location of the second tap, so the second tap is "moved" to its new location *backward in time* after the third tap.

This is akin to the well-known "McGurk effect." In the classic experiment carried out by Harry McGurk and John MacDonald (1976), the sound input to the subject is a sequence "ba ba ba," and the simultaneous visual input is the video of that person saying "ga ga ga." What the subject hears is da da da. The b in "ba" is pronounced at the front of the mouth with the lips, the g in "ga" is pronounced at the back of the mouth, and the d in "da" is pronounced in between. The brain *creates* a sound *da* that is *not said* but is "heard" between what is said (ba) and what is seen (ga). A slightly different version was performed by Lawrence Rosenblum (2005) and colleagues and is illustrated in a BBC video on YouTube at https://www.youtube.com/watch?v =G-lN8vWm3m0. In this version, the subject is again present with a speaker who is saying "ba ba ba," but this time the subject is shown a video of the speaker saying "fa fa fa." In this case, the subject *hears* not the *spoken* "ba ba ba" but rather the *seen* "fa fa fa." The explanation given is that f is pronounced with the lower teeth seen against the upper lip and that this is so noticeable (salient) that it overrides and changes the actual sound.

Note that in the Shimojo beeps-and-flashes case, the number of *seen* flashes is determined by the *sound* of the beeps. In the Rosenblum version of the McGurk effect, the reverse occurs: the *sound* is determined by what is *seen*. In the original McGurk case and in the Shimojo jumping rabbit case, the result is intermediate, fitting a linear order pattern.

THE POINT

We are going to be discussing complex thought and language, not just flashes of light, filled-in colors, taps on the arm, and McGurk cases. But many of the same phenomena occur with thought and language. It takes time for the brain's neural circuitry to activate a complex word meaning once a word is heard. There has to be circuitry connecting the word sound (or spelling) to the meaning, and if the meaning is complex, the circuitry will have to branch to different brain regions. Yet, the meaning will commonly be *consciously understood all at once*, as if no signal passed to distant complex branches over time. The time course will be somewhere in the hundred millisecond range. But to conscious perception, it is as if there were *no process* and *no time course*. The *result* of the activation and the changes is all that is consciously perceived.

CONSCIOUSNESS IS A PRODUCT OF UNCONSCIOUS INTEGRATION

There is no general theory of what consciousness is. And we don't offer any in this book. In fact, first-person subjective experience still poses considerable mysteries to current neuroscience. But from results cited above and what we hypothesize below, there seem to be some extremely basic conclusions about some of the properties of consciousness despite the lack of a general theory:

- Consciousness is highly structured (by its unconscious structure before, during, and after integration).
- Consciousness is unconsciously integrated; its structure and integration mechanisms are not consciously experienced.
- Conscious thought, like much unconscious thought, is in the result of an integration process that extends across multiple, often widely separated, areas of the brain, linking multiple areas of embodied circuitry. This clearly includes the cortex but also importantly many subcortical, brain stem, and peripheral regions.
- *Thus, there is no single brain location for a conscious thought.*
- Rather, maximal activation showing up in fMRIs may be integration points where complex circuitry converges, sometimes called **convergence zones**.
- There are time constraints on what can be conscious. Minimal conscious experiences seem to be at or around the 100 millisecond range. Some unconscious integration can take up to 400–500 milliseconds to become conscious. The most radical experiments on the time difference are in the domain of intention and action. Using fMRI, Bode et al. (2011) tracked the unconscious generation of free decisions using ultrahigh field fMRI. They were able to decode the outcome of free decisions several seconds (2 to 4 seconds) *before the subjects were consciously aware of the decision!*

The point of these examples is that unconscious neural integration exists. Although we don't model these particular visual phenomena in this book, the idea of unconscious neural integration is central to all of our hypothesized neural cascades for thought and language.

Thought—both unconscious and conscious thought—is real, physically real in the neural systems of our brains and experientially real in that we really do have conscious experiences of thinking. Conscious experience in itself matters enormously. What we consciously experience affects much of what we do, and so conscious experience can bring about real events. That is, conscious thought is causal, just as unconscious thought is causal. Because *conscious experience is an integrated form of the neural unconscious*, the functioning of the neural unconscious regularly has a causal effect on how our behavior changes the world.[21]

The neural unconscious is thus powerful in its ability to change reality to fit a pattern. Understanding how neural systems think is a deep scientific enterprise. But it is also urgent, because *the neural unconscious, working socially and politically, is constantly creating a reality that can be very scary and needs to be understood.* This book is about science, but its ramifications are about power.

Language has power. Qualitative experience has power. The cognitive unconscious has power. These are real sources of power whose effects in the world are underappreciated. The detailed mechanisms of thought are important in creating our reality, and they influence the social and political world we live in. Please keep that in mind as you read this book, especially in the more technically detailed sections. These technical results are crucial in order to understand the relationship between the mechanisms of mind and the sources of social and political power.

The Takeaways

EMBODIMENT: A TRANSFORMATIVE IDEA

It is worth stopping at this point to integrate and make sense of the basic points of this chapter. The results we have just cited fly in the face of a huge amount that most of us have been taught about thought and language, taught by our public culture, reinforced by popular ideas, and officially taught in our schools and research universities. To make total sense of the embodiment of thought and language is to greatly transform our understanding of ourselves and our cultural and intellectual institutions.

To begin to see why, let us review some of the new ideas in small steps.

THE BODY IS NEURAL

We generally think of the body as flesh and blood, bone and muscle, heart, stomach, guts, and so on. Yet, the body is neural through and through, with connections between the somatosensory cortex and all of your skin and every place you can feel, between motor cortex and every muscle in the body, between each flexor and its corresponding extensor muscle, between the pain center of your brain and every place where you can feel pain, and between the insula and your visceral organs. The idea of neural embodiment starts with these basic facts about the vast network of neural connections between brain and body.

THOUGHT CANNOT BE LOCALIZED IN THE BRAIN ALONE

*If all thought just occurred in the brain without connections to the body, there would be nothing to think **about**.* Without connections to the eyes, there could be no seeing. Without connections to the ears and mouth, there would be no speech or hearing. Without connections to the muscles, there would be no movement. Without connections to the somatosensory cortex, there would be no feeling. And without anything in the external world to think about, there could be no concepts. A brain isolated in a vat could not think, because it would have nothing to think *about!*

THE BODY STRUCTURES THE BRAIN

The brain is structured to optimize connections to the body. The attentional and control centers are toward the front—the prefrontal cortex—and the connections to the body are peripheral, extending outward from the front of the brain to the periphery.

TOPOGRAPHIC MAPS

Topographic maps are an example of neural embodiment.

A topographic map in the brain is a two-dimensional region whose neurons receive connections from other regions of the body, some of which are three-dimensional. What makes the regions topographic is that they preserve closeness of something, often physical closeness.

The primary visual cortex, V1, is a topographic map of the retina. Two *adjacent* connections leaving the retina are *adjacent* in V1. The motor cortex is a topographic map of the body, with *adjacent cortical locations connected to adjacent muscles*. For example, *the neurons controlling two adjacent fingers are adjacent in the motor cortex. The somatosensory cortex is a topographical map of connections to the skin: if adjacent spots on the skin are touched, adjacent positions in the somatosensory cortex are activated.*

Those are large extended maps. There are also small two-dimensional regions that are topographic maps of motion that is seen, where adjacent points on the path of motion are adjacent in the cortical map. Studies of phantom limbs have revealed that there are topographic maps along the arms and legs. Someone who has lost part of an arm and still has feelings as if he had fingers have been shown to have topographic maps in the arm linking the brain to fingers. Their role, we hypothesize, is to make sure the adjacent brain regions connect to the corresponding adjacent fingers.

REPURPOSING THROUGH EVOLUTION

Human neural circuitry has been inherited from animals via evolution and "repurposed" for thought. Human thought and language use the same basic circuitry types that are used in the perceptual-motor systems of both humans and other animals.

LEARNING = NEURAL RECRUITMENT + SYNAPTIC STRENGTHENING

We are born with between 85 billion and 100 billion neurons in our brains and with between 1,000 and 10,000 connections for each neuron. That translates into hundreds of trillions of connections, with each connection between neurons actually being realized not as a physical connection but rather a chemical connection established in the chemical flow between neurons. Many of those connections in chemical flow are randomly distributed throughout the neural system, throughout the body and the brain. What we conceptualize as a "circuit" comes into being when that random fluid distribution allows for a meaningful action, such as kicking, turning over, and thumb-sucking in the womb.

When those occur and are repeated, synaptic strengthening occurs, and circuits form through use. When this happens, we say that a useful random circuit has been *recruited* to perform a meaningful action, and that repeated use results in synaptic strengthening, which occurs via chemical action. If sufficiently repeated, a fixed circuit will be formed. All neural learning works like this, including the learning of systems of thought and language. This is a natural selection mechanism, sometimes referred to as neural Darwinism.

WHY ARE THERE CONCEPTUAL AND LINGUISTIC GENERALIZATIONS?

Because of the similarities in human bodies, human neural processes, and human environments around the world, the same general types of conceptual systems and types of linguistic systems have evolved.

We hypothesize that complexes of embodied primitive concepts (containment, goal-oriented motion, etc.) form the basis for the formation of general concepts.

Because neural circuitry is hierarchical and optimized, general (higher-level) circuits branch into multiple other (special case) circuits. Each special case circuit makes use of the general circuit and adds specific constraints to that general circuit. In neural recruitment and learning, it takes less energy and shorter connections to extend an existing general circuit via branching to special case circuits than to form entirely new circuits from scratch for each special case. The **optimization of energy use** is why we have generalizations in thought and language.

THE BRAIN IS MULTIMODAL

A modality is a way to interact with the world. Our modalities include vision, movement, touch, hearing, imagining, emotions, etc. The brain has connections across modalities so that you can move on the basis of what you see, and your emotions (your feelings) are tied to embodied experiences, such as hot and cold, fine motor control or the lack of it, clear or unclear vision, undergoing or exerting force, visceral sensations, and so on.

Much of thought has a multimodal basis. What are called "basic-level concepts" are defined relative to three modalities—**mental imagery, gestalt perception, and motor programs,** as we shall see below. *Multimodal concepts require neural circuits that extend across the brain from one area to another. They cannot all be located in any one area.* This is true of the range of human ideas and meaningful language in general. In short, multimodality is normal in the human brain. Thus, reason—conceptual thought—is not located in any one region in the brain. *There can be no such thing as a local region, a "module," in the brain responsible for all human reason.*

Basic-level concepts are defined by *the correlation of mental imagery, gestalt perception, and motor control*. Why should there be such a correlation? As Martha Farah (1989) discovered, mental imagery uses the entire visual system down to V1. There is a common circuitry for gestalt perception and mental imagery. Mirror neuron experiments show that there is common circuitry linking the visual and motor systems: when you see someone else move, neural regions in the brain mirror those movements by providing neural correlations to the corresponding regions in your brain. In short, mental imagery, gestalt perception, and motor control use common circuitry. That is why they come together in basic-level concepts.

Basic-level concepts also occur at a level of generality *midway between* the superordinate (higher-level) concepts and the subordinate (lower-level) concepts. Research on the LPFC indicates that the most general concepts occur toward the front (rostral), with the more specific concepts toward the back (caudal), *with the apex of a hierarchy in between the front and back, in the mid-LPFC*. This suggests that the mid-LPFC may be the locus of basic-level concepts.

GLOBALITY: NO WAREHOUSE THEORY

The neural mind is global, not local. The old warehouse theory that ideas are localized in brain modules is false. That is, the neural circuitry for an idea does not reside in one place in the brain that "lights up" on an fMRI when you have that idea. Although neural control for an idea may be localized, just about every idea of any complexity is characterized by complex circuitry that extends across brain regions beyond the locus of control.

We see this in neural binding, where two neural ensembles separated by a distance fire in sync. The effect is that the synchronously firing ensembles are taken as characterizing the same entity. The reason is that *other circuitry in the brain cannot tell the difference between the firing of one and the firing of the other*. Synchronous firing can only occur when there are connections going in both directions linking the two ensembles so that the firing of each tends to make the other fire.

Such neural binding requires a gate—a gating node—that, when activated, turns the flow of activation between the ensembles on or off. Other neural circuits have control nodes that determine whether the whole circuit, often extending across the brain, is turned on or off. Those gating nodes and control nodes can themselves be turned on or off via cascade circuits that extend across brain regions. Gating, controlling, and cascading are all global neural phenomena.

CONVERGENCE ZONES

Convergence zones are where cascade circuits crisscross as they extend across the brain. There is a large amount of evidence from patients with brain lesions for a great number of convergence zones in the brain. This evidence destroys the warehouse metaphor for memory and thought. Under the extremely common warehouse metaphor, memories are stored in the brain, in some local region. In another version of the warehouse metaphor, each bit of knowledge is stored in some specific location in the brain. In still another version, the lexicon and the grammar of a language are each stored in some brain location. And fMRI activation is commonly thought of as indicating the place in the brain where a memory or piece of knowledge is stored. The brain does not work according to the warehouse metaphor.

SOUND SYMBOLISM

Now consider sound symbolism, where image schemas (types of universal embodied primitive concepts, such as containment and motion toward a goal). In sound symbolism, minimally meaningful sounds have semantic content, structuring the mouth as it pronounces sound sequences. For that to happen, there must be cascade circuits linking the relevant parts of the mouth with circuitry for image schemas. Only then can an image schema impose its structure on the mouth. A common example is the *str-* sound for long, thin things, such as straight lines, strings, strips (strips of paper, air strips at airports, etc.), stripes, the process of extruding, and so on. Another example is the *-ip* sound for short paths to a sudden end, as in drip, slip, clip, sip, blip, grip, dip, whip, snip, tip, and so on.

WHAT DOES NEURAL PLASTICITY DEPEND ON?

Neural plasticity is the capacity for neural change that leads to a change of function. The degree and kind of plasticity depends on

- the number of synapses to be changed,
- the connectivity among the synapses to be changed,

- the physical distance between one synapse and another that is undergoing changes,
- the chemical composition (proteins, neurotransmitters, ions) of a synapse,
- the sum of the changes needed per synapse, and
- the regularity of the firing of those synapses.

First, an existing circuit can become part of a new network; hence, we tend to use what we already know. Why, in terms of the physics of neural networks? Because of conservation of energy. It takes less overall energy to use existing knowledge for many purposes than to create new neural circuitry for each new purpose.

Second, the closer the new network is to the already existing network, the more likely recruitment will occur, because there will be fewer synapses to change. Third, the more the network has been used, the greater its overall synaptic strength will be already, and the less there will be to change in that direction. And fourth, the less inhibition encountered by the network to be changed, the more the regularity of its firing, and the more likely the change will be.

THE WHAT AND WHERE PATHWAYS

In the primary visual cortex at the rear of the brain, there is a pathway split with information about entities (e.g., color, shape, solidity) being computed along the what pathway in one direction (to the side and down). Information about motion, location, etc. is computed along the where pathway, a pathway up and to the middle. Both pathways go to the front of the brain, the prefrontal cortex. There are interconnections between the two pathways along the way as well.

DUAL MODELING: THE BRIDGE

Modeling of neural systems by computational researchers is done with formal mathematical structures computing with numbers. These numerical structures map onto important aspects of what is being modeled. Take, for example, computational weather modeling. It begins with prior meteorological weather modeling that theorized that the important sequences of numbers in a weather model are sequences for temperature, barometric pressure, humidity, wind velocity, and wind direction for given times and local areas.

Computer models start with that meteorological model, divide areas up into square miles, and do computational weather modeling based on what those numbers are at given times.

The computational model, given the output of the meteorological models, is a theory of the weather. The model, of course, is a model, not the weather itself. The weather can get you wet; the model cannot.

We are using is a two-level dual model: a model that models two different phenomena simultaneously and thereby forms links between them.

The two different phenomena modeled are

- thought and language within cognitive linguistics, and
- functional neural structures for embodied neural processes.

Dual modeling is a computational system capable of doing both modeling jobs at once. A dual model is a single computational system whose symbols can each be meaningfully mapped onto two distinct kinds of phenomena:

- For thought and language, the computational system must be able to model meaning (semantics, such as frames and conceptual metaphors) as well as grammar and the lexicon (words and morphemes as well as hierarchical and linear structures and their meanings).
- For the neural system, the computational system must be based on models of neurons and neural circuitry, including such things as neural connections, synaptic strength, firing thresholds, input activation and inhibition, output activation and inhibition, types of circuit structures, gating, binding, and so on.

For this to happen, there must be two mappings: (1) a mapping between the computational system and the cognitive linguistic model of thought and language and (2) a mapping between the computational system and the functional model of the neural system. These mappings must provide a theory of how the details of thought and language are carried out via the details of neural circuitry.

BEYOND THE DUAL MODEL: NEUROANATOMY AND THE USE OF THOUGHT AND LANGUAGE

Neuroscience is, of course, concerned with neuroanatomy. Here we depend on the expertise of neuroscientists who investigate and discover the anatomical and functional connectivity of the brain. The neuroanatomy supplements the bridging theory in important ways. It matters where the *prefrontal* cortex is located in the brain relative to *peripheral* brain connections to the body. For example, we are guided in our modeling by the discovery in neuroscience that more general circuitry is more to the front (rostral) area of the brain, while the details of special cases (caudal) are

- in the details of corticothalamic loops and the basal ganglia,
- in the links between the acoustic cortex and the articulatory cortex in phonetics, and
- in the existence and nature of convergence zones.

We believe that neuroanatomical research into such details must fit with our bridging theory to yield an account of how thought and language are carried out anatomically. Research in both areas—neuroscience and corresponding thought—is necessary but is rarely done at present.

A neural understanding of thought and language matters for social and political reasons. George Lakoff for two decades has been applying versions of the neural mind theory to social and political issues, charting the details of the two worldviews now dividing America and many other nations. The issue is this: Are generalizations about social and political views reflected in generalizations about neural circuitry? For example, are there generalizable brain differences between liberals and conservatives? It's an important and interesting empirical question.

THE CREATION OF OUR EXPERIENCE OF REALITY

We started this chapter where embodied cognition is most powerful, with the role that our neural systems play in the creation of our experience of reality and hence our understanding of what reality is.

We saw a result that is initially shocking to many people: that there is no color in the external world. Objects reflect multiple isolated wavelengths of light, and those wavelengths are converted into what we see as color via color cones in the retina attached to neural circuitry in our brains. We also saw that the operation of color cones is affected by X chromosomes and that since typically, females have two X chromosomes and males have one X chromosome, fine differences in color creation in the body and the brain often vary with gender. In other words, men and women looking at the same object may actually see different colors because the chromosomal differences in their color cones create colors differently for them. And even more dramatically, some women have four color cones and can distinguish roughly a million more shades of color than men can.

We ended the chapter with Shinsuke Shimojo's remarkable survey of experiments where perceptual inputs to vision, touch, and hearing are systematically changed by our neural systems so as to fit existing neural structures. This is not simply the *creation* of otherwise nonexisting external reality (as with color) but rather the *change of the perception of existing external reality to fit neural reality*.

NEUROPHILOSOPHY

Much of the passion for the research in this book was inspired by Patricia Churchland's (1986) great book *Neurophilosophy: Toward a Unified Science of the Mind-Brain*. Churchland pointed out that taking neuroscience seriously would utterly change not just professional academic philosophy but also our ordinary folk theories of the mind (pp. 481–82). She presciently observed that "political and entrepreneurial abuses of scientific knowledge" have led to "a catastrophic disregard for humanistic principles" (p. 482), which has unfortunately become all too obvious over time.

Churchland also observed that taking up such philosophical issues as the nature of thought is necessary for the development of neuroscience "because ongoing research must have a synoptic vision within which the immediate research goals make sense" (1986, p. 482).

Churchland pointed out that "neuroscience may even teach us a substantial thing or two about how science and mathematics are possible for our species" (1986, p. 482). That idea was followed up on and significantly realized in *Where Mathematics Comes From* (Lakoff and Núñez 2000), which took inspiration from neuroscience research by Stanislas Dehaene (1997) in *The Number Sense*.

As Churchland concludes, "So it is that the brain investigates the brain, theorizing about what brains do when they theorize, finding out what brains do when they find out, and being changed forever by the knowledge" (1986, p. 482).

2

How Thought Works

From Motor Control to Thought

Motor control is as embodied as any human neural mechanism: it controls everything we can do by moving a body part. The discovery that motor control is central to thought was not unexpected, given other evidence of the embodiment of mind, but it was certainly dramatic.

The neural circuitry for human motor control—especially the goal-directed coordination of complex motor actions—is like that of other higher animals. Via evolution, human beings have repurposed (exapted) the neural circuitry for motor control. It now serves two human purposes:

- understanding actions and events in the world conceptually in terms of what we can do with our bodies, and
- allowing us to reason and structure arguments and to provide a structure for language.

These ideas have come out of the Neural Theory of Language (NTL) project at the International Computer Science Institute at the University of California, Berkeley, over the past two and a half decades. Both of the present authors were members of this research group originally founded by George Lakoff and Jerome Feldman. The idea of using computation to study the relation of sensorimotor control to thought and language came from Feldman (2006).

Feldman and his then graduate students in the NTL group wanted to construct a computational neural learning program that could learn the verbs of basic hand motion from a model of how people moved their hands in performing certain tasks. They worked with a computational model of the body, with the bones, joints, and muscles. The model was named Jack and was designed at the University of Pennsylvania by Norm Badler and his group. Unfortunately, Jack didn't come with a program to make it move via verbal interaction, and one was needed.

The problem was formidable. Here is what the NTL group had to consider. First, there are lots of verbs of one-hand motion: seize, snatch, grab, grasp, pick up, hold, grip, clutch, put, place, lay, drop, release, pull, push, shove, yank, slide, flick, tug, nudge, lift, raise, lower, lob, toss, fling, tap, rap, slap, press, poke, punch, rub, shake, pry, turn, flip, rotate, spin, twirl, squeeze, pinch, twist, bounce, stroke, wave, caress, stack, salute, and many, many more.

Second, languages differ:

- In Tamil, the terms *thallu* and *ilu* correspond to the English terms "push" and "pull" except that they connote a sudden action as opposed to a smooth continuous force. The continuous reading can be obtained by adding a directional suffix, but there is no purely lexical way to indicate smooth pushing or pulling in an arbitrary direction.
- In Farsi, the term *zadan* refers to a large number of object manipulations involving quick motions. The prototypical term *zadan* is a hitting action, although it can also mean to snatch (*ghaap zadan*) or to strum a guitar or play any other musical instrument.
- In Cantonese, the term *meet* covers both pinching and tearing and connotes forceful manipulation using the two-finger posture but is also acceptable for tearing larger items when two full grasps are used. Cantonese has no distinct word equivalent to the English term "drop"; there is a word meaning "release," but it applies whether or not the object is supported.
- In Spanish, there are two separate words for different senses of the English verb "push." The Spanish word *pulsar* corresponds to pushing a button, and *presionar* covers most of the other uses.

The meanings of these verbs are concepts, concepts of hand action based on those aspects of hand action that can be conceptualized. There are six dimensions to take into account:

- There are parameters that we can be or become aware of, such as which body part is moving and how (e.g., shoulder, arm, elbow, whole hand, fingers): its degree of force, its direction of motion, its speed, whether it is instantaneous or prolonged, one's body posture, whether a motion is repeated or not, and so on.
- There are sequencing of parts of a complete action: preconditions to be met, beginning, middle, ending, final state; changes of parameters at a given stage (e.g., whether the degree of force changes); whether movements are concurrent (e.g., in grasping, the arm moves to the object as the hand sequentially forms a grip), etc. Moreover, movements can be paused and resumed or abandoned midcourse. Some movements are short and instantaneous, such as *flick* and *punch*. Others are iterative, such as *shake*. Still others are continuous, such as *slide*. Most are purposeful, but *drop* is typically not purposeful, although it can be purposeful in a wide range of contexts. Some movements are directional relative to the body, such as *push* (away from the body) and *pull* (toward the body). Some involve rotating motions such as *turn*, *flip*, *rotate*, *spin*, *twirl*, and *twist*. Some are forceful such as *yank* and *tug*. Others are gentle, such as *caress*. In short, there are lots of parameters to consider.
- You need to take account of the kind of context Jack is in. For instance, what objects in the world around Jack that might be in the way and what Jack's current arm position is.
- There are aspects of actual hand actions that do *not* form part of a hand action concept. These are subtle muscular neural mechanisms that are triggered by the above considerations but are not subject to awareness even by experts such as professional athletes. In the motor control literature, they are called **functional muscle synergies**, defined as *patterns of coactivation recruited by a single neural command signal*. One muscle can be part of multiple synergies, and one synergy can activate multiple muscles.
- Individual synergies are controlled by the primary motor cortex.[1] They are simple, such as opening and closing the hand, opening and closing the elbow, raising and lowering the shoulder, and so on. To perform a hand action, synergies have to be "choreographed," that is, smoothly combined in sequence and in parallel, and parameters and contexts must be brought in. Such choreography is performed by the premotor and supplementary motor cortices via neural connections to the motor cortex.
- The program for Jack's movements could not use a global clock, such

as standard sequential programs on a computer. Because the program was a model of neural functioning, it had to be dependent on distributed and semiautonomous techniques for local control of limbs and joints based on neural firing or spiking.

In 1992, we undertook the job of creating a neural computational modeling program that could perform simple hand actions. It had to meet the above specifications plus one more: it had to be biologically plausible, that is, consistent with known biology.

Results from biological motor control theory suggest that motor control is hierarchical and has the following features:

- Specific motor actions are carried out by functional circuits that integrate sensory and motor information to achieve specific goals.
- In the literature, these functional circuits are referred to as motor schemas. Examples include basic actions such as pushing and pulling and also more complex actions such as stacking blocks and pouring liquids. One way to think about this is to imagine what a mime would do to suggest such actions.
- Complex motor control schemas form a hierarchy. The schemas in the hierarchy bring together primitive motor schemas, or functional motor synergies. The internal action structure of those synergies is not available for direct control by the higher schemas. An individual synergy may, however, have parameters whose values result in specific behavior. For example, the action of reaching is a motor synergy, but reaching can be made fast or slow. The choice of speed can affect the synergies used.
- There is a distributed network of brain regions including prefrontal, motor, premotor, and parietal areas that, acting together, may implement a single motor schema. Motor schemas are constituted and carried out by neural circuits that occur distributed over various regions in the brain.
- Many motor schemas may be active at any given time. The triggering and activation of motor schemas takes place without a centralized clock, or controller, and only via the amount of accumulated neural activation controlling each motor schema.
- Multiple schemas may be triggered for achieving a single goal. Often there is a dynamic interaction between schemas, including competing schemas mediated by what are called winner-takes-all networks.

- Schemas perform complex coordination and control functions. Some types of coordination mechanisms include central pattern generators, homeostatic control circuits, and complex action sequencing.
- Higher-level motor schemas that put together lower-level schemas are learned via neural learning mechanisms, such as reward-modulated reinforcement, Hebbian learning, and spike-timing-dependent plasticity (STDP).

In constructing a biologically plausible system for controlling Jack's movements, these were some of the constraints the NTL group had to meet. The task of control was to build coordination networks that could combine synergies in complex hierarchical and temporal arrangements to achieve specific goals with given resources.

The NTL group succeeded in building a program that could model and learn verbs in various languages, which resulted in the PhD thesis of David Bailey. Bailey did fieldwork comparing the outputs of his program to the judgments of native speakers of a number of languages. The program did relatively well, and the results are described in Bailey (1997).

DETAILS

When building the action models, we noticed something that surprised us. The neural computational models for all the diverse cases of hand action in various languages (and for such leg actions as walking, running, stumbling, falling, and so on) *all looked essentially the same at the highest level in the action hierarchy!*

The neural computational structures we discovered in modeling motor control are called **executing networks (X-Nets)**. X-Nets have the structure of purposeful actions in general. The elements of X-Nets are actions and states (shown in Table 3). Each action sequence has a structure: preconditions, action, and relevant consequence.

For example, to take a drink, the usual precondition is that you are thirsty and near a container of liquid, say, a glass. The starting action is to reach out, grasp the glass, and raise it to your mouth. The central action is to take a drink of the liquid. The test for achieved purpose is whether you are still thirsty; if so, keeping drinking; if not, put the glass down. The relevant consequence afterward would be that you are no longer thirsty.

There are variations. At any point, there can be a pause and resumption, or one can quit. Moreover, there are additional variations. If the action does not

TABLE 3. The structure of actions

PRECONDITION	ACTION	CONSEQUENT STATE
activate	prepare	ready
ready	start	started
started	act	ongoing
ongoing and goal realized	finish	done and goal satisfied
ongoing and goal not realized	continue act	ongoing

have a purpose, there is no purpose satisfaction test. If the action goes on and does not end, there is no final state.

This structure is given by the names listed. They are names for phases of an action that *we* can understand. The motor program has a network structure and operates in time, doing things. Computer scientists use the term "execute" for carrying out an action, hence X-Nets. The X-Nets themselves just model the flow of activation in a neural circuit.

The X-Net structure is general and could apply to any action or can be adapted to structure any understood event. The result is a two-stage model: (1) the general structure given above and (2) lower-level particular actions carrying out the general structure. The particular structures consist of the hierarchies and sequences of synergies (with parameters) needed to carry out particular actions.

The Big Surprise: Aspect Is Motor Control

When we presented these results at the regular NTL meeting, linguists in the room pointed out that at the highest level of the motor hierarchy, there was a general structure that was essentially the same for all cases and that characterized a familiar linguistic structure known as "aspect" in linguistics. Aspect, conceptually, refers to the common conceptual structures of all actions and events in every known language in the world, although "aspect" is expressed differently in linguistic form from language to language.

The grammar of aspect is defined relative to X-Nets. In English, aspect is marked by *about to* (the stage after precondition, just before start); *start to* (the start stage); *be+ing*, as in *She is sipping the wine* (the central action); the purpose test stage for purposeful actions (e.g., drinking enough to be satisfied or drinking all the wine); if purpose is unsatisfied, iterate as in *She sipped and sipped*; the finishing-up stage (e.g., putting down the wine glass); the final state, as in *She is no longer sipping the wine*; and the relevant consequence, as in *She has sipped the wine*.

At the time of this discovery, the logic of aspect had never been accurately worked out by logicians. After realizing that aspect has the structure of motor control, we could then demonstrate that for each typical physical action, the hypothesized X-Net correctly computed the logic of aspect for physical action and event cases, namely at any given point in an event or action, what happened earlier is logically inferred, and what happens later is not yet logically inferred.

This works for physical events. But what about abstract events?

The subsequent work reported by Srini Narayanan (1997a) in his PhD dissertation showed how X-Nets characterize the logic of abstract events. The computational model used the Lakoff-Johnson theory of conceptual metaphor to characterize abstract concepts, using metaphors mapping from physical events to abstract events: metaphorical structures such as understanding reasoning to a conclusion in terms of moving to a goal. The metaphor is "actions are motions," with purposes as destinations. In doing so, we were able to create a computational neural theory of conceptual metaphor. Narayanan's dissertation demonstrated that the resulting theory of both physical and abstract actions and their inferences worked for a variety of simple and complex range of cases. Subsequent experimental work by Teenie Matlock (2002) at the University of California, Merced, and Ben Bergen (2001) at the University of California, San Diego, demonstrated the validity of the model and its predictions on multiple abstract domains of human discourse, including politics, criminal justice, and economics.

STRUCTURED NEURAL COMPUTATION AND THE BRIDGE

Structured neural computation was developed by Jerome Feldman in the 1980s and formed the basis of research in the NTL group at Berkeley's International Computer Science Institute. The development of X-Nets, that is, networks that carried out actions, took structured neural computation to a new level of sophistication. In the decades since then, structured neural computation using X-Nets has been developed so much further that it now functions as the bridge in our theory, as discussed in the section "Dual Models" in chapter 1 above and in detail in chapter 3.

X-Nets are structured neural computational models of processes in general. There are various specific types of X-Net activations that are relevant to linguistic aspect:

- Completed processes have final states and relevant consequences, which are called **perfective** in the field of linguistics.
- Uncompleted processes are those that continue, but no completion is indicated. These are called **imperfective**.
- Each event in an X-Net action can be either instantaneous, durative (i.e., extended), or iterative. This corresponds to the internal aspect of what has been called the "Aksionsart" of a verb (as discussed by Zeno Vendler [1957] and David Dowty [1979]).
- At each stage in an X-Net, there can be either a brief pause, a wait and a continuation, or a stop with no continuation.
- At each stage, there can be a branch into two or more coordinated processes, which can be coordinated actions, experiences, or processes in general.
- X-Nets can be combined to form sequences; to form embeddings, each with a whole X-Net embedded within a single stage of a higher-level X-Net, or can be interspersed, with the stages of one X-Net occurring at the pauses of another X-Net.
- X-Nets have a sequential part-whole structure, with the stages as parts forming a whole.
- X-Nets can combine with other sequential part-whole structures: for example, the purpose schema with a desirer, a desired state, resources, an action sequence, possible difficulties, and an achieved state. In a failure, the X-Net ends with difficulties, and no achieved state is reached.

A general X-Net thus has the structure of a schema, which can be made specific in various ways by choosing among the above choices of properties. The general X-Net schema has no filled-in specific properties. One can imagine or perceive a specific X-Net as a simulation of the entailed action with the choices among its properties filled in, sometimes with default values and other times with specific knowledge of the action and objects acted upon.

Importantly, the general concept of a purposeful action is characterized via X-Nets. **Purposes** have the following structure:

- There is a desirer and a desired state or object.
- The desirer has resources (e.g., energy, instruments, other people or animals, institutions).
- The desirer acts with the intention of bringing the desired state or object into being.
- There can be difficulties (e.g., blockages, features of the terrain such as mountains or swamps, strong opponents).
- The resources are used in the face of difficulties.
- The difficulties must be overcome or avoided for the desired state to come into being.

In the X-Net model, there is a test for the desired state right after the central action. If the purpose is satisfied, the X-Net goes on to the finishing action. If not, more of the central action is performed, either by iteration or by extended action. If the difficulty cannot be overcome or avoided, then the desired state cannot be reached. This is **failure**.

More complex cases have been worked out: for example, the logic of putting together time and aspect as well as the composition of multiple aspectual schemas in context.[2]

What is remarkable is that the same neural circuitry used to run our bodies physically can also structure our reasoning processes about *all* events and actions, not just physical ones but abstract actions and events as well. Moreover, these structures are represented as types of aspect in the grammars of languages throughout the world.

This neural computational model of hierarchical motor control appears to characterize motor control for animals as well as human beings. In order to characterize motor concepts, inferences, aspect, purpose and causation, we hypothesize that the neural circuitry for motor control in humans has been exapted, that is, repurposed over the course of evolution from the circuitry used for purposeful movement by animals. Moreover, results by Martha Farah (1989) and a subsequent study by Graybiel and Smith (2014), show that *the same circuitry is used for both acting and imagining acting*. We conclude that the same circuit types—X-Nets—are used for both actual physically carried-out motor control and a *conceptual understanding* of actions and events. That is, imagining moving uses neural circuits involved in actually moving.

In short, if our hypothesis holds up, X-Nets play a major role in thought, especially about ideas of events, events with consequences and purposeful actions. First, X-Nets structure and carry out mental inferences for all action and event concepts, both physical and abstract. Second, X-Nets are used in **simulations** to carry out *the action of reasoning itself!* This is important. *Reasoning is an action*, an action that has the same general conceptual structure as other actions.

Physical action requires movement. Thinking is a kind of mental movement, understood via the metaphor "thinking as moving"; thinking, like movement, is done sequentially in time. Examples of the common conceptual metaphor of thinking as moving include "I'm stuck," "You've gone off on a tangent," "I'm not getting anywhere on this problem," and so on. It would not be surprising if the neural circuitry for guiding physical movement were to be used as well for the action of thinking.

Image Schemas: Starting with Kant

There are types of triangles. They differ in angle size and the length of their sides:

- Right triangle: A 90-degree angle
- Acute triangle: All angles less than 90 degrees
- Obtuse triangle: One angle more than 90 degrees
- Equilateral triangle: All sides the same length
- Isosceles triangle: Two sides the same length
- Scalene triangle: No sides the same length

Examples of each type can be drawn, and the drawings can be perceived.

What makes them all triangles is what Immanuel Kant called a triangle "schema." The schema generalizes over all types but cannot be drawn. Each drawing constitutes an image. But the general case is not an image. It is an image schema. It is understood but cannot be perceived or even explicitly formed in mental imagery.

As Kant (1965) writes in *Critique of Pure Reason*, "No image could ever be adequate to the concept of a triangle in general. It would never attain the universality of the concept which renders it valid of all triangles, whether right-angles, acute-angles, or obtuse-angles; it would always be limited to a part only of this sphere. The schema of the triangle can exist nowhere but in thought" (p. 182).

The schema consists of three points (not in a line) and three lines, each connecting two of the points. The points are vertices; two lines (the sides, or rays) extend from each point, forming an angle. You can think of this as a rule or program for forming a triangle: first pick three points, then draw the connecting lines in sequence. Or you can think of it as a static part-whole structure: a gestalt in which the whole is the triangle and the parts are the points, the lines, and the relations between them.

We are concerned in this book with certain details, in particular with the detailed nature and structure of ideas and with the functional neural circuitry needed to characterize the nature and structure of those ideas. In order to point out the details that have to be accounted for, we have adopted the following notation commonplace in cognitive science for spelling out such details.

We have started using this notation with the example of a triangle schema simply because the details of the triangle schema are easy to understand, and so the notation should be easy to understand as well.

Schema: Triangle

Parts
- Vertex 1 is a point.
- Vertex 2 is a point.
- Vertex 3 is a point.
- Side 1 is a line.
- Side 2 is a line.
- Side 3 is a line.

Relations
- Side 1 connects vertices 1 and 2.
- Side 2 connects vertices 2 and 3.
- Side 3 connects vertices 3 and 1.
- Angle 1 is formed at vertex 1 by sides 3 and 1.
- Angle 2 is formed at vertex 2 by sides 1 and 2.
- Angle 3 is formed at vertex 3 by sides 2 and 3.

The special cases of triangle types can be characterized by adding conditions to the triangle schema:

- Right triangle: Angle 1 equals 90 degrees.
- Acute triangle: All angles are less than 90 degrees.
- Obtuse triangle: Angle 1 is more than 90 degrees.
- Equilateral triangle: Length of line 1 equals length of line 2 equals length of line 3.
- Isosceles triangle: Length of line 1 equals length of line 2.
- Scalene triangle: No lines have the same length.

Here you can see precisely how image schemas form generalizations over their special-case images and how the special-case images "contain" the schema within them and add properties to the schema to form each image.

This property of schemas is important. *Schemas are general, so general they cannot be made concrete at that level of generality.* The specific cases all activate the general schema and add properties or relations. This means that if you activate the circuitry for any specific schema, you will also activate the circuitry for the general schema. But the general schema does not necessarily activate any specific schema. This characterizes an internal logic of schemas: the specific cases imply the general case, but the general case does not necessarily imply any specific case.

PROPORTIONALITY

The triangle schema has a further logic:

- For any triangle in a flat plane, the ratio between the lengths of each two sides is the same as the ratio between sizes of the corresponding angles for those two sides.

Triangles may be bigger or smaller, but once the overall size is fixed, the side–to–opposite-angle ratio is fixed. This also holds for triangles on the surface of a sphere.

Given this, further consequences hold. Consider the lines that bisect each angle and its opposite side. Those three lines will intersect at a single point inside the triangle. And each such bisecting line will divide the triangle into two equal areas.

These inferences arise from the general triangle schema independent of any specific triangle. What is interesting is that these inferences occur even though you can never perceive the general triangle schema. These inferences are invariants that apply to all triangles by virtue of arising from the general schema, which exists even though you cannot perceive it.

The triangle schema has two kinds of logics:

- There is a general case that holds for every specific case.
- There are entailments that hold for all cases.

As we shall see, this is true not just for the triangle schema but also for many other schemas.

On Schemas

PRIMARY SCHEMAS

The general triangle schema is an instance of what we call a **primary schema**. Primary schemas are relatively simple schemas that, so far as we know, are common to all human beings. They are all embodied. General X-Nets are primary schemas, as are at least two other common types: image schemas and force schemas.

GESTALT CIRCUITS AND PART-WHOLE SCHEMAS

Some simple circuits arise readily via neural recruitment. One of the simplest is the gestalt circuit, which characterizes a whole structure (the gestalt) and the parts (elements that play "roles" in the gestalt).

The gestalt circuit has three or more nodes: a gestalt node and at least two role nodes. For example, a movement schema has a source of movement, a path, and a goal (the endpoint of the movement). A gestalt circuit for a movement would have a gestalt node for the whole movement and three role nodes for the source, the path, and the goal of the movement. The key properties of gestalt circuits are as follows:

- Connectivity: The gestalt node is connected to each role node. Each role node is connected to the gestalt node.
- Thresholds and input activation: Each node has a firing threshold. If

the input activation is above threshold, the node fires. Otherwise, it doesn't. The details as to how this works will be discussed in chapter 3.

Gestalt circuits are so simple that they should form randomly in relatively large numbers and should be easily recruited. The reason is that the constraints on the connectivity, activation, inhibition, and threshold values should be relatively easy to meet when there are hundreds of trillions of neurons in the brain.

As we shall see, gestalt circuits play a major role in our theory of the neural mind. They provide the neural infrastructure for schemas and frames.

IMAGE-SCHEMA COMPOSITIONALITY

In the 1970s, linguist Leonard Talmy at the University of California, Berkeley, made a remarkable discovery. He was studying the language of spatial relations in various languages and found that even closely related languages had very different words or morphemes for spatial relations. But he also found that each word or morpheme could be decomposed into primitive schemas, primitives in the sense that they could not be further decomposed. In addition, he discovered that the primitives were the same across the languages he studied. Talmy's primitives were mostly what we now call image schemas!

In the 1980s, Talmy studied types of force interactions in the semantics of various languages. Again, he found that they could all be decomposed into primitives that are the same across languages, and those primitives are all schemas: force schemas! As Talmy observed, force schemas and image schemas operate together.

As a linguist looking at various languages, Talmy studied words and morphemes, and to find their meanings—their semantic structures—he "decomposed"—that is, analyzed into components—the meanings of the individual words and morphemes. For example, Talmy observed that the English word "on" in *The cup is on the table* can be decomposed into the components above, contact, and support. In *The lamp is on the wall*, "on" just has the components contact and support. But if you understand what support means, you know that it requires contact, and if you know how tables provide support, you know that the cup has to be above and in contact with the table. Above, contact, and support are all primitive image schemas, but above and contact in these examples can be inferred from support plus contextual knowledge.

One can see the centrality of support in the meaning of "on" in sentences such as *Harry depends on his brother*. This occurrence of "on" makes use of the widespread conceptual metaphor in which emotional support is understood metaphorically in terms of physical support, as in sentences such as *Lean on me*, which can be an offer of emotional support.

Talmy decomposed other spatial terms as well. He pointed out that "across" decomposes into a side schema and a motion schema:

- The side schema has two regions ("sides") separated by a boundary.
- The motion schema has a mover and a source, a path, an endpoint of motion, and distance traveled. It can apply to the motion of the whole body from one place to another or to any motion: a movement of the arm, a turning of the head, or the falling of a leaf from a tree.

The meaning of "across" is formed by bindings of these schemas:

- The source of the motion is bound to one of the regions of the side schema.
- The endpoint of the motion is bound to the other region of the side schema.
- The boundary of the side schemas is bound to some portion of the path of the motion schema.

In the sentence *The quarterback dove a yard and a half across the goal line*, the goal line is the boundary between two regions: goal territory (the end zone) and nongoal territory. The quarterback is the mover in the motion schema and begins his dive in the nongoal region and *ends his dive in the goal region*. The distance traveled in the motion schema *is a yard and a half*. These two schemas are integrated by three bindings. The first identifies the source of the diving motion with a location in the nongoal region, the second identifies a point on the path of motion within the goal line (the boundary in the side schema), and the third identifies the endpoint of motion with a location in the goal region.

From the perspective of the linguistic analyst, this is a decomposition of "across" into two schemas and three bindings. From the perspective of a thinker, this sentence involves compositionality, integrating the side schema and the motion schema, which together provide structure to the meaning of "dove across" through the use of three bindings.

Such examples also work metaphorically, as we will see in more detail below. Here's a simple example: *It's still two weeks before the election, and the leading candidate is within sight of a touchdown but still has to move the ball a short distance across the goal line.* The conceptual metaphor is "achieving a purpose is reaching a destination." The purpose is winning the election. In the football metaphor, winning the election is scoring a touchdown, which requires moving the ball the required distance across the goal line. Or take the academic metaphor "He's close to finishing his dissertation and should be *across* the finish line in another week."

The master of viewpoint studies was the great linguist Charles Fillmore. The classic text is Fillmore (1977). Fillmore called the study of viewpoint in language "deixis."

The deixis schema consists of a viewer located at a viewpoint (called a **deictic center**), a region, and entities in that region. The *come/go* distinction is based on the deixis schema. "Come" refers to motion *to* the deictic center, while "go" refers to motion in other directions.

For example, upon seeing the bus approaching you, you could report, "The bus is coming here" but not "The bus is going here." Suppose a friend rings your doorbell. From your viewpoint you can say "Please come in," with you being in control of her entrance. If you say "Please go in," you are taking her viewpoint outside the house, with her in control of whether she enters or not.

Goals that one is trying to reach are often conceptualized as deictic centers. Suppose you are driving somewhere and you have to stop at an intersection that you didn't know in advance was there. You might report that by saying "I came to an intersection," not "I went to an intersection," which can be used only if you knew the intersection was there and purposely drove there. If you are applying the brakes of the car with the goal of stopping, you could say "I came to a stop," with the *goal* of the action as the deictic center, and not "I went to a stop."

The same is true of metaphorical uses of "come" and "go": *the deictic center is shifted from the state you are in to the state you want to be in or naturally would be in.* As you are finishing an argument, you can say "I'm coming to my main point," not "I'm going to my main point." This is also true of natural states, as in "Come to your senses," not "Go to your senses." It also works for natural states of development: "As I grew older, I came to appreciate my parents," not "As I grew older, I went to appreciate my parents."

In the 1980s, Leonard Talmy made another advance in the study of schemas: force and force dynamics. A force schema consists of

- an exerter of force,
- an object of force (what the force is applied to),
- a force,
- a degree of force,
- an instrument of force, and
- a result.

A force-dynamic schema is a force interaction between two exerters of force on each other with a single result. It can be written as the composition of two force schemas, as defined by three neural bindings:

- Exerter 1 = Object 2
- Exerter 2 = Object 1
- Result 1 = Result 2

A common conceptual metaphor for force is that a force is an object transfer, with the exerter as transferer and the object as transferee. For example, if a punch is conceptualized metaphorically as an object that is transferred, the exerter can "give" the object a punch, with the object "taking" the punch. The "punch" example is an instantaneous transfer with the force as an object, but there can also be a metaphorical durative transfer of force as a fluid. Thus, the exerter can "pour it on," and the object can be "inundated" and "overwhelmed." Further metaphors can apply. Thus, you can be inundated by verbal attacks or can turn up the heat in your criticisms (with "force" of the criticism as metaphorical heat).

The durative and repetitive versions of the force schema extend over time via the process schema (carried out by an X-Net). The release of force occurs when the degree of force is reduced, perhaps to zero, in which the exerter "eases up on" or "lets go of" the object. A supporting force is durative and keeps the object up. Metaphorical emotional support can come from below (you lift someone up) or from above (someone is hanging onto you). Metaphorical support is released when you let someone down. An interesting example of the force-dynamic schema is the expression "to hang in there." It uses the metaphor that states are locations. To stay in the same state is to stay in the same location: for example, when there is a force to be resisted by exerting a force to hang onto a source of support to keep you in the same location (metaphorically, the same state).

Talmy, in his discussion of force dynamics, pointed out one of the most important conceptual metaphors: causes are forces, where the result of force application is the effect of a cause. The reason why this is important is that each specific force application has a different logic with different inferences. The "causes are forces" metaphor shows that there is no one logic of causation but rather many depending on the specific force. The "causes are forces" metaphor is thus a metaphor schema whose metaphorical source is the force schema. Different types of forces (e.g., supporting, pushing, pulling, releasing, instantaneous, durative) are mapped by the "causes are forces" metaphor schema onto correspondingly different types of causes.

This analysis fits together with the X-Net metaphor for causation: an X-Net that carries out a complex action in which the X-Net controls bodily movements that exert forces. The fact that schemas are experienced in the body brings together X-Nets, forces, and causes, as we have seen. The neural mechanisms for this will be covered below in our discussion of how conceptual metaphors are acquired naturally and mostly unconsciously.

VERY BASIC SCHEMAS

The very basic schemas are as follows:

- **Trajector-landmark schema**. Trajectors are entities that are located or are moving; landmarks are entities that they are located or moving with respect to. *There is a tree (trajector) in front of the house (landmark).* This sentence looks simple, but there is a complexity: the landmark (the house) defines a region in front of the house, and the trajector is bound to a location in that region. There is a region extending from the front of the house, and the tree is located in that region.

    In a sentence such as *Under the bed is where I keep my slippers*, the bed is a landmark, and *under the bed* refers to a region under the bed. That region is an entity and functions like an entity grammatically in sentences such as *Under the bed is known to be where he keeps his cash.* Here the region *under the bed* is an entity, and moreover it is the subject of the sentence.

- **Event schema (carried out by X-Nets)**. The parts of the schema are an event, a time, a place, circumstances, and impediments: *The plane arrived (event) at O'Hare (place) at midnight (time) despite a thunderstorm (impediment) when we thought it would be late (circumstance).* Negating the place, time, or circumstance indicates that the event did not happen.

  - Negating time: *Never have I been to Paris* (I haven't been there).
  - Negating place: *Nowhere could I find him* (I couldn't find him).
  - Negating circumstances: *Under no circumstances would I accept a ride from Count Dracula* (I would not accept a ride from Count Dracula).

- **Action schema.** The parts of the action schema are an agent, an action, a patient, and other roles: *Harry (agent) sent (action) the book (patient) to me (recipient) as a present (benefit).*

Many embodied schemas have structures with reversals. Take the embodied schemas of forward motion versus reverse motion. Reverse motion is not just about walking or running or driving somewhere and then going back. You can move your arms in a forward direction or move them back. You can extend your fingers forward from a fist position and move them back into a fist. You can stick your tongue out, moving it forward, or retract it, moving it back. The general motor schema governing all of this is extension and retraction. It is a general schema, applying to all such physical cases.

The extension and retraction schema also works in metaphorical cases. You can put forth a claim and then retract it. You can make an offer and then take it back. You can make a plan and then reconsider it. You can reason to a conclusion and then backtrack to see how you got there. This is a case of repurposing motor control using metaphor, applying the same neural circuitry used in motor control to nonphysical cases.

Such cases support our views on exaptation: via evolution, neural circuitry for motor control in animals has been repurposed for use in conceptual thought by human beings.

The into schema is a composite embodied schema with three identity constraints, or what are called **binding constraints**: (1) the source of motion is exterior, (2) the goal of motion is interior, and (3) the path of motion crosses the boundary, typically overlapping a portal. Each binding constraint binds some part of the whole motion schema to some part of the whole container schema.

We hypothesize that such bindings are carried out by part-to-part binding circuits across schemas and that *all* conceptual bindings in composite concepts are of this character.

The three binding constraints, taken together, can occur as part of an integrating structure that creates a single concept: into.

The composition mechanism of integrating multiple bindings creates an integrated composite structured experience of moving into some interior, or perceiving or imagining an into motion. Moreover, the into concept is general, fitting all sorts of motion and containers: walking, running, skipping and hopping into rooms; pouring coffee into cups, popcorn into a popper, and sand into a cement mixer; and as putting sushi pieces into your mouth.

PRIMARY SCHEMAS AND BLINDNESS

It appears that people who were born blind or who have gone blind nonetheless have image schemas.

One such primary schema has been studied from this perspective—the rotation schema—which has been shown to exist in people who are blind (Carpenter and Eisenberg 1978). A reliable nonexperimental observation has been made by Leonard Talmy, who first discovered the primary embodied schemas and many of their properties. Talmy, as a graduate student in linguistics and cognitive science at the University of California, Berkeley, was going blind and knew he would eventually be completely blind. He decided to prepare himself. He became the world's expert on the structuring of space across languages, that is, independently of any specific language. He did this through the linguistic study of primary schemas. Talmy (2000a, 2000b) has written extensively on the subject, has done keynote lectures at conferences around the world, and spent over two decades as chair of the Cognitive Science Program at SUNY Buffalo. While in Berkeley, he spent time in the Cognitive Science Research program and at the NTL Lab at the University of California, Berkeley.

The study of how primary schemas are retained despite blindness might be a good place to extend such neuroscience research beyond the rotation schema. But the question to ask is not just what primary schemas are retained but also what circuitry carries them out.

Many embodied schemas are topological in character (Talmy 2000a, 2000b). Thus, the container schema is **topological** and hence independent of the **geometrical** properties of size, location, and shape. A container can be of any size, from an atom to the universe: an atom contains its nucleus and electrons, and the universe contains everything in the universe. Just as all containers are instances of the container schema, so too all motions are instances of the motion schema, all forms of contact are instances of the contact schema, and so on.

An unanswered question remains: What kind of neural circuitry that can apply across modalities (e.g., vision and action) in a wide range of very different cases can constitute a topological property for a schema, say, the into schema?

A reasonable first guess would be neural structures that are topological in character: topographic maps. These are two-dimensional arrays of neurons that have internal and external connections. What makes them topological is that they tend to preserve closeness. In the primary visual cortex, the closeness preserved is closeness in the retina. In the primary motor cortex, the closeness preserved is closeness of functionally related body parts. For example, in the motor cortex, the neurons controlling fingers that are next to one another in a hand are in brain locations that are next to one another in the brain, as close in the brain as they can be. Such topographic maps are not just in the brain but also run throughout the body.[3] For this reason, it does not seem strange that certain schemas carried out by the body are topological in character.

However, it is not known for sure whether topographic maps are the source of the topological properties of embodied schemas. And it is not known just what kind of circuitry accounts for the kinds of embodied schemas that exist. This is a research project for neuroscience.[4]

MECHANISTIC CONSTRAINTS ON COMPOSITION

The circuitry carrying out bindings, integrations, and cascades allows for enormous flexibility in combining schemas. Since schemas can be combined with both great precision and great speed, we suggest the following mechanisms:

- **Gating.** All composite-forming circuits are "gated" by a neural ensemble that can immediately turn on or off a composite schema formed by multiple bindings, such as those for into and across. Gating reduces the enormous control problems for combining schemas correctly. The combining of schemas is reduced to the problem of combining their gates.

- **Control by disinhibition**. One mechanism works as follows. the gates of a circuit normally inhibit firing. The circuits fire when the gates are disinhibited, that is, when the inhibition is inhibited.

What control by disinhibition offers is selective activation and realistic timing. Attentional mechanisms use disinhibition to perform recall of these stored memories. Disinhibition mechanisms work very quickly, allowing the formation of very complex conceptual thought within milliseconds. It takes about 80–100 milliseconds in simple cases for unconscious complex concepts to become integrated into conscious thought. Given the number of steps between complex language structure and complex grounded conceptual structure, the link between language and meaningful thought must take place very quickly. The disinhibition of inhibitive gates would allow for such speed.

EXPLANATORY POWER

The study of embodied schemas is not just about embodied schemas. It is also about the embodied basis of a huge range of conceptual thought. Embodied schemas have an explanatory power far in excess of the primary concepts they express. The reason is that they can be combined to form complex schemas, and they are used, as we shall see, in other kinds of conceptual structures: frames, conceptual metaphors, and conceptual integration. In each case, embodied schemas play a necessary role in embodied meaning.

Schema Embodiment

Mark Johnson in two remarkable books, *The Body in the Mind* (Johnson 1987) and *The Meaning of the Body* (Johnson 2007), has given us a sense of image schemas as forms of embodied experience, gestalts (wholes with parts) that reside in and structure bodily experience.

Certain cases are obvious. We experience motion and the lack of it. We experience certain types of motions (plural), movements of the whole body, of the arm and legs, of the mouth, head, eyelids. But do we experience movement in general, the motion schema we have described? An interesting test is the instruction "Don't move—anything!"; that is, no path of motion by the body as a whole or an arm or leg, the head, etc. Or "He's dead—he can't move." Do you understand the schematic meaning of the verb "move" by itself? The answer "yes" seems clear. You don't have to give a list of all the kinds of movement that "move" covers.

You experience the part-whole schema, which characterizes the general case of parts that constitute wholes and wholes made up of parts and their relations to one another. This starts with your whole body and your body parts. The part-whole schema structures the body itself, the parts of the hand: the fingers, the palm, the back, etc. versus the hand as a whole.

You experience the balance schema in your body, and you know when you are off balance and could fall. You know what it means to regain your balance after you start to fall.

You experience the container schema in your body. When you drink or eat, you know there is something *in* your body, and when you excrete it or spit it out, you know that something is no longer in your body's container. You also know that your body has more than one container. Some thing can be *in* your mouth, *in* your ear, or *in* your nose. You experience them all in terms of containment, though they are all different instances. You also experience the full-empty schema in your body. When you eat and/or drink you sense when you are full. When you can't hold any more in your mouth, your mouth is full.

You experience the contact schema in your body. You can feel contact on your skin and in your muscles.

And you can experience contact separate from balance, balance separate from motion, motion separate from part-whole, and so on.

How is this possible? Conceptual schemas are used to conceptualize the world and are expressed in language. How can these same schemas actually structure embodied experience?

Remember that via connections between body and brain, the body is mapped onto the brain. Every muscle we can move is mapped onto the motor cortex. The choreography of complex movements is carried out by circuitry in the premotor and supplementary motor regions. Everything you can feel is mapped topographically onto the somatosensory cortex. Whatever is activated in the retina is mapped topographically onto the visual cortex. The viscera are mapped onto the insula (but not topologically).

This leads to a reasonable hypothesis: embodied experience of schemas is structured via the brain. Wherever and however schemas are characterized in the brain, they seem to be able to recognize and impose schema structure in those regions of the brain that are maps of the body. Remember that although you experience pain as being in your leg, the pain center is actually in the brain.

With your index finger you can trace a triangle on your arm, your leg, your face, or your stomach, a triangle you can not only make but also feel. Hold out your right arm. Now rotate it to make a circle in the air. Now with an open flat palm, move it back and forth to form a line in the air. Now trace a circle in the air. In order to do these things, you need X-schema circuitry in your brain, neurally bound to triangle schema circuitry, to move your index finger in a triangle shape on your arm.

Although we experience the schemas as being on and in the body, the neural activation that produces the bodily experience of those schemas occurs not in the body itself but rather in the brain.

Thinking with Schemas

EMBODIED SCHEMAS HAVE LOGICS

One of the most important properties of embodied schemas is that they give rise to inferences. The inferences are mostly carried out unconsciously and effortlessly and go unnoticed. But we use them all day, every day. Here's a simple and obvious example, based on the motion schema:

- Consider distinct locations A, B, and C. If a mover has moved from A to B and from B to C, then the mover has moved from A to C.

If you go from the kitchen to the front door, open the door, and then go from the front door to the garage, you have gone from the kitchen to the garage. Obvious. But there has to be neural circuitry to carry out the inference.

Do we have special circuitry for every such inference, with entirely different circuitry for different inferences? It appears not. Our hypothesis is that these inferences are carried out through the image schematic generalizations (such as source-path-goal in this case). There are inferences based on other embodied schemas that work in the same way. Here are three more, although there are more than three:

- **In**: If A is in B and B is in C, then A is in C.
- **Behind**: If A is behind B and B is behind C, the A is behind C.
- **Above**: If A is above B and B is above C, then A is above C.

There is a generalization across all of these, and neural systems naturally learn generalizations that occur regularly.

We will be using the symbol S to generalize the specific embodied schemas in these cases. The general schema logic can be notated as follows:

- If A S B and B S C, then A S C.

The special cases occur when S = motion from-to, in, behind, above, and so on.

But this simplified notation can be misleading.[5] We hypothesize that such general cases are neurally learned, that there is a circuit for them, and that the special case circuitry *adds additional structure to the general case* in each specific case. In other words, we hypothesize that *general circuitry is physically part of each specific case of neural circuitry.*

To visualize this, we need a somewhat more detailed notation. We will notate a general schema, two of its roles, and the referents that fill these roles as follows:

SCHEMA	ROLE 1:	ROLE 2:
	REFERENT 1	REFERENT 2

Think of this as what is meant by the single symbol S as used above. The special case "A IN B" is represented by In in the Schema slot, Content in the Role 1 slot, Interior in the Role 2 slot, A in the Referent 1 slot, and B in the Referent 2 slot. That is, the second table has the structure of the general schema shown in the first table and *adds* what is in the boxes. Both the general structure above and the specific added details below are present at once.

IN	TRAJECTOR	LANDMARK
	A	B

Similarly, A Behind B and A Above B as in the following:

BEHIND	TRAJECTOR	LANDMARK
	A	B

ABOVE	TRAJECTOR	LANDMARK
	A	B

We will continue to use the simplified notation A ABOVE B, given this understanding that A is the trajector and B is the landmark.

X-NETS CAN MODEL SCHEMA INFERENCES

Recall that X-Nets are computational networks that "execute," that is, do things.

Our computational models of schema inferences will, of course, be using computational models of neural circuitry to carry out such inferences. Within those computational models of neural circuitry, X-Nets play two important roles:

- X-Nets in computers can, using computational techniques, model the functional neural circuitry that controls bodily movements in the performance of purposeful actions.
- X-Nets also have the right structure to model image schematic reasoning, as we are about to see.

Bear in mind that the computer models are different kinds of things than what they are modeling. As an example, a computer can model the flow of a river, but the river can make you wet, while the computer model cannot.

We justify this use of X-Nets with a major hypothesis: Over the course of evolution, neural circuitry has been repurposed (exapted) from controlling motor movements to carrying out reasoning. To show how this might happen, we show how X-Nets for modeling motor control can carry out logical reasoning.

We begin with simple logical inferences from premises to conclusions. Such logical inferences can be modeled computationally by a composition of X-Nets, schemas, and bindings. The result is a conceptual metaphor for understanding logical reasoning via neural networks in the brain in terms of X-Nets designed for the computational modeling of motor control.

THE LOGIC X-NET METAPHOR: A COMPUTER X-NET STRUCTURE METAPHORICALLY UNDERSTOOD AS PERFORMING INFERENCES VIA NEURAL NETWORKS IN THE BRAIN

Logical reasoning moves from premises to conclusions. Here is how this is done using computer models:

1. The portion of the X-Net from the precondition up to just before the final state constitutes the premises.

2. The conclusion of the action performed by the X-Net is the logical conclusion from the premises, given any bindings linking premises and the conclusion.

Consider the structure A is IN B, D is IN C, and B is bound to D. A neural binding between B and D would make B and D neurally indistinguishable. This is equivalent to the notation that if A is IN B and B is IN C, then A is IN C.

It seems simple. We do reasoning such as this unconsciously and effortlessly every day. The logic X-Net structure is a part-whole schema with two parts, what Aristotle called a "premise" and a "conclusion," with the premise preceding the conclusion. The premise has two parts—a major premise and a minor premise, each a statement of a content in a container—plus a binding that brings together the major and minor premises. Here the container B in the major premise is the content B in the minor premise. The same letter, B, is a notation indicating a neural binding.

In honor of Aristotle, we will be using his names for the parts of the reasoning process: premise, conclusion, major premise, and minor premise. Those names are arbitrary; they could be called, correspondingly, Part 1, Part 2, Part 3, and Part 4. But Aristotle's names make it easier to keep in mind the roles played by those parts.

The following shows how the parts of X-Net are integrated with the schema (IN), its roles (content, interior), and their referents (A, B).

- Whole: logic schema for IN
- Parts: premise, conclusion
- Relation: premise before conclusion
- Premise:
    - Major premise: content A is IN interior of B
    - Minor premise: content B is IN interior of C
    - Binding: major container B = minor content B
- Conclusion: A is IN C

One might well ask why say that we do reasoning of this sort? Why not just superimpose the two schemas and read off the relation A is IN C? This superimposition is what we are doing by binding D and B and calling them both B.

This is a form of reasoning performed unconsciously by most people every day. It has to be carried out by neural circuitry. The neural circuitry must be capable of characterizing a part-whole structure and a neural binding of a part in one part-whole structure to a part in another part-whole structure. In short, there must be part-whole circuits and binding circuits as well as what we are calling an "integration circuit," which determines how the parts of the logic schema fit together in a neural system.

There is a single general logic schema with the general schema S of the following form. Note that all logic schemas have a binding linking the major premise to the minor premise.

- Whole: logic schema
- Parts: premise, conclusion
- Relation: premise before conclusion
- Premise:
  ○ Major premise: A S B
  ○ Minor premise: B S C
- Binding: major premise B = minor premise B
- Conclusion: A S C

SPECIAL CASES: S = IN, BEHIND, ABOVE, MOTION START-END

Here we have the same general logic schema with four different special cases determined by plugging in four different image schemas, all notated by S. This is a class, S, of image schemas to which the same logic schema applies.

CONTAINMENT: NEUTRAL, IN, AND OUT

There is a neutral container schema with roles: content, boundary, interior, and exterior. There are two basic subcases, a schema for IN and a schema for OUT. They both have the container schema with an interior, a boundary, an exterior, and (potential) content.

- The IN schema adds the binding: Content is bound to interior.
- The OUT schema adds the binding: Content is bound to exterior.

Note that although these are special cases of the general containment schema, they too are schemas, not particular images. Note also that the general containment schema does not come with a logic schema. Since the containment schema lacks a binding, no logic is defined.

We have just seen that image schemas are built into embodied experience. As such, they are separate. They are wholes with parts, and yet they are experienced bodily throughout everyday life. They are not abstract. They are physical, and as embodied, they have a physical existence, not just an abstract existence in the mind.

Although we are using a formalism to name and characterize the structure of image schemas, the formalism is in our theory, not in the neural system itself. Our model here is a computational abstraction of the neural circuitry that results in these inferences.

We are hypothesizing that this is where so-called abstract reason really comes from: neural circuitry extending between the brain and the body. In short, abstract reason is not abstract; it is physical. It uses the neural circuitry of embodied schemas, which can be modeled computationally by X-Nets, that is, by computer networks that "execute" (carry out computer operations).

DOES SIMULATION USE LOGIC SCHEMAS?

Simulation is central to everyday life. For example, someone deciding where to keep a valuable piece of jewelry mentally simulates—whether consciously or unconsciously—where she might keep the jewelry. One possibility is in a dresser drawer. She might imagine the drawer in the dresser and imagine pulling out the drawer, inserting the jewelry, and closing the drawer and remembering later that the drawer is in the dresser and the jewelry is in the drawer; therefore, the jewelry is in the dresser. Moreover, the dresser in in the bedroom, so the jewelry is in the bedroom. The reasoning is typically unconscious, effortless, and unnoticed. This simulation makes use of the logic of containment. A logic schema is being used as a part of mental simulation. This is not surprising. Simulation is coherent; that is, the parts fit together into an integrated whole in which logic schemas make the simulation useful for reasoning. Logic schemas are what make simulations coherent and useful in everyday life.[6]

WHAT DOES "LOGIC SCHEMA" MEAN?

We are hypothesizing logic schemas of the following form in the cognitive linguistics dimension of our bridge theory. This form is meant to map onto an X-Net via our neural computational bridge and then onto the functional neural circuitry for carrying out that X-Net. The X-Net circuitry has been adapted from motor control to reasoning.

- Whole: logic schema for [schema name]
- Parts: premises, conclusion
- Relation: conclusion after premises
- Premises:
    ◦ Major premise
    ◦ Minor premise
- Binding
- Conclusion

X-Nets have a part-whole schema structure. In these X-Nets, the whole is the whole logic schema, the parts are (1) the premises, that is, the beginning up to the final state; and (2) the conclusion, that is, the relevant conclusion resulting from the application of the X-Net. The relation between premises and conclusion is that the conclusion follows the premises. The premises come in two parts, major and minor, as observed by Aristotle.

There is a systematic relation between (1) the structures of major and minor premises and (2) the structure of the conclusion. In general, the trajector of the major premise is the trajector of the conclusion, while the landmark of the minor premise is the landmark of the conclusion. The binding indicates that the landmark of the major premise is the same entity as the trajector of the minor premise. The job of the binding is to show how the major and minor premises fit together logically. Once made into a single connecting element by the neural binding, that entity, having done its job, drops out of the conclusion.

LOGIC SCHEMAS ARE RECRUITED VIA SIMULATION

Mental simulations of real or imagined situations are constructed in our imaginations, usually in real time. They are commonly used in moment-by-moment decisions, such as whether to turn left or right when someone is walking toward you. They can also be used in making longer-term decisions. Or they can be used in understanding what is about to happen or what just happened. The point here is that we are simulating constantly, mostly unconsciously, throughout every day.

We hypothesize that logic schemas of the sort we have been discussing are recruited via repeated simulations. Each time the neural structure is used in simulations, the connections in that neural structure get stronger. Inferences are necessary in simulations. Children begin simulating very early in life. For that to happen, the right combinations of neurons has to be in the right places to constitute a circuit for a logic schema, and for an embodied schema that is relevant in this situation. Once recruited through repeated use, those neural circuits keep firing in appropriate situations, getting stronger and stronger until they are permanent. That is how their content is learned.

Frames

WHAT IS A FRAME?

A frame is a mental structure used for conceptualizing a type of situation, real or imagined. A frame is technically a type of schema. As a schema, the frame generalizes over special cases. That is, the frame fits all the (nonmetaphorical) special cases; it has a part-whole structure with relations among the parts and gives rise to inferences. The frame is a general mental structure and, like the triangle schema, cannot fit real situations in the world without additions to create special cases that fit the general frame structure.

Frames differ from the primary schemas (image schemas, force schemas, and process schemas) in that they are socially, culturally, and institutionally dependent. Frames characterize social and cultural entities, practices, and institutions that we regularly encounter in our experience or imagination, such as a general type of situation (e.g., a theft), a physical structure (e.g., a church), a social activity (e.g., folk dancing), a life form (e.g., a butterfly), a kind of person (e.g., a fireman), and a professional (e.g., a lawyer). As such, frames tend to be much more complex than primary schemas.

A frame is an embodied mental structure. In our model, a frame is constituted by a type of neural circuit, which we will be calling a **frame circuit** when we discuss it in later chapters.

Frames incorporate primary schemas in their structure and derive much (some researchers think all) of their inferential power from the logic of primary schemas. Many frames make use of conceptual metaphors.

Complex frames form hierarchical systems, with a hierarchy of successively more general to more specific frames.

Like other schemas, frames combine via bindings, integrations, and cascades to form larger frames.

And like other schemas, frames are mostly used unconsciously, automatically, and effortlessly and so tend to go unnoticed. On the other hand, frames can be consciously imposed on a situation through the use of language.

Charles Fillmore, one of the discoverers of frames, observed that all words and morphemes in a natural language are defined relative to frames. Since frames are special cases of schemas, all words and morphemes are defined relative to schemas (Fillmore 1976).

Are frames embodied? Many of them are. Take the tennis frame, in which the proper swing of a tennis racket is an embodied part of the game, as are the motor skills of serving, hitting a backhand, and so on. Special motor programs, typically with the use of equipment, constitute part of the embodiment of an athletic frame. Such forms of embodiment occur in many professions and avocations, such as surgery, welding, gardening, carpentry, surfing, texting, and driving.

Frames can be small, with few roles and little structure (e.g., the week frame, with roles Sunday, Monday, etc.). Or a frame can be very extensive, with lots of roles and subframes. The hospital frame, for example, has people who function in the roles of patient, doctor, nurse, orderly, visitor, receptionist, and so on. In the hospital frame there are place roles, areas that function as operating rooms, recovery rooms, reception areas, and so on. The hospital frame will also have object roles, with objects such as gurneys, scalpels, defibrillators, ambulances, and so on. In addition, the hospital frame will include happenings, things that normally happen in hospitals: doctors examine and operate on patients, nurses assist doctors and attend to the needs of patients, visitors register with receptionists, and so on. Of course, there are types of entities and happenings that are *not* in the hospital frame: herds of elephants, fighter planes, ski instructors, investment counselors, operations performed by patients on doctors at reception desks, and so on. If one of these entities or happenings appeared in a hospital, we would speak of it as "breaking the frame."[7] Indeed, a way to explore the boundaries of a frame is to figure out the minimal cases needed to break the frame.

All words are defined relative to frames. For example, "surgeon" is defined relative to a surgery frame, "spoon" is defined relative to an eating utensils frame and an eating frame, and "price" is defined relative to a commercial event frame. Saying or hearing a word evokes the corresponding frame, which, in the brain, involves activating the frame circuit that neurally constitutes that frame. Just hearing and understanding a word has the consequence that its neural frame circuit will be activated and therefore strengthened. Negating a frame cannot happen without activating that frame. George

Lakoff's (2004) book title *Don't Think of an Elephant!* makes this point clearly: the title makes you think of an elephant. When Richard Nixon during the Watergate investigation said "I am not a crook," he led people to think of him as a crook by negating and thus activating the crook frame.

As we shall see shortly, frames are structured by primary embodied schemas and often by a sequence of frames hierarchically structured and experientially grounded via primary embodied concepts. A house, for example, is structured by a container schema and has an interior, an exterior, and a door as a portal. In intermediate framing, a house is also a physical enclosure for human beings, as is a prison, a school, and an office building.

Human conceptual systems are very rich, as we have seen. There are hundreds of embodied primary schemas, thousands of complexes made up of embodied primary schemas (e.g., those for *into*, *across*, and *along*), and thousands of frames (characterizing the tens of thousands of words we know).

THE MODERN ORIGINS OF FRAME SEMANTICS

The contemporary notion of a frame arose from research performed independently at roughly the same time in the early to mid-1970s by three major intellectual leaders in three different fields: Erving Goffman in sociology, Charles Fillmore in linguistics, and Marvin Minsky in computer science. Goffman conducted sociological research on institutions, famously taking jobs and working in those institutions for a significant period of time, such as working in a mental hospital and in a Las Vegas casino. He noticed that people in their private lives often change when functioning in an institution as if they were playing a role in a play, with certain lines to be spoken, clothes to be worn, body postures to be adopted, interpersonal relations to be maintained, and acts to be performed. Metaphorically, he thought of life as a play. As in a play, he saw the institution defined independently of the people. People were functioning in what he called "frames": the collection of roles, defined as in a play by clothes, postures, relationships, language, and actions.

Goffman saw frames as lived by people in institutions and cultures. They could be described by conceptual structures, but those conceptual structures were understood as implicit guides for living. Croupiers in a casino and nurses in a hospital were seen by Goffman as living out different frames.

Fillmore, as a linguist concerned with word meaning and use, had studied the European tradition of semantic fields: sets of related words such as "knife, fork, spoon," "Sunday, Monday, Tuesday, . . . ," and "buy, sell, goods, cost, price, . . ."; minimal contrasts such as "on land/at sea/in the air," and "beach" versus "shore"; background-foreground differences such as "accuse" versus "blame"; and differences in deixis such as "go" versus "come." To account for these semantic relationships, Fillmore postulated "frames"—wholes containing parts called "semantic roles," where the roles were defined by certain properties (e.g., person, animal, object), relations were characterized by other frames, and "scenarios" included actions and events among the roles in a scenario frame.

Marvin Minsky, a founder of the field of artificial intelligence, was concerned with knowledge structures as well as postulated frame structures characterizing areas of knowledge, structures that could be used in computer programs. Minsky's former student Terry Winograd, along with colleagues Daniel Bobrow and Donald Norman, consulted with Fillmore and his Berkeley colleague George Lakoff in 1974 in designing the early knowledge representation language for computer scientists, KL-ONE, that became a model for computational research on knowledge structure. In all cases, frames were considered conceptual in nature, part of the nature of thought. In artificial intelligence, what were called "frames" had a formal, mathematical structure.

It was Fillmore who systematically turned the study of conceptual frames into a science over nearly four decades of research, with students and colleagues working on the frame semantics of languages around world. So far, Fillmore's hypothesis that all words are defined by semantic frames has held up. His research group, called FrameNet, was located at the International Computer Science Institute at the University of California, Berkeley, a few feet away from the NTL research group run by Jerome Feldman, Srini Narayanan, and George Lakoff.[8]

Beginning in the mid-1990s, Lakoff took the sophisticated understanding of frames that had developed over two decades of FrameNet research and started applying the science of frame semantics to the study of political understanding and political discourse (Lakoff 1996, 2004, 2006a, 2006b, 2008, 2014a; Lakoff and Wehling 2012). Since then, the awareness of conceptual framing in politics has spread throughout the political world. TV political commentators regularly discuss how issues are "framed" by contemporary political actors and how they could be framed otherwise with different political consequences. The most widely read text in this area is Lakoff's (2004) book *Don't Think of an Elephant! Know Your Values and Frame the Debate.* On television, the 2022 CNN miniseries *Reframed: Marilyn Monroe* showed Marilyn Monroe as being very different from the way she has been typically portrayed.

In our neural embodiment hypothesis, primary image schemas, force-dynamic schemas, and X-Nets are part of a universal collection of primary concepts that are embodied in pretty much the same way around the world. Frames, on the other hand, tend to be cultural in nature and also tend to be complex, getting semantic superstructure via complexes of primary image schemas, force schemas, X-Nets, and so on.

Thus, frames have two kinds of structure. First, there are universally embodied structures consisting of embodied primitives. These include embodied concepts such as actions (with agents, patients, instruments, and so on), causation of many kinds defined metaphorically (via the causes are forces metaphor) by kinds of forces, aspectual concepts (types of events and actions), purposive structures (desires, goals, resources, means, difficulties) represented via X-Nets, evaluative concepts deriving from "reward" circuitry that neurally embodies what we experience as well-being and ill-being, linear scales, and so on. Second, there are culturally specific experiences such as eating with knife and fork versus chopsticks, religious rituals, cultural institutions, skilled practices, and so forth.

A classic example of a cultural frame is the commercial event frame. Fillmore considered sentences such as

- *Harry **sold** the book to Bill for ten dollars.*
- *Bill **bought** the book from Harry for ten dollars.*
- *The book **cost** Bill ten dollars.*
- *Ten dollars was the **price** of the book.*

He noted that words such as "buy," "sell," "cost," and "price" all made use of the same frame, in which there are four semantic roles: a buyer, a seller, goods, and money. These roles are integrated by a scenario:

- First stage: The buyer wants the goods and has the money. The seller has the goods and wants the money.
- Second stage: There is a mutual transfer. The buyer transfers the money to the seller, and the seller transfers the goods to the buyer.
- Final stage: The buyer has the goods, and the seller has the money.

Though the frame is the same, there are different *perspectives* on the frame, different points of view expressed via grammar. In buying, the buyer is the agent. In selling, the seller is the agent. Each perspective is given by a basic embodied experience, say the experience of being an agent performing an action. The action of buying is different from the action of selling. Buyers shop and compare prices; sellers set prices, market their goods, and try to convince buyers to buy. Buyers want bargains. Sellers want profits. Each perspective gives rise to different instances of the general commercial event frame. In getting a loan from a bank, the seller is the bank, the goods sold is money itself, and the price is also money (the interest).

THE NEURALLY EMBODIED PRIMARY STRUCTURE WITHIN FRAMES

Frames have the structure of schemas: a part-whole structure, with the whole frame and semantic roles as the parts. As with other schemas, frames are gestalts, with each role node activating the whole frame (the gestalt) and the activation of the whole frame (the gestalt) activating all the roles (the parts constituting the whole). Frames also have an X-Net structure: the "scenarios" characterizing the actions and events that give to frames dynamic structures playing out over time.

Frames occur in hierarchies. X-Nets, force schemas, and image schemas can be seen in the commercial event hierarchy (Chang et al. 2002). A prototypical commercial event is a physical exchange of money for goods. An exchange is a mutual transfer. A physical transfer involves the application of a force to a possessed entity, which then moves to a recipient, who becomes the new possessor. A commercial event involves two such forced movements involving the same participants:

- In buying, the transferer is the buyer, the recipient is the seller, and the transferred entity is the money.
- In selling, the transferer is the seller, the recipient is the buyer, and the transferred entity is the goods.

There are bindings across the two transfers, integrating the selling and buying transfers: the transferrer of goods in the selling frame is bound to the recipient of money in the buying frame and so on. Money is understood as a special kind of resource, that is, a resource for obtaining other resources, which is depleted by buying and made less able to buy goods by inflation.

Such forced movements from seller to buyer and buyer to seller make use of embodied schemas such as the application of force, motion, and possession. To form each transfer, there must be a force exerter applying force to a possession, moving it to a recipient, which makes it the recipient's possession.

All of these are primary embodied schemas and are part of what you know about transfers, mutual transfers, and exchanges. In understanding *Harry bought a book for ten dollars,* all of this is unconsciously, implicitly understood, so quickly and effortlessly that it all goes unnoticed.

That is how the commercial event frame is given meaning. It is linked to some of the most basic forms of embodied cognition: forced movement and change of possession.

The two transfers are each further structured by an X-Net with a purpose schema, whose roles are means, cost (as resource used), and benefit (as purpose achieved). For the buyer, the benefit is the goods, the cost is the money, and the means is the buying transaction. For the seller, the benefit is the money, the cost is the goods, and the means is the selling transaction. The general-purpose schema is deeply embodied and is used a great many times every day.

These everyday frames are all characterized by fixed embodied schemas, acting unconsciously, effortlessly, and automatically and deriving their meaning via embodiment and their logics via simulation.

Finally, Fillmore's frames are intended to be active, with each scenario evolving over time. Each scenario makes use of an X-Net: a network characterizing actions that are executed, hence the "X" in X-Net.

Frames also have schema structure in that they are gestalts in which the semantic roles are defined relative to the frame as a whole. There is no selling without buying, and there are no sellers without buyers, goods, and money and no buyers without sellers, goods, and money.

When you think in terms of frames, bear in mind that a great deal of what gives a frame its meaning is the structure of fixed embodied primitives (e.g., goods, money, and agents) that come together to structure the frame.

FRAMES IN CONCEPTUAL SYSTEMS

Frames are not isolated. They take part in whole systems of thought. As we have just seen, frames are composed in part of primary embodied schemas and complexes of them. Moreover, frames occur at various levels of generality. There are subtypes of commercial events, with special cases of roles, such as sales by computer, sales to and from corporations, sales of mineral rights or insurance, and sales via intermediaries (e.g., eBay, Craigslist).

There are also complex commercial endeavors such as restaurants, where the serving-of-food frame and a host-guest institution frame are integrated with a business frame. In such cases, there are bindings. In a restaurant, the customer (business frame) is the eater (food service frame) and the guest (host-guest frame), the maître d' is the host, and the food is the goods.

There are special cases of restaurants (e.g., dim sum restaurants, fast-food restaurants, small-plate restaurants, bars serving food, food carts, caterers, pop-ups, and so on).

And then there are commercial transactions based on metaphors. For example, in economic theory, employment is the sale of one's labor to an employer within a "labor market," the employee is the seller, the employer is the buyer, and the employee's labor is the goods. In a loan, the money loaned is the goods, and the interest is the money paid for the goods.

All such conceptual complexities provide challenges for any account of neural circuitry for composite frames. Composite frames are formed by bindings that identify roles across frames, integrations that bring together multiple bindings or perform multiple activations and inhibitions, and cascades that extend across brain regions controlling complex composites of activations, inhibitions, bindings, and integrations, as we will see.

METONYMY IS FRAME-BASED

Metonymy is a cognitive process in which one role in a frame stands for another role in that frame or for the whole frame. The stands-for relation is characterized cognitively by a role-to-role mapping within a frame. Given that they occur with frames, metonymies can be general, just as frames are. And they have lots of specific instances, just as frames do.

We recognize general metonymies by their instances: *Populists are attacking Wall Street. Let's not let Iraq become another Vietnam. Selma was a turning point in the struggle for civil rights. Hollywood is threatened by Silicon Valley.* The general metonymy governing these cases is that places stand for well-known events occurring there. This is based on a simple general frame for occurrences in which events occur at times, in places, and due to circumstances. Not only can places stand for events, but so can times. *9/11 gave new power to American conservatives. New Year's Eve is always crazy in Times Square. The French celebrate July 14th.*

Just as frames can have special cases, so can the corresponding metonymies. There is the production frame, with producers and products as roles.

In general, producers can stand for their products. Special cases include the following:

- Authors stand for their writings (*I'm reading* **Hegel**. **Chomsky** *takes up a foot on my bookshelf*).
- Artists stand for their artworks (*I've got a* **Picasso** *on my wall.* **Rothko** *is selling well in auctions. Every* **Vermeer** *is exquisite*).

Given that frames are neurally constituted by frame circuits, a metonymy uses a neural frame circuit and adds to it a neural subcircuit in which one node (in the subcircuit) activates another node (in the subcircuit).

The 2,500-Year-Old Theory

When we were in high school, we were taught a false theory of metaphor, a theory that arose with the ancient Greeks. In many places, it is still being taught.

More than 2,500 years ago Aristotle proposed a theory of metaphor to fit his philosophy. That philosophy required that meaning was objective and resided in the external world. He saw the world as structured by categories defined by essences, putative collections of properties seen as inherent in things in the world.[9] Aristotle proposed three kinds of essences: substance, form, and patterns of change. A tree, for example, has substance (wood), form (roots, trunk, branches, leaves), and a pattern of change (the oak sprouts from an acorn, grows to maturity, dies, and falls). Each essence (collection of general properties) was itself seen by Aristotle as an entity in the world.

Aristotle saw essences as causal: the natural behavior of things, he believed, followed from their essences. For example, you can cut up trees because they are made of wood (substance). You can climb trees because they have a trunk and branches (form). You can grow orchards because trees sprout from seeds (pattern of change).

Interestingly, this was not just a strange 2,500-year-old philosophical claim. A lot of ordinary people these days also believe in essences. Intelligence is seen as an essence that defines intelligent people, leading them to act intelligently, allowing for intelligence tests, and making possible the idea that computers, which can do computations that intelligent people can do, can have intelligence as an essence (artificial intelligence) and therefore can be used to make "intelligent" decisions and do "intelligent" reasoning.

From the perspective of frame semantics, many people have a common-place essence frame, with roles pretty much as Aristotle described them: substance, form, process of change, and natural behavior following from the essence. The essence frame is a conceptual frame that has practical utility in many cases, both for natural entities such as trees and in building artifacts to fit a given frame.

Aristotle mistook the frame for the world. And it led him to take a commonplace metaphor literally, the metaphor that "understanding is grasping," as in *He grasped the idea* and *That idea went over my head* (couldn't be "grasped").

Aristotle believed that we can have knowledge of the world because the mind can reach out and grasp the essences of the world. This, he believed, would allow us to understand natural causation and thereby get control of nature. The mind, he thought, was simply a mirror of nature. That is, he believed that we think in terms of categories defined by necessary and sufficient conditions on the external world. For Aristotle, it was therefore vital to understand similarity, since he understood similarity between entities as the sharing of essences. Understanding similarity, he believed, allowed you to understand essences and their relation to one another.

Words, for Aristotle, were therefore defined in terms of essences in the world: lists of properties defined by necessary and sufficient conditions on the world. Metaphor, for Aristotle, could only reside in words. There could be no metaphorical *thought*, since be believed that thought was carried out in terms of essences of things in the external world as "grasped" by the mind. Metaphor was for Aristotle a matter of language, not thought. Moreover, he saw metaphor as residing not in the ordinary language of reason but rather in myth, poetry, and political rhetoric. He believed that when metaphor had meaning, it had to be literal meaning. His theory was that metaphorical language worked by similarity, that is, the sharing of properties between essences of things in the world. He saw linguistic metaphor, according to his theory, as a great thing, since one could gain knowledge of the essences of things in the world through similarity, that is, the partial identity of the properties of things in the world.

That Aristotelian theory stood for more than two thousand years. It has been taught and is still being taught in schools and colleges around the world. There are philosophers and scientists who still think of the world as Aristotle did.

But Aristotle's metaphor theory is a false view of the world, drastically and tragically false. Philosophers such as Spinoza, Nietzsche, I. A. Richards, and a few others have sensed the fallacy in Aristotle's account of metaphor, but it was only in 1980 that a serious scientific study of conceptual metaphor began. This late-coming study of conceptual metaphor is tragic, because those lost 2,500 years have denied us for so long one of the deepest insights into the nature of human thought and of human behavior based on that thought. We ordinarily think using metaphorical thought. Like almost all thought, metaphorical thought is mostly unconscious. Like all thought, metaphorical thought is carried out via brain circuitry and so can affect how we conceptualize the world and how we behave. We live much of our lives according to largely unconscious conceptual metaphor.

Conceptual Metaphor

THE INDEPENDENT DISCOVERIES

The modern discovery of conceptual metaphor was made independently, about six months apart, in 1977 by Michael Reddy (1979) and in 1978 by George Lakoff (Lakoff and Johnson 1980). Both were professors at major universities: Reddy at Columbia University and Lakoff at the University of California, Berkeley. Both were linguists. They were friends but hadn't communicated for some years. They were both using the usual linguistic methodology of looking for generalizations over linguistic expressions (words, idioms, etc.) and semantic inferences (general patterns of thought from which the language follows). Both knew formal logic and were trained in the use of logic with the goal of characterizing linguistic meaning.

Reddy and Lakoff were both well aware of the Aristotelian theory that metaphors were supposed to be a matter of words and mean similarity (shared properties of the literal meanings of individual words). Both happened to be looking at the use of language in a significant domain of thought, Reddy in communication and Lakoff with love relationships. And both made the same discovery: the generalization governing the use of language in those domains did not fit Aristotle's theory.

Reddy was looking at cases such as *You're not getting through to him* and *That poem is densely packed with meaning*. Both examples make use of a single conceptual metaphor for communication in which the language is about the transfer of objects in packages, while the subject matter is communication. Lakoff was looking at cases such as *The marriage is on the rocks* and *We're spin-*

ning our wheels in this relationship, where the language is about the difficulty of travel to a destination, while the subject matter is difficulty in long-term love relationships.

Moreover, in both the Reddy and Lakoff cases, frame semantics characterizes both the subject matter (communication, love with difficulties) and the literal meaning of the words (transfer of objects in packages, difficult journeys). In both cases, the form of reasoning (the "logic") used in the domain of literal word meaning (object transfer, difficult journeys) was being applied to the subject-matter domains (communication, love). In short, people are reasoning about subject-matter domains in terms of generalizations over literal word-meaning domains (object transfer, difficulties in reaching a common destination) and using that reasoning via metaphorical thought.

Both Reddy and Lakoff looked at a wide range of examples and independently concluded that people are *thinking metaphorically* about important subject matters, that is, reasoning about communication as if it were object transfer and about love difficulties as travel difficulties. And people are not aware that they are thinking metaphorically!

The first book on the subject was Lakoff and Johnson's (1980) *Metaphors We Live By*. Since then, a wide-ranging science of metaphorical thought and language has developed, and we will go through the main results in this section. The following terminology has developed: The subject-matter domain is called the **target domain**. The word-meaning domain is called the **source domain**. The idea behind the terminology is that the inference structure of the source is used to understand the target. The relation between the source and the target is called a **mapping** from source to target. Since both source and target have a frame structure, a conceptual metaphor is understood as a *mapping from a source frame to a target frame*.

As we shall see, the primary embodied schemas of the source (motion, containment, etc.) are imposed on the target via the mapping.[10] For example, in the journey metaphor for a long-term love relationship, the beginning of the journey is mapped to the beginning of the relationship, the path of the journey to the course of the relationship, the goal of the journey to the goal of the relationship, and the difficulties in the relationship journey to the difficulties in travel along the path. Each metaphor is given a name with the following form: *Target is source* (e.g., love is a journey). It should be borne in mind that *this is just a name for the conceptual metaphor. The conceptual metaphor itself is the mapping from the elements in the source frame to the corresponding elements in the target frame.*

The advances made since 1980 are huge. Since we think using neural circuitry, every conceptual metaphor uses neural circuitry. And Narayanan has shown how neural mappings may be accomplished, which we will discuss in chapter 3.

One of the most important developments in conceptual metaphor theory is what is called a **primary metaphor**, a conceptual mapping from one primary embodied schema to another primary embodied schema. The primary metaphors are the main linkages across and among the embodied schemas. These primary metaphors are what ground the entire metaphor system.

We now turn to details, looking at primary metaphors and how they work.

THE CONDUIT METAPHOR

Michael Reddy had found that the abstract concepts of communication and ideas are understood via a conceptual metaphor that he dubbed the conduit metaphor:

- Ideas are objects.
- Language is a container for idea objects.
- Communication is sending idea objects in language containers.
- The means of communication includes a "conduit."

The "conduit" can be a telephone hard line, an internet transmission, a literary text, or the space between two people or between a speaker or writer and an audience.

The above notation, having the form "A is B" or "Target Element (Object) Is Corresponding Source Element (Idea)," characterizes a conceptual mapping from a source domain frame for sending objects in containers to a target domain frame for communicating ideas via language (Lakoff and Johnson 1980).

Reddy (1979) found over 100 classes of linguistic expressions for the conduit metaphor. Examples include, among others, the following:

- *You finally **got through to** him.*
- *The meaning is right there **in the words**.*
- ***Put** your thoughts **into** clear language.*
- *Your words are **hollow**.*
- *The ideas are **buried in** dense paragraphs.*
- *You haven't **given** me any idea of what you mean.*
- ***Put** those ideas **down** on paper.*

- *Those ideas have been **floating around** for ages.*
- *When you have a good idea, try to **capture it** immediately **in** words.*
- *Emily Dickinson **crams** an incredible amount of meaning **into** her poems.*
- *The introduction has a great deal of thought **content**.*

Reddy's point was that the generalization covering such linguistic metaphors was not in language but rather in the metaphorical *concept* of communication as putting ideas into words and sending idea objects in language containers.

Reddy furthermore pointed out that the conduit metaphor created an important inference about communication: the speaker is primarily responsible for its success. If you put an object in a container and send it, the receiver will find the *same* object inside. Reddy observes that in real communication the hearer has as much responsibility as the speaker and that what the hearer hears is very often not what the speaker intends. The metaphor, however misleading, is often taken literally, as if it were true. In contemporary politics, Republican operative Frank Luntz (2007) has adopted the slogan "It's not what you say, it's what they hear."

What is implicit in these examples is a distinction between *linguistic* and *conceptual* metaphors. Linguistic metaphors are just expressions in a language, using words. The words in linguistic metaphors get their meaning via conceptual metaphors that can occur within a system of frames and mappings from frame to frame.

METAPHOR SYSTEMS

A crucial idea in the science of metaphor is that a *domain of thought* can be characterized by a *conceptual metaphor system*. This idea was first worked out by Eve Sweetser and Alan Schwartz,[11] who observed that there is a domain of mind (a metaphorical target) that is understood via a very general metaphor that is in turn split into four subcases, each associated with a separate source domain. The general metaphor is the following conceptual mapping:

- The mind is a body.
- Mental functioning is bodily functioning.
- Ideas are objects of bodily functioning.

The four special cases of conceptual metaphors are as follows:

1. Thinking is moving, ideas are locations, **communicating is leading**, and understanding is following.
2. Understanding is seeing, ideas are things seen, and **communicating is showing**.
3. Thinking is object manipulation, ideas are objects, **communication is sending**, and understanding is grasping.
4. Thinking is eating, ideas are food, **communication is feeding**, and understanding is digesting.

Their point is that Reddy's conduit metaphor for communication is a special case of one of four conceptual metaphors that are in turn special cases of one general metaphor for thinking. *The main insight is that conceptual metaphors occur in systems.*

There are many linguistic examples of each of the above conceptual metaphors, such as the following:

- Moving: *reach a conclusion; go off on a tangent; do you follow me; go step-by-step.*
- Seeing: *see what I mean; point of view; shed light on; clear; brilliant.*
- Manipulating: *turn it over in your mind; toss ideas around; I gave him that idea.*
- Eating: *food for thought; raw facts; half-baked ideas; digest; he won't swallow that; it smells fishy.*

What seems to define these domains are embodied brain regions or structures significantly involved in performing these functions. There are two questions raised by this analysis: *What is a "domain" in the brain? What defines a metaphor system and a domain neurally?*

Domains seem to be characterized by hierarchically structured frames. A frame is a complex schema, a mental structure that organizes knowledge. Each frame makes use of primitive embodied schemas and may make use of prior conceptual metaphors. The elements of a frame are called "semantic roles." For example, the semantic roles of the seeing frame are the viewpoint, the viewer, eyes, light, the directing of the eyes, the act of seeing, things seen, the gaze (the line from the eyes to the thing seen), and degree of clarity. There is also knowledge about seeing: You need enough light to see, light has a source, the gaze must extend from the eyes to the thing seen in order to see, things look different from different viewpoints, and so on.

The concept of a gaze is itself metaphorical, based on the metaphor of seeing is touching. The gaze is like an outstretched limb. The eyes can in many cases stand metonymically for the metaphorical outstretched limbs. Examples include *Their eyes met. My eyes **picked out** every detail of the pattern. I **ran** my eyes **over** the wall. She **undressed** me with her eyes. From Berkeley I can see all the way to San Francisco. On a clear day, my gaze **stretches to** Mount Diablo.*

A crucial thing we learn from this is that important abstract concepts are understood not merely via one conceptual metaphor but instead via multiple conceptual metaphors that provide different understandings of the concepts. For example, communication is not just sending but it is also leading (when thinking is moving), showing (when understanding is seeing clearly), and feeding (when thinking is eating). Ideas, metaphorically, can be not only manipulable objects but also locations and food.

Other metaphors for the mind include the following:

- Thinking is addition: *That just doesn't add up. Let me sum it up for you. I just put two and two together. That doesn't count!*
- Thought is language: *Do I have to spell it out for you? It's Greek to me. Liberals and conservatives don't speak the same language. The argument is abbreviated. He's computer literate.*
- The mind is a machine: *The wheels are turning now! I'm a little rusty. He had a mental breakdown.*

Lakoff and Johnson (1980, 1999) have shown that important concepts such as events, actions, causation, the mind, the self, morality, and being are each defined via multiple conceptual metaphors, sometimes between a dozen and two dozen. Each metaphor supplies a distinct way to understand the concept, which includes an inference pattern for reasoning.

LOVE

Lakoff's initial discovery of conceptual metaphors was similar to Reddy's (1979). Lakoff had found that the abstract concept of love is commonly understood in terms of a journey. There are lots of linguistic expressions of this sort, such as the following: *Our relationship hit a dead-end street. The marriage is on the rocks. We're getting nowhere in this relationship. We're going in different directions. We're at a crossroads in our relationship. We're spinning our wheels in this relationship.* The generalization over these cases is not in the linguistic expressions but rather in a conceptual mapping (which is indicated here by the notation "→" to be read as "maps to"):

- Travelers → Lovers
- Vehicle → Relationship
- Common Destinations → Common Life Goals
- Impediments to Travel → Relationship Difficulties

METAPHOR AND THE MEANING OF IDIOMS

The earliest examples that Lakoff and Johnson looked at led them to the study of idioms. The traditional theory of idioms held that idioms had arbitrary meanings, that a given idiom could mean anything at all. They discovered that the meanings of a huge range of idioms were anything but arbitrary. Those idioms made use of conceptual metaphors but not in any obvious way.

The first idiom Lakoff looked at was *We're spinning our wheels in this relationship*. It has a conventional image with knowledge about the image: The wheels are on a car. The car is stuck with the wheels spinning (in sand, in mud, on ice, etc.). The car isn't moving. They people in the relationship are putting a lot of effort into getting it moving, but it won't move. They are frustrated.

The love is a journey mapping applies to the conceptual knowledge about the image. The car (a vehicle) is the relationship, the travelers are lovers, and they are not making progress toward common destinations (compatible life goals). They feel frustrated.

That is what it means to be spinning your wheels in a relationship. *The conceptual metaphor applies to knowledge about the image, yielding the meaning of the idiom!*

But although the love is a journey metaphor applies systematically in understanding this idiom, the literal meanings of the words in the idiom ("spinning" and "wheels") are not mapped by this metaphor. *Those words activate a conventional mental image with associated knowledge that is commonplace.* There is a system of conceptual metaphors fixed in the mind that applies naturally, automatically, very quickly, and unconsciously to such *knowledge (and not necessarily the words)*, thus *linking the knowledge of the image via conceptual metaphor to the meaning of the idiom.*

There are a huge number of idioms like these. Consider *The marriage is on the rocks*. The marriage (the relationship) is a boat (a vehicle). A boat on the rocks is not moving forward. The couple in the boat is not progressing toward their common destination (compatible life goals). The boat is likely to be harmed in some way. Even if it gets free of the rocks, it may not be able to continue on the journey. That is, even if the marriage survives, the couple may still split up. And when the boat hits the rocks, the passengers may be hurt physically. Given the metaphor that psychological harm is physical harm, the couple may be psychologically harmed.

If you have that image for the idiom and that knowledge about the image, then the metaphorical mapping tells you what the idiom means metaphorically from the knowledge of the image. That same love is a journey metaphor, applied to a different image and knowledge, yields a different meaning. Being on the rocks has different inferences than going in different directions.

These constitute a special class of idioms: they are both imageable and metaphorical. New ones are being created all the time (Lakoff 1987b, Case Study 2).[12]

Note that metaphorical mappings occur at a certain level of generalization. In the love is a journey metaphor, the relationship is a generalized vehicle. There are special cases of vehicles: cars, boats, planes (*We may have to bail out*), rockets (*We've just taken off*), trains (*We're off track*). It's important to recognize the general level of the conceptual metaphor. Encountering *The marriage is on the rocks*, you should *not* conclude that the conceptual metaphor is love is a boat but rather that love is journey (with the special case of a journey by boat). This example is important to remember. The right level of generalization is crucial in understanding metaphorical thought.

THE DISCOVERY OF PRIMARY METAPHORS

In everyday life, achieving purposes often requires getting to a destination. If you want a cold beer, you'll have to go to the refrigerator or the ice chest. In American culture, people are expected to have goals in life, and a couple in a long-term love relationship is expected to have compatible life goals. If love is a journey, that means having common destinations.

This allows us to explain why a love relationship is conceptualized as a metaphorical vehicle. There are three reasons:

- A vehicle is a means of getting to a destination.
- A vehicle is a *container*. In general, relationships are understood in terms of containers: you are *in* a relationship; you can *enter* or *leave* or be *trapped in* a relationship.
- Intimacy is understood metaphorically in terms of closeness, as can be seen in such expressions as *We're very close* and *We're drifting apart*.

Thus, a relationship is conceptualized as a container you are in where you are close, and the container is a means for reaching destinations. Cars, boats, and planes fit this conception.

Lakoff and Johnson (1980) then asked themselves further questions and reasoned as follows:

- Why is intimacy understood metaphorically as closeness? Because intimacy requires being physically close.
- Why is a relationship a container? Because when you are growing up, you tend to live in the same enclosed space as your relatives.
- Why is a long-term love relationship conceptualized as travel to common destinations? Because goals are conceptualized as destinations, and long-term love relationships require having compatible life goals.
- Why are purposes characterized metaphorically as destinations? Because over and over again, action is conceptualized as motion, and to achieve a purpose you have to move to a specific location.

There are two general principles:

- Complex conceptual metaphors are made up of primary conceptual metaphors.
- Primary conceptual metaphors arise when there are regular correlations in real-world embodied experience.

Those principles lead to two further questions:

- What are primary conceptual metaphors neurally?
- Why should regular correlations in real-world experience lead to primary conceptual metaphors?

These questions are answered by our proposal for a neural theory of primary conceptual metaphor, which we turn to in the section "The Neural Theory of Metaphor."

Scales and Their Logic

Linear scales are central to thought and language around the world. We use them for measurement, for comparison, for estimation, for defining goals, for measuring progress and regress, and in running machines and other equipment (thermostats, radios, speedometers). It may not be obvious, but linear scales are metaphorical.

Traveling along a route to a destination fits the motion schema, but it is more than mere motion: it is a special case in which the whole body moves with a purpose (reaching a given destination) and is moving not just along a path of motion but also along a prescribed route. It is an image schema in that it is general (not specific), has parts and relations among the parts, and so on.

THE LINEAR SCALE METAPHOR

There is a conceptual metaphor that linear scales are routes of travel. For example, in talking about your grades as a student or your income as a businessman, you can say "I'm *ahead* of where I was last year" or "I'm *behind* where I was last year." In talking about a scale of success, a new CEO of a corporation can say "I've *come far* in my career." The linear scale schema can be used for comparisons on a scale: "John is *far* more intelligent than Bill" and "John's intelligence *goes beyond* Bill's." The subject matter is the comparative locations of John and Bill on the intelligence scale, but you can see the travel metaphor in the use of motion words such as "ahead of," "behind," "come" "far," "go," and "beyond."

A further conceptual metaphor can be added: more is up and less is down, to make the linear scale vertical. The businessman can say "My income has risen this year" if he has made more money and "My income has fallen this year" if he has made less. What makes the linear scale schema a schema is that it can be about any measurable subject matter, just as the triangle schema can be about any type of triangle and the container schema can be about any kind of container.

The linear scale metaphor applies to the travel schema as follows:

- The route of travel maps to the scale.
- The start of travel maps to the zero point on the scale.
- The travel destination maps to the endpoint on the scale.
- Travel maps to a quality (Q) on the scale.
- The traveler maps to an entity measured for values of quality (Q).
- Locations reached during travel map to values measured on the scale.
- The greatest distance traveled maps to the greatest value measured.
- Lesser distances traveled map to lesser values measured.

Here is an example of the logic of travel along a route.

Suppose you start walking on a route that is two miles long, and you walk one mile. Then you have walked a quarter of a mile, a half a mile, three-quarters of a mile, and all distances less than a mile but no distances on the route longer than a mile.

Now consider an example of the corresponding logic of linear scales. The scale measures the values of your wealth in a bank account.

Suppose you have $1,000 in the bank account. Then you also have $800, $500, $200, and every lower amount less than $1,000 and greater than zero in the account. But you do not have $1,200, $1,500, $2,000 and every amount greater than $1,000 in the account.

Here is another example:

Suppose the scale measures the height of a tree that is not being pruned as it grows in a yard. Then, if the height is of the tree is 25 feet, it has previously grown to 10 feet, 20 feet, and all heights less than 25 feet but no heights greater than 25 feet.

In short, the linear scale metaphor maps the logic of travel to the logic of linear scales, preserving the structure of the logic of travel. Here is how it works.

The travel schema comes with a travel logic schema:

- A traveler is traveling on the route from a starting point A to a destination B.
- D is the distance traveled.
- Then, on that trip, the traveler has been at all distances from A less than D.

The linear scale metaphor maps the travel logic schema to the linear scale logic schema, preserving the structure of the travel logic schema.

The linear scale logic schema looks like this:

- An entity is assigned values for a quality between the zero point and the endpoint of the scale.
- The maximum value assigned is V.
- Then, all values less than V also hold for the entity on the scale.

The inference that follows from this metaphorical mapping is as follows:

- All of the values between zero and V on the scale hold.
- None of the values between V and the endpoint of the scale hold.

CONCEPTUAL METAPHORS ARE LOGIC-PRESERVING

In general, conceptual metaphors that apply to schemas preserve the logics of those schemas. This applies in all conceptual metaphors, and it is the mechanism by which metaphorical reasoning works. This was discovered in the early 1980s. Here are some examples:

- Knowing is seeing: *I see what you're getting at* means I know what you're getting at. *I don't see what you're getting at* means I don't know what you're getting at.
- The conduit metaphor: *You got that idea across* means you communicated that idea. *You didn't get that idea across* means you didn't communicate that idea.
- More is up: *The stock price went up* means the value of the stock increased. *The stock price went down* means the value of the stock decreased.
- Groups are containers for members: *I'm in the group* means I am a member of the group. *I've left the group* means I am no longer a member of the group.

Linear scales for opposite concepts have reverse directions (e.g., rich vs. poor):

- *Harry is far poorer than he used to be* means Harry has much less money than he used to have.
- *Harry is far richer than he used to be* means Harry has much more money than he used to have.

Examples such as these are everywhere. They reflect the fact that conceptual metaphors preserve the logics of their source domains.

Primary Embodied Metaphor

The theory of primary metaphor is one of the deepest and most important results in scientific study of conceptual metaphor. The primary metaphor theory explains the existence of the most prized of human conceptual thought: imaginative thought, the mode of thought that lies behind philosophical theories, moral theories, poetry, cultural practices, religion, and even higher mathematics and science. The theory explains why imaginative metaphorical thought is a human universal, why it arises naturally in children just from living in the everyday world, and also explains why imaginative metaphorical thought is mostly unconscious and why it could have gone unnoticed for 2,500 years.

The primary metaphor theory is a neural theory, suggesting that the circuitry for primary conceptual metaphor crisscrosses the brain, linking embodied circuitry in one brain region (a source domain) to embodied circuitry in another brain region (a target domain) in such a way as to extend embodied reason from a source schema to a target schema, enriching each target schema with an additional mode of thought from the source. This is what metaphorical thought is about: conceptual richness and the ability to understand one domain (the target) in terms of another domain (the source). Hundreds of such primary conceptual metaphors have been found so far, placing conceptual metaphor at the center of imaginative thought.

Those primary conceptual metaphors are joined together by binding circuitry with other primary metaphors and frames to form complex conceptual metaphors. Further bindings link those complex conceptual metaphors together, forming cascades. Those cascades of conceptual metaphors form the imaginative conceptual products of culture, from narratives to philosophies to forms of politics to businesses to mathematics.

BACKGROUND TO THE THEORY OF PRIMARY METAPHORS

The theory begins with a basic observation about conscious thought. The division between "concrete" and "abstract" thought is based on what can be observed consciously from the outside. Physical entities, properties, and activities are "concrete." What is not visible or touchable is called "abstract": emotions, purposes, ideas, and understandings of other nonvisible things (freedom, time, social organization, systems of thought, and so on).

But from the perspective of the brain, each of those "abstract" concepts *is* physical because all thought and understanding is physical, carried out by neural circuitry. This puts "concrete" and "abstract" ideas on the same physical basis in the brain. Whereas early conceptual metaphor theorists (e.g., Lakoff, Johnson, Grady) saw conceptual metaphor as conceptualizing the abstract in terms of the concrete, neural metaphor theory links physical neural circuitry to other physical neural circuitry, allowing for a uniform theory. Neural metaphor theory also explains why early theorists thought that conceptual metaphor went from concrete to abstract.

The idea of primary metaphor was first developed by Lakoff and Johnson in *Metaphors We Live By*. They noticed that the conceptual metaphors "happy is up" and "sad is down" (as in *I'm feeling up today* and *I'm down in the dumps*) are motivated by the relationship between emotions and bodily musculature.

Lakoff and Johnson also noticed that the conceptual metaphors "more is up" and "less is down" are motivated by a repeated everyday experience in the world that we all have: every time you pour more liquid into a container or pile more objects on a table, the level goes up. They reasoned that *somehow this regular correlation in real-word experience leads to a primary metaphor.*

Joseph Grady (1997) in his dissertation did a survey of primary metaphors up to that time, analyzing about 100 of them. Lakoff and Johnson (1999) in *Philosophy in the Flesh* showed how primary metaphors constituted the basis of our most central "abstract" concepts, those not only central to Western philosophical thought but also present in cultures throughout the world: time, events, causation, the mind, the self, and morality.

CORRELATIONS IN EXPERIENCE

The fact that there are repeated correlations in everyday experience lies behind the theory of primary metaphor.

Here is a commonplace example. A common occurrence in everyday life is that one has to *go to a specific location* in order *to achieve a given purpose*. If you want a cold beer, you have go to the refrigerator where the beer is kept. If you want to brush your teeth, you have to go the bathroom where the toothbrush and toothpaste are kept and there is a sink. Even infants, to feel secure, have to crawl over to where their favorite toy animal or their blanket is lying. These experiences requiring you to go to a destination to achieve a purpose give rise to the primary metaphor "purposes are destinations," which is widespread around the world. This primary metaphor maps the motion schema onto the purposeful action schema, as follows:

- The mover maps to the actor.
- The motion maps to the action.
- The motion source maps to the action precondition.
- The motion goal maps to the purpose.
- The motion path maps to the means.
- Resources for motion (e.g., a vehicle, energy) maps to resources for achieving the purpose.

- An impediment to motion maps to a difficulty in achieving the purpose.
- Reaching the destination maps to achieving the purpose.

Each of these is a submapping; the whole collection of submappings jointly constitutes the metaphor mapping.

This mapping is not arbitrary or fanciful; it reflects a real-world fact. In the repeated experiences of going to a location to achieve a purpose, the elements of the motion schema correspond to the elements of the purposeful action schema. That is, the actor *is* the mover, the action *is* the motion, and achieving the purpose *is* reaching the destination.

THE NEURAL THEORY OF METAPHOR

Earlier in the book we were left with two questions:

- What are primary conceptual metaphors neurally?
- Why should regular correlations in real-world experience lead to primary conceptual metaphors?

In answering these questions, we reasoned as follows:

- Whether or not one consciously notices such experiential correspondences, from early childhood our brains notice them; that is, each experience is registered in the brain, each in a different brain region, with the neural circuitry in the two regions *firing at the same time*.
- The nodes in each region that regularly fire together get stronger (via Hebbian learning) with regular firing.
- Such correspondences in the real world cannot be noticed even unconsciously by the brain unless there is neural circuitry across the brain regions linking the right corresponding circuits; that is, motion: circuits linked to action circuits, destination circuits linked to purpose circuits, and so on.
- The neural activation spreads out from each of the two neural ensembles along existing pathways, creating neural links that get stronger as regular firing continues. The spreading from two brain regions keeps extending and strengthening until such persisting circuitry for regular correlations in the world is formed. Only with such linking

circuitry can the correlations be "recognized," at least unconsciously, by the brain.

- Eventually a shortest pathway is reached in each direction. The pathway is strengthened by regular firing, and a cross-region circuit is formed linking the circuitry across the two distinct regions to form a correlation circuit.
- Between the two brain regions there will be neural connections going in opposite directions, sharing the synapses along that shortest pathway.
- This creates the condition for timing-based synaptic learning mechanisms such as STDP. In STDP, the synaptic connections of the neurons that regularly spike first in a temporally correlated manner are strengthened in its direction, while the synaptic connections in the opposite direction from the other later-firing neurons are weakened.
- This asymmetric pattern of strengthening in one direction and weakening in the opposite direction occurs all along the pathway of the two-way correlation circuits.
- The result is an asymmetric activation pattern, with activation going from one region (which we call the source) to the other region (which we call the target).
- The directionality of activation flow is determined by the neurons all along the pathway that regularly spike first. This in turn depends on which of the circuits at the ends of the pathway receive more overall activation.

This raises the question of why neurons in one region regularly spike in repeated correlated patterns before the neurons in the other region. The answer is simple: neurons in one region spike in the direction from which the most correlated activation comes regularly. That will be the metaphorical source region. The other will be the metaphorical target region.

But now that raises a further question: Why should neurons in one region regularly get more activation than neurons in another region?

When we look at examples, we get two answers, which provide a hypothesis. Here are some examples:

- More is up, less is down: *Stock prices went up. Turn the radio down.* Here verticality is the source, and quantity is the target. Why? Because the brain is always computing verticality both internally (even when you are sleeping) and externally (based on visual stimuli), but the brain is not always computing quantity. Thus, there is more activation regularly flowing from the verticality region to the quantity region, which will lead to first-spiking in the verticality-to-quantity direction.

- Affection is warm, disaffection is cold: *He's a warm person. She's cold as ice.* Here temperature is the source domain, and affection is the target. Why? The brain is always monitoring/computing/maintaining temperature but is not always computing affection. Thus, there is more correlated early activation regularly flowing from the temperature region to the affection region, which will lead to first-spiking in the temperature-to-affection direction.
- Purposes are destinations: *There's nothing standing in my way. I hit a roadblock on this project. He has reached his goal.* This is, of course, another obvious example. Not all of our motions are purposeful. We do a lot of aimless moving. For this reason, the motion schema will be activated more than the purposeful action schema, resulting in first-spiking occurring in the motion-to-action direction and therefore predicting that motion will be the metaphorical source and that purposeful action will be the metaphorical target.

So far, all the cases we have thought through are explained either by (1) activation for a longer time or (2) activation from more sources.

As a result, we hypothesize that the sources and targets of primary metaphors *might well be* predictable by the Hebbian theory (such as STDP) of neural learning. To be able to predict the directionality of primary metaphors on the basis of low-level neural learning via STDP would be a truly remarkable result if it holds up. At present, this is a theoretical hypothesis; there is no *experimental* evidence at all for (or against) this hypothesis.

CHIMPANZEE CONCEPTUAL METAPHOR

Dahl and Adachi (2013) reported in their paper "Conceptual Metaphorical Mapping in Chimpanzees (*Pan troglodytes*)" that they experimentally showed the effects of a conceptual metaphor well known in English and other human languages that apparently also occurs in chimpanzees: high rank is up, low rank is down. They presented pictures of high- and low-ranked chimpanzees at the top and bottom of a computer screen in two conditions: high-ranked on top and low-ranked on bottom versus low-ranked on top and high-ranked on bottom:

We found a modulation of response latencies by the rank of the presented individual and the position on the display: *a high-ranked individual presented in the higher and a low-ranked individual in the lower position led to quicker identity discrimination than a high-ranked individual in the lower and a low-ranked individual in the higher position.* Such a spatial representation of dominance hierarchy in chimpanzees suggests that a natural tendency to systematically map an abstract dimension exists in the common ancestor of humans and chimpanzees. (Dahl and Adachi 2013, p. 1, emphasis added)

Dahl and Adachi conclude that these results indicate the existence of the high rank is up, low rank is down metaphor in chimpanzee cognition.

This result is made sensible in the research by Ravignani and Sonnweber (2017) in their paper "Chimpanzees Process Structural Isomorphisms across Sensory Modalities." They tested chimpanzees across the visual and sound modalities and used two patterns: symmetric (A B A) versus edge (A B B). In the visual modality A and B were shapes, and in the sound modality they were sounds. The chimpanzees correctly matched the symmetric shapes with the symmetric sounds and the edge shapes with the edge sounds.

Sounds and shapes are, of course, activated in different brain regions. The chimpanzees could match the shape patterns with the corresponding sound patterns across the brain regions.

From the neural mind perspective, this requires neural circuitry linking these structural correlations across brain regions. This is just what is required of the correlation stage in our theory of primary metaphor.

If we assume that the chimpanzees studied by Dahl and Adachi also have this capacity, then they should have circuitry linking height and social rank. Since verticality is activated throughout chimpanzees' habitat whether or not social rank is an issue, verticality should get more regular activation than rank, and hence, via STDP, the directionality of the circuits linking verticality and rank should get the verticality to rank directionality, creating high rank is up, low rank is down as a conceptual metaphor.

Although it may seem remarkable and strange that chimpanzees should have conceptual metaphor, these considerations indicate that it is entirely natural that this should happen.

We would like to encourage other animal cognition researchers to read the literature on conceptual metaphor as an inspiration to future research on conceptual metaphor in animals.

There are hundreds of primary metaphors structuring our conceptual system. They are learned via neural learning mechanisms early in life, usually before language, just via our functioning in the everyday world.

Each primary metaphor neurally maps one primitive schema onto another via an asymmetric circuit linking them. But each primitive schema (e.g., motion, purpose) can also occur independently of any metaphor circuitry linking them. For example, motion can be performed without purpose, and purposes can be achieved without moving from one place to another.

Since the source and the target can occur independently, we conclude that the metaphor circuitry is not always active and so must be "gated," with a node called a "gate" controlling when the metaphor circuitry is active and when it is not. We hypothesize that there is a gating node normally inhibiting the operation of the metaphor. When the gating node is disinhibited, the metaphor circuit is turned on. To be more precise, when the gating node is activated, an inhibiting connection to the neural mapping (the conceptual metaphor) stops the mapping. When the gating node is itself inhibited (turned off), it can no longer stop the metaphor (the cross-region neural mapping) from proceeding.

Each submapping has a gate. In the whole mapping, the gates work together. How? The theory hypothesizes that this is done by a simple gestalt circuit, which characterizes a part-whole structure. Each submapping in the metaphor is a part of the metaphor. When all parts are inhibited, the neural mapping is inhibited. This happens when the gestalt node for the whole circuit is activated and when it sends inhibiting connections to the parts. As before, part-whole circuits have easy-to-learn combinations of activation strengths and threshold strengths.

The gating nodes allow metaphors to be turned on and off. There are many thousands of cases of complex metaphors that are combinations of primary metaphors. We hypothesize that there are "cascade circuits" that control which metaphors are turned on in particular cases as well as which binding circuits are turned on to link the individual metaphors into a coherent whole.

The cascade circuitry linking the metaphors controls the activation of both primary metaphors and complex metaphors, thus allowing the cascade to control complex metaphorical inferences.

Each primary metaphor allows reasoning characterized by the circuitry in one region of the brain to apply to circuitry in another region of the brain. Primary metaphor thus *makes reasoning patterns for individual schemas that are active locally into reasoning patterns that can apply globally.* As we saw in the section "Integrative Bindings," the container schema characterizes an inherent container logic: If A is in B and B is in C, then A is in C. And if X is outside of Y and Y is outside of Z, then X is outside of Z.

Via the primary metaphor that categories are containers (for their members and subcategories), we get the neural equivalent of the Aristotelian syllogisms. For example, "all men are mortal" is the category statement that the category "man" is in the category "mortal." *Under the metaphor that categories are containers, the category-container man is inside the category-container mortal.* The statement "Socrates is a man" places Socrates in the inner container, which in turn places him in the mortal category. The logic of containers thus becomes the classical syllogism, given the primary metaphor that categories are containers. In general, *the standard principles of formal logic arise* **biologically** *from the logic of* **embodied** *primary schemas.*

Primary metaphors provide the elements of the overall inferential structure to our system of concepts. The neural binding of primary metaphors together to form complex metaphors results in complex inference patterns. The main point here is that *reasoning is neural and therefore physical rather than being abstract and somehow floating in air.*

THE RESULTING THEORY

From this point, the following theory emerges:

- The human brain is structured by thousands of embodied metaphor mapping circuits that create an extraordinary richness within the human conceptual system. Like most thought, they largely function unconsciously.
- These mapping circuits arise from connected regions in the brain including multimodal association areas, where there are initial connections between sensorimotor (homeostatic control and social cognition) circuits such as temperature and space (distance, velocity) and more abstract and subjective circuits for quantity, affection, and purpose.

- These mapping circuits asymmetrically link distinct brain regions, allowing reasoning patterns from one brain region to apply to another brain region.
- Although metaphorical in content, the circuits reflect a reality. Real correspondences in real-world physical and social experiences starting in infancy give rise to primary metaphors, which provide the meaningful basis for complex metaphors, and complex metaphorical reasoning.
- Where the basic experiences are essentially the same across cultures, the primary metaphor mappings tend to be the same. They appear to be learned by experience via neural learning. The asymmetry of the mappings appears to arise via STDP, from which metaphor sources and targets can (we believe) be predicted.
- Simple metaphorical thought is learned prior to and independent of language, just through living in the world and having a neural system that learns via neural learning mechanisms.
- Complex metaphorical thought is formed via neural binding, integration of regions linked via neural bindings (as discussed in chapter 3), and cascades.
- Certain metaphor complexes arise cross-culturally via experiences in the everyday world.
- Complex metaphorical thought shows up not just in language but also in gesture, imagery (paintings, movies, dance, etc.), mathematics, science, and moral and political ideology.
- The compositional properties of language, not surprisingly, lead to an unbounded range of complex metaphorical thought expressed linguistically.
- All imaginative and poetic metaphor uses the primary metaphor system and conventional complex metaphors.

Complex metaphorical thought is ubiquitous in our conceptual systems but is hard to notice in many cases. The reason is that (1) each conceptual metaphor in the complex has a different source domain, and (2) the source-to-target mappings are *integrated* in a complex cascade and are therefore less obvious and harder to tease apart. When we form complexes of four, five, and six metaphors, the sources domains become so integrated that they are difficult to see separately. Nonetheless, there are techniques for doing so. We will be discussing such complex metaphors starting in chapter 3.

The theory of primary metaphor is an explanatory theory and a powerful one. But it is nice to have external experimental evidence as well. In the section "What Is Structured Neural Computational Modeling" in chapter 3, we will survey a range of such cases. But after a theoretical discussion of the present depth, it is sweet to have a simple, straightforward example of a confirming experiment.

The two sentences (1) *She looked at him sweetly* versus (2) *She looked at him kindly* were studied in a functional magnetic resonance imaging (fMRI) experiment by Francesca Citron of the Free University of Berlin and Adele Goldberg of Princeton University that compared brain activations for the two sentences (Citron and Goldberg 2014). "Sweetly" expresses roughly the same abstract meaning as "kindly" in the sentence, but the abstract meaning is expressed via a neurally constituted conceptual metaphor and therefore activates additional brain regions: gustatory and emotional (the amygdala).

"Kindly" indicates a strong positive emotion. That emotion showed up on the fMRIs for sentence 1. "Sweetly" is more complex because of a conceptual metaphor, namely "kindness is sweetness." In sentence 2, kindness is expressed metaphorically via sweetness. In the fMRIs for sentence 2, the same strong emotion showed up as in sentence 1, but there was activation is the gustatory region as well. There were two important findings in this result.

- The word "sweet" was triggering noticeable brain activity on the basis of its meaning. But its meaning in this case was not literally about flavor but instead was metaphorical about kindness via the conceptual metaphor "kindness is sweetness." For this to happen, that metaphor had to be characterized in the brain via brain circuitry: since activating sweetness in the gustatory region activated the idea of kindness in emotional regions, there had to be a circuit linking them. In the brain, that circuit *constitutes* the metaphor.
- The adverb "sweetly" is usually not used to talk about flavor. We don't usually say, for example, "The fruit sweetly flavored the yogurt," although we could understand it. But the fact that presence of the word "sweet" in "sweetly" shows that "sweetly" does not just literally mean only kindness but also that "sweetly" makes use of the gustatory region of the brain and so characterizes the mode of thought in the conceptual metaphor "kindness is sweetness."

To summarize, the gustatory meaning of "sweet" is embodied: it activates the region in the brain with connections to taste buds in the body. Via circuitry linking sweetness to kindness, the same emotional regions of the brain are activated as with "kindly." As predicted, two distinct and separate regions are activated. We hypothesize that the two regions are connected by circuitry that carries out in the brain what we understand as a metaphorical thought: kindness is sweetness.

What can we conclude about embodiment and thought? The meaning of "sweet" is embodied via the gustatory region of the brain, which uses embodied circuitry connecting that region to our taste buds. The meaning of "She greeted him sweetly" is arrived at neurally in the brain via activation of the gustatory region of the brain. Thus, *the embodiment of sweet taste is used in understanding the thought about kindness expressed in the sentence.* And the emotional correlates of sweetness are thereby associated with kindness in the sentence.

EMBODIMENT AND EMOTION METAPHORS

The Goldberg-Citron experiment provides straightforward neuroscience evidence for the role of neural embodiment in conceptual metaphor. We now turn to a dramatic case of physiological evidence.

In the early 1980s, Zoltán Kövecses (2000) and George Lakoff (1987b) discovered that systems of emotion metaphors arise from the physiology of emotions. For example, Paul Ekman and his colleagues (1983) found that *when one is angry, skin temperature rises, blood pressure increases, and there is interference with accurate visual perception and fine motor control.* That is why we get such linguistic metaphorical expressions as *boiling mad, he exploded, blind with rage, hopping mad,* and many more (see Lakoff 1987b, Case Study 1).

Antonio Damasio (1996) has observed that such bodily experiences have correlates in the brain's somatosensory system, which are registered and can be seen via neuroimaging in the ventromedial prefrontal cortex as "somatic markers" that play an important role in decision-making. This raises the possibility that emotions *are constituted by* the bodily effects that are registered in brain during emotional experience. Thus, *it would be natural for emotions to be metaphorically conceptualized as those bodily effects,* as Kövecses and Lakoff observed. This accords with the theoretical model of Lindeman and Abramson (2008) of the causal mechanisms of depression. They hypothesize that "(a) the inability to alter events is conceptualized metaphorically as motor incapacity; (b) as part of this conceptualization, the experience of motor incapacity is mentally simulated; and (c) this simulation leads to both feelings of lethargy and peripheral physiological changes consistent with motor incapacity" (p. 228).

These ideas, together with our emotion metaphor research, raise the possibility that one can get insight into emotional states via neuroscience and the study of linguistic metaphors for physical states.

CONCEPTUAL METAPHOR IN LANGUAGE

A whole field of metaphor science developed after 1980, including research on the role of conceptual metaphor in grammar. The first major paper on construction grammar came out in 1987, a 100+ page study of there constructions that demonstrated the importance of conceptual metaphor in grammar (Lakoff 1987b, Case Study 3). Since then in book-length studies, Adele Goldberg (1995) and Ellen Dodge and Abby Wright (2002) have demonstrated how conceptual metaphors work in grammatical constructions. Following those insights, Karen Sullivan (2007) has since provided the first general theory of how conceptual metaphor structures grammatical constructions. Dancygier and Sweetser (2014) provide an outstanding introductory text on these matters.

Why does research on metaphor in grammar matter for an understanding of conceptual thought? This research matters because grammar is widely seen as purely formal, meaningless ways of organizing linguistic form: ordering words and morphemes, putting them together into hierarchical and linear structures, and using "markers" for case, tense, number, gender, etc. according to presumably purely formal rules. As such, grammar would appear to be impervious to metaphor.

The research just cited demonstrates dramatically that conceptual metaphor enters into grammar in a variety of systematic ways, bringing embodiment to grammar.

FOR NEUROSCIENTISTS DOING METAPHOR RESEARCH: A WARNING

Neuroscientists who study metaphor by looking at idioms are subject to making mistakes.

When a neuroscientist is using an idiom in metaphor research where there is averaging over a number of subjects, it is important to make sure that all the subjects use the same metaphor in understanding the idiom. That is not easy to do. Moreover, the metaphor may apply systematically not to the words "spin" and "wheels," or to the words "on" and "rocks," but rather to the concepts in the way the whole image is understood—if it is understood at all!

Some idioms are completely arbitrary, that is, you cannot figure out the meaning from the words. Take "by and large." It was originally a nautical term from the days of sailing ships. To sail "by" meant close by the wind, whereas to sail "large" meant with the wind fully behind you filling the sails (making them large). If a ship sailed well both "by and large," then it sailed well under most conditions. Via the commonplace metaphor that Action Is Motion, with sailing as a special case of motion, sailing by and large came metaphorically to mean action by and large, that is, under most conditions. With the complete loss of "by and large" in its nautical meaning, the meaning of "by and large" kept the meaning of "mostly," but the systematic metaphorical relationship to the words was lost.

Some neuroscientists choose to study idioms with body part names like *hand* or with words for what body parts do, like *kick* or *bite*. The point is to see if the relevant body part word activates the brain region in the topographic map of the body in the motor cortex. But such idioms vary in their degree of arbitrariness and directness. There is a commonplace conceptual metaphor, Control Is Control by the Hands. It occurs in the understanding of idioms such as *It's in your **hands** now, He's got the whole world in his **hands**, They **handed** over the company to the Mafia*. In these cases there is a relatively direct metaphorical connection between hands and control. But that particular metaphor is not present in the understanding of *He's an old **hand** at phonological analysis, Tax cuts are **hand**outs to the wealthy, Don't bite the **hand** that feeds you*.

The idiom *kick the bucket* has been used in some neuroscience experiments to see if there is activation in the foot region of the motor cortex. What would one expect? Not much. First, there is a lot of variation across speakers. For many speakers it is an arbitrary idiom, with the meaning of kick playing no role at all in the meaning. For some there is a weak mental image. Here is mine:

The bucket is upright. There is some but not much liquid in it. It is weakly kicked over and what liquid there is spills out, and it is empty and on its side after the kick. There is a common conceptual metaphor that seems to be applying here: Life is a Fluid in the Body, as in sentences like *The life drained out of him; He's full of life; He's brimming with life*. The spilling out of the fluid from the bucket means death. But since there was not much fluid in it in the first place, it suggests a particular kind of death—death when there is not much life left, as with an old person expected to die soon. You won't say *She kicked the bucket* of a child run over by a car or a young woman who died in childbirth.

Incidentally like mine appears in a prominent place in two popular movies. In *It's a Mad, Mad, Mad, Mad World,* Jimmy Durante plays an old man who dies of a heart attack on a mountain. As rigor mortis sets in, his leg goes out and kicks over a bucket that tumbles down the mountain. In *Young Frankenstein,* the man soon to become the monster dies and, in rigor mortis, kicks over a slop bucket at the edge of the bed. The kicking of the bucket is a comic way of indicating death, a visual pun in two slapstick movies.

But for many speakers, *kick the bucket* is an arbitrary idiom, with no mental image of kicking. Even in the best of cases, one shouldn't expect much by way of foot activation in the motor cortex. The kicking is only indirectly connected to the death, and then only via a conceptual metaphor that has nothing directly to do with kicking. In addition, the bucket may be a container, like the body, but that's a weak connection. And for most speakers, there is no connection at all.

The morals for neuroscientists: Be aware of what kind of idioms you are using in your experiments and what their cognitive analysis is. Always list the idioms you are using in any write-up of your experiment. And test your subjects for the images they may or may not associate with the idioms. (Lakoff 2014b)

Incidentally, mental images are often unconscious. Nonetheless, subjects can often answer detailed questions about their unconscious images. Lakoff (1987b, Case Study 2) covers the many uses of the word "over"), such as *Do it over, but don't overdo it.*

THE METAPHOR SYSTEM, PHILOSOPHY, AND MATHEMATICS

The metaphor system is anything but trivial. It structures everyday thought from the mundane to the deepest of philosophical thought. *Philosophy in the Flesh* (Lakoff and Johnson 1999) surveys the system of primary and complex conceptual metaphors central to Western philosophical thought, showing that each major philosopher discussed—the pre-Socratics, Plato, Aristotle, Descartes, Kant (his moral theory), and twentieth-century Anglo-American philosophers—used a small set of conceptual metaphors, took them as literally true, and scrupulously worked out the entailments of those metaphors, seeing them as philosophical truths. Many of them are used today in everyday thought by ordinary people. A commonplace example is the mind as computer metaphor recently reinforced by Open AI's ChatGPT-4, which searches the web for data and simulates human thought and language in an eerily convincing way.

Lakoff and Núñez (2000), in *Where Mathematics Comes From*, show that mathematics, at both the basic and higher levels, is also understood unconsciously and intuitively in terms of embodied conceptual metaphors with special properties such as precision, symbolizability, and computability that make the metaphors suitable for mathematics. When metaphor is piled upon metaphor, the result is "abstract, higher" mathematics, as metaphors for space and number add more and more metaphors to yield negative numbers, set theory, mathematical logic, logarithms, trigonometry, abstract algebra, calculus, complex numbers, infinite series, space-filling curves, infinite numbers, and all forms of infinity in general. The general point is that even mathematics is a creation of the human mind, using exaptation from the sensorimotor systems that we share with animals—in particular, X-Nets, embodied image schemas, frames, bindings, and conceptual metaphors—just like the rest of human thought.

Conceptual Integration and Consciousness

Contemporary theories of consciousness tend to agree that consciousness is integrated. Unconscious conceptual thought, as we have shown repeatedly, is highly fragmented, with many brain regions activated and linked indirectly by cascades of connections across those regions. Given that thought is embodied, the reason for the fragmentation of unconscious thought is obvious: the brain connects to the body in many different ways from many different brain locations. Those diverse brain-body connections are what make conceptual thought not merely meaningful but also meaningful in complex overlapping ways. Meaning is far from unitary, as we have seen. Meaning requires many embodied schemas and many ways of creating neural composites of them to form complex schemas, logic schemas, conceptual frames, and conceptual metaphors, all of which contribute to the complexity of meaningful thought.

But for consciousness, fragmented unconscious thought has to become integrated. That is, composites of fragmentary thought have to form composites that fit together as integrated visual scenes, integrated sounds, and integrated ideas. Neurally, visual circuitry separates colors from shapes, and shapes are put together via cascades of separate parts. But in conscious vision, colors and shapes are not separate things floating before us but instead form integrated wholes: colored objects with colored shapes.

The neural change from unconscious fragmented chunks to conscious integrated wholes is not simple. As Shimojo (2014) showed (see chapter 1, "Creating Perception," in this volume), what is unconsciously fragmented seems to have to be changed in order to become conscious. In the experiments Shimojo cites, disjoint percepts are changed to already integrated, well-established patterns before they can be seen, felt, or heard consciously. We don't necessarily see, feel, or hear certain fragmented material that is presented to our senses. In those experimental cases, principles of best fit (that is, least energy) actively change what impinges on our sensory organs to fit prior well-integrated patterns already in the brain.

Conceptual integration seems to involve at least two kinds of cases:

- Sensory or conceptual inputs that do not fit existing integrated patterns are neurally changed to fit existing integrated patterns as closely as possible.
- Cultural constructions that go beyond natural experience—such as myths, stories, cultural practices, and institutions—have been conceptually (and neurally, of course) prefabricated to create a kind of sense outside of natural experience, a sense that better fits our emotional, imaginative, social, cultural, and even political lives.

Stories are necessary for this development, and the elements of such stories show up around the world in integrated forms in myths and other forms of imagination.

In chapter 4, we will discuss a simple neural computational model of learned conceptual integration structures. Briefly, our theory looks like this:

- There is in the brain a collection of control nodes and gates that govern the activation or inhibition of the fixed circuits for schemas, frames, metaphors, and bindings.
- Integration circuits create wholes by activating or inhibiting combinations of control nodes and gates.

In short, integration circuits turn combinations of other fixed circuits on or off as needed to avoid neural contradictions. The result creates entities such as flying horses (Pegasus); gods such as Zeus, who can turn himself into a swan; a walking, talking skeleton dressed in black who comes to your door, knocks, and leads you off to death; and cartoons with talking ducks and wise-cracking rabbits.

The structure of an integration circuit contains nodes and connections that activate or inhibit those control nodes and gates. Learning which of your existing circuits have to be activated or inhibited in a given case is carried out via the recruitment of an integration circuit that does the job. When the integration node is activated, the integration circuit forms an integrated whole. When it is not activated, the fragmented unconscious circuits are free to form other combinations.

Let us start with a well-known example.

THE GRIM REAPER

There is a very general personification metaphor that has many special cases:

- A phenomenon is a person causing that phenomenon.

Thus, Fortune brings about good or bad fortune, and Envy makes people envious, and, of course, Death brings about one's death. That's Death, as in John Donne's holy sonnet *Death, be not proud. . . .*

Perhaps the best-known personification of Death is the Grim Reaper, a mythological figure dating back to the fifteenth century, the time of the Black Death in Europe. Grim Reaper, a personification of death, appears in the form of a skeleton wearing a black cloak and hood and carrying a scythe. He comes to the door of a person in the prime of life and says you are coming with him. You have no choice but to go, forever.

George Lakoff and Mark Turner (1989) observed that the concept of the Grim Reaper was a combination of existing conceptual metaphors and metonymies. Let us start with these metaphors. Each of them occurs independently of the myth. *This means that there must be an integration circuit that brings the metaphors all together in just the right way when the circuit is turned on but leaves the metaphors to act separately and independently when the integration circuit is turned off.*

- A phenomenon is a person causing the phenomenon. Special case:
  Death is a person causing death.

- Death is departure. The prematurely dying person must leave, escorted by Death.
- People are plants with respect to the life cycle. Special case: Grain is cut down at the prime of life by a person wielding a scythe.

There have to be bindings across the metaphors so that they act together:

- The person Death, causing death, is the same person, Death, coming to the door escorting the newly dead person away.
- The person wielding a scythe and cutting down grain at the prime of life is the same as the person Death wielding a scythe, coming to the door, and causing death at the prime of life.

There is a metonymy: The skeleton that a person becomes after death stands for Death, the metaphorical person who causes death. This is the skeleton who comes to the door with the scythe.

All the metaphors, the bindings across them, and the skeleton metonymy have to be activated. But because the skeleton has to come to the door and walk and talk, the fact that skeletons do not walk and talk has to be inhibited.

There are also two metaphorical metonymies:

- Light stands for life, since you can see light when alive.
- Darkness stands for death, since a person cannot see when dead.

The metaphor "life is light, death is darkness" is activated. So are two metonymies:

- The black of the boils accompanying the Black Death maps onto the darkness of death in the metaphor.
- The black cloak—the clothing of a monk presiding over a funeral of a Black Death victim—stands metonymically for the clothing of the skeleton coming to the door.

In order to form the myth, all these elements have to be brought together by a conceptual integration circuit:

- The integration circuit has to activate all the metaphors via their gating nodes by disinhibition. That is, the integration circuit must inhibit each gating node that inhibits the corresponding metaphor, thus activating the metaphor enabling activation to flow from source to target.

- The integration circuit has to activate all the bindings going across the metaphors, linking the metaphors together into a single complex circuit.
- The integration circuit has to activate all the metonymic mappings—for example, mapping the skeleton onto Death—and must also inhibit the knowledge that skeletons don't walk and talk.

When the integration circuit performs all of these tasks of activating and inhibiting just the right control nodes and gates, what results is an integrated myth, which is conscious and can only be conscious because it is an integrated whole.

Integration does not guarantee consciousness. We all live by unconscious myths, as cognitive therapists well understand (e.g., books on the Cinderella syndrome in which women wait for their prince to come).

GENERAL PATTERNS OF INTEGRATION CIRCUITRY

Much of imagination is scripted at a general high level, with original variations at a specific low level. Fauconnier and Turner (2002) cite the case of Dante's account of Bertran de Born, who is encountered in hell. De Born plotted for his king to be murdered by the king's son. Metaphorically, such a crime is the removal of the head of state. De Born's punishment in hell is to have his own head removed and to walk around holding his head by the hair, with the head explaining what he had done to earn such a punishment.

There are integration patterns that constitute rules for myths that define what has been called "poetic justice."

- Rule 1: There is a metaphor defining a punishment that constitutes poetic justice for committing a crime. In the Bertran de Born case, the metaphor is the king is the head of state. The crime is treason: plotting the murder of the king, which is cutting off the head of the head of state. The punishment is to have the traitor's own head cut off.
- Rule 2: The metaphors, metonymies, and integration must require that something happen in the myth that we know cannot happen in the world. That knowledge must be inhibited by the integration. We know that heads that have been cut off cannot talk. That knowledge is inhibited in the case of Bertran de Born.

- Rule 3: Suppose a character in a myth is defined by a property associated with an animal. The character in the myth can then be half-person, half-animal with that property. Examples are centaurs, mermaids, and the Minotaur. Centaurs, half-man, half-horse, are hunters. Mermaids are voluptuous women who live in the sea and are half-fish; their sexuality and beautiful singing lures sailors to their death. The Minotaur is half-man, half-bull. He lives in a maze and will kill you if he runs across you in the maze. The Minotaur in the maze characterizes the fear of confusion that can lead to disaster or death.

Conceptual integrations do a lot more than just create mythological figures. They characterize concepts that require the partial activation and partial inhibition of parts of other concepts.

As Seana Coulson (2001) observes, a pet fish is a fish small enough to be domesticated and treated as a pet. The compound *pet fish* activates both concepts of pet and fish and binds them together, providing that the fish is small enough. The pet fish category inhibits the category members from being too large. A great white shark is not a good example of a pet fish even if you happen to keep one in your swimming pool.

A fake gun looks like a gun and can lead people to think it is a gun, but it doesn't fire bullets. Fauconnier and Turner (2002) is replete with such examples. These are all examples with both activating and inhibitory bindings. "Fake" as a modifier of "gun" inhibits the function of a gun in that it cannot shoot while also adding the frame of deception to the inherent nonfunctionality of the gun. You use a fake gun when you don't have a real one but are trying to deceive certain other people into thinking it is real.

CONCEPTUAL INTEGRATION AND CONSCIOUSNESS

Consciousness is integrated. Unconscious thought may be made up of multiple separate conceptual parts. But if a thought becomes conscious, it must be integrated.

The reverse is not true. Not every conceptual integration is conscious!

The examples are cases in which an integration circuit is required to perform a partial activation and partial inhibition to unconsciously and automatically arrive at the understanding of a phrase or a concept.

Consider a case discussed earlier. In discussing a relationship, one person says "We're spinning our wheels." There is an image of a car stuck and not moving while the wheels are spinning. Unconscious, implicit metaphors are activated and apply to knowledge of the image. Among them are the metaphors: a relationship is a container (the car), purposes are destinations, and difficulties are impediments to motion. They are activated together and integrated unconsciously, along with their inferences. The metaphors map the car as a vehicle, the fact that it is not moving; the travelers are expending effort trying to get it to move, and they are frustrated. But what is *not* mapped are the wheels and the spinning of the wheels, even though the sentence is "We're spinning our wheels." The metaphors apply to knowledge about the image, but an integration circuit has to explicitly indicate just which metaphors apply and exactly what they apply to and what they do not apply to in order to provide an integrated meaning for the idiom. The pairing of imageable idioms with their meanings requires such integration circuits.

Conceptual integration is also at the heart of the formation of complex words. Compare "oversee" and "overlook," as in a sentence such as "It is the job of a police commissioner to *oversee* the police force and not to *overlook* violations of law by police." Or compare "look over" and "overlook" in a sentence such as "*Look over* the manuscript, and don't *overlook* any typos." Here the same words "over" and "look" have their meanings integrated in very different ways depending on their differences in grammar. A circuit has to indicate exactly how the integrated meanings of "over" and "look" fit together for "overlook" and "look over."

Once you notice it, conceptual integration is everywhere in language and thought. We owe Fauconnier and Turner (2002) a debt of gratitude for pointing this out. Further examples will be discussed in chapter 4.

Categories

CATEGORIES ARE MEANINGFUL

"Don't categorize me!" It's a plea to be treated as an individual, not a type—or a stereotype—not limited by properties defining membership in some group. The plea comes from a recognition that categories have meaning and come with inferences about the members of the categories. One might like being categorized favorably—intelligent, good-looking, kind, thoughtful, dependable, a friend—or dislike being categorized with the opposite qualities. Either way, categorization matters and is inescapable.

Categorization is a natural neural phenomenon. Our neural systems categorize automatically. It cannot be helped. You have no choice.

Categorization is also a form of attribution: It has a causal effect. It creates a mental understanding (real or not) of what or who is categorized, an understanding that is active, that leads us to imaginatively simulate how someone might behave.

In short, our neural systems engage—naturally, automatically, effortlessly, and mostly unconsciously—in forming *categories that have a causal effect* by creating an implicit mental understanding of what is categorized rightly or wrongly, justified or not, but a real-world effect via the categorization.

The plea *Don't categorize me!* is a response to the reality that being categorized is hard to avoid. But it is a reality that is unlikely to change. The best we can do is to understand categorization itself.

There are, of course, cases of relatively neutral categorization. The goods in a supermarket have to be arranged in some way via categories: frozen foods, canned vegetables, soups, international foods, organic produce, and so on. But as Michael Pollan (in a personal communication) has claimed, even supermarkets put the high-caloric foods in the center of the store and the vegetables and fruits at the periphery, an unhealthful categorization. This is probably done unconsciously, not according to Pollan's observation but instead according to patterns of sales in which unhealthy foods just sell better than healthy foods. Conscious or unconscious, the categorization is real, and that reality is reflected in the way products are arranged in a store.

WHY CATEGORIZATION WORKS THIS WAY

The kinds of categories we create are complex and make use of all of the basic mechanisms of thought: embodied conceptual schemas, frames, metaphors, bindings, integrations, and cascades. Categories thus use all the neural mechanisms that constitute those mechanisms of thought. They are neurally natural, whether conscious or not. The purpose of this section is to point out how these neurally natural mechanisms of thought give rise to the remarkably complex types of categories.

CATEGORIZATION IS COMPLEX

George Lakoff's (1987b) book *Women, Fire, and Dangerous Things* is the standard text on categorization. It reveals the enormous complexity of category phenomena: Aristotelian categories, stereotypes, prototypes, basic-level categories, natural kinds, cluster categories, radial categories, salient exemplars, and so on.

Categorization is not simple, not just a matter of checking off items on a list such as the one given above. Because categorization matters, it is important to understand what kinds of categories there are and what makes them meaningful.

Meaning, as we have seen, derives from embodied experience. Meaningful ideas have structure: primary embodied schemas, complex schemas, logic schemas, frames, metaphors, metonymies, and conceptual integrations; the mechanisms for accessing them, namely control nodes and gates; and the mechanisms for combining them, namely bindings, integrations, and cascades. The question we ask in this section is how these mechanisms come together to create the complexities of forms of categorization.

CATEGORIES AND MEMBERS

The most basic fact about categories is that they have members, whether real or just conceptualized. There are two widespread metaphors for conceptualizing the category-member relationship. The first is the **category-as-container metaphor**: Categories are containers; members are contents in the interior of the container.

Using this metaphor, we view subcategories as containers inside other containers.

This is accomplished by neural binding, which identifies a content of the category-as-container metaphor with a second container. This allows the larger container to contain the smaller container, which has its own contents.

The basic container inference works here: *If container A contains container B and container B contains X, then container A contains X.* Here X can be either another container or another kind of entity.

Under the category-as-container metaphor, this inference applies to subcategories: *If a higher category contains a subcategory, then the members of the subcategory are members of the higher category.* This is automatic, because metaphors naturally map source domain inferences onto target domain inferences (unless inhibited by overriding constraints on the target domain).

In case you were wondering why the term "higher category" is used, the reason is the commonplace metaphor "more is up." Higher categories are called "higher" because they contain *more* members than each of their subcategories.

Before picking up this book, you could tell that desk chairs are chairs and that any particular desk chair is a chair. You could do this because, consciously or more likely unconsciously, you already knew that *subcategories* are contained in *higher* categories and that members of *subcategories* are members of the *higher* categories.

Eleanor Rosch (1977) has observed that categories can be defined by their **prototypical** members, which are understood as **central** to the category, while **nonprototypical** members are seen as **peripheral**. For example, sparrows and robins are prototypical birds, while pelicans and ostriches are nonprototypical birds. Such categories are called "radial" in that their prototypical members are conceptualized as central, while nonprototypical members are conceptualized as peripheral.

ANOTHER METAPHOR DEFINING CATEGORIES

In addition to the category-as-container metaphor, there is another commonplace metaphor characterizing categories, which is the **part-whole metaphor**: Categories are wholes; their members are their parts.

What about subcategories? Each subcategory is a category and hence a whole with parts. Each subcategory is also a part of a larger whole category. Neural binding accomplishes this, identifying the subcategory as a part of the more inclusive whole category.

A basic part-whole inference is at work here: *If whole B is a part of whole A, then the parts of B are also parts of A.* For example, since your hand is a part of your arm and your arm is part of your body, it follows that your hand is a part of your body.

THE GRAMMAR OF THE PART-WHOLE CATEGORY METAPHOR

The part-whole metaphor for categories and their members has a grammatical consequence in English. English grammar has a "partitive" construction, in which the whole is marked with the preposition "of," and the part-whole relationship is grammatically expressed as a noun phrase of the form "part of whole," as in *the hood of the car, the tail of the pig, the top of the can* and, correspondingly, in cases extended by other metaphors, such as *members of the club, citizens of the country,* and *officers of the corporation* as well as measures of a substance: *teaspoons of vanilla, cups of water, a tank of gas.*

Those are basic category-member relations, although they get much more complex. In fluid measurement, a "cup" is defined as 8 ounces, a "quart" is defined as 4 cups (32 ounces), and a "gallon" is defined as 4 quarts. Thus, we use the grammatical structure part-of-whole to speak of a quart *of* milk, a gallon *of* gas, etc.

Entities are naturally understood in terms of their properties, such as size, shape, color, and function. Particular properties tend to appear grammatically as adjectives: the *green* apple, the *big* rock, the *round* window, and a *washing* machine, with function tending to appear in nouns and noun compounds, as in *computer* (*an object that computes*), *underwear* (*clothing worn under other clothing*), a *drink* (*a liquid that is drunk*), *lawn mower* (*an object for mowing lawns*), *water bottle* (*a bottle for carrying water*), *dishwasher* (*a machine for washing dishes*), and so on. **General properties** (such as size, color, shape, and function) *are conceptualized as parts*, and the **partitive construction** is used: *the size **of** the bottle, the color **of** the wall, the shape **of** the patio, the function **of** the corkscrew.*

Other general properties have to do with date and place of origin or optimal use: for example, when and where the wine was produced, when the can of beans or the medicine should be sold or used by, and a person's birth date and place of birth. The same applies for a person's role in a significant event or function, as in *She took part in the free speech movement at Berkeley* and *He was campaign manager for Barack Obama.*

Leibniz observed that *no two entities have exactly the same properties*; even identical twins differ in time of birth. So, from the perspective of mathematical logic, there is *a one-to-one correspondence between any single entity, including a person, and the set of its properties.* Leibnitz in his philosophy took this principle as a rationale for *identifying an entity with the set of its properties.*

An interesting consequence of this unique correspondence of an entity with its properties is exploited in today's notion of *entities in a database*. As far as the database is concerned, *you are the list of your properties in the database*: your name, social security number, birth date, place of birth, mother's maiden name, what you bought, what you owe, what you paid, etc. And in many cases this could include the URLs of every website you have clicked on, which is how companies get to send ads to your computer for items you have searched for. *Your identity can then be computed as a unique property-vector, an ordered n-tuple combining all the properties and values in your database entry.* For companies keeping track of their customers by the data on what they use or consume, *you are your data*, and no two people are defined by the same data.

Comparing two database entities (yours and someone else's) would then boil down to comparing values along the dimensions given by those properties.

Aristotle, the founder of logic, distinguished between two kinds of properties: incidental and essential. Every entity and every person was assumed to have incidental properties (e.g., Did you happen to have coffee with your lunch today?) and essential properties (e.g., Do you have two arms? two eyes? perfect pitch? Are you intelligent?). Aristotle assumed that essential properties defined natural categories in the world and determined one's natural behavior.

That idea is still with us. Many categorizations are defined by what are taken as essential properties (called "essences"), as when someone is described as *intelligent* and *trustworthy* or *stupid* and *lazy*. Someone categorized as intelligent is expected to act intelligently, and someone categorized as lazy is expected not to put out a lot of effort. These forms of reasoning arise from a fundamental conceptual metaphor:

- The essence metaphor: All entities have intrinsic properties (essences) that determine their natural behavior and the natural categories they belong to.
- Examples include laziness and intelligence. The natural behavior of a lazy person is not to put out much effort. The natural behavior of an intelligent person is to act intelligently and not normally do stupid things (actions ignoring relevant knowledge).

Such categories are defined by necessary and sufficient conditions on their essences. The use of the essence metaphor is commonly used implicitly in academic fields such as mathematics, computer science, and philosophy.

Given this understanding of essences, two further metaphors use this notion of an essence to characterize forms of identity:

- The essential identity metaphor: The collection of an entity's essences defines its identity.
- Dedication to a kind of activity is an essence that defines a kind of entity.
- Incidental (nonessential) properties that an entity happens to have do not characterize the natural kind of an entity.

For example, if you happen to have a glass of milk on your kitchen table, the milk's being in a glass on your kitchen table is an incidental (not essential) property of that milk, of your kitchen, or of you.

A vast number of special cases exist, some about people, some about institutions, and some, via metonymy, about buildings and other spaces where institutions are located. Here are some common examples:

- He is an athlete, a teacher, a poet, an artist, a politician, a businessman, so on.
- It is a gym, a library, a school, a store, a boutique, a restaurant, and so on.

The words "athlete" and "poet" define a kind of person who is inherently dedicated to athletics, poetry, or both. A library is an institution intrinsically dedicated to acquiring, storing, and loaning out books to be read, and the word "library" can refer to a building housing such an institution. *Intrinsic, inherent dedication* is here seen as an *essence characterizing the natural behavior of a person or institution.*

THE CATEGORIES "ESSENTIAL" AND "INCIDENTAL"

Is essential versus incidental an all-or-nothing distinction? As it happens, there are cases in which a scale of essentialness enters. Although one cannot be a little bit pregnant, one can be very intelligent, pretty trustworthy, somewhat lazy, not very athletic, and more than a little bit wary of politicians.

Linear scales are extremely important in category structure, and we turn to them next.

LINEAR SCALES IN CATEGORIZATION

During the 1970s Eleanor Rosch, while teaching at Berkeley, opened up the field of categorization research, bringing in such ideas as prototypes and basic-level concepts.

BEST EXAMPLES OF A CATEGORY

For example, Mervis, Catlin, and Rosch (1976) introduced the concept of "degree of goodness of example" for a category and showed that there were consistent results across a number of experiments. She found, not surprisingly, that robins, sparrows, and other *well-known small songbirds* were the best examples of the category "bird," with goodness of example ratings going downward along a scale for *predators such as eagles and hawks* → *flying water fowl such as ducks and geese* → *nonflying barnyard fowl such as chickens* → *large big-beaked fish-eating fowl, namely pelicans* → *large long-necked head-in-the-ground ostriches* → *and nonflying, swimming, fish-eating, cold-weather penguins.*

The experiments include response times to sentences of the form A____ is a bird, with robin and sparrow fastest, eagle a little slower, chicken slower, and then pelican, ostrich, and penguin. When asked to produce a list of examples of birds, the fastest were robins and sparrows, then the rest in pretty much that order, and so on for various other experiments.

All the examples were in the category "bird." This scale did not rule out fixed boundaries and necessary conditions for birds: laying eggs, two legs, having wings and feathers. All birds fit those conditions.

The best examples of birds appeared to be (1) small, (2) songbirds, (3) flying, (4) eating seeds and bugs, and (5) making nests; for example, sparrows, robins, towhees, and so on. Deviations from this list appeared to determine lesser degrees of goodness of example:

- Birds that are predators: eagles, falcons, hawks, herons, egrets, vultures, ospreys, owls, etc.
- Water fowl: ducks and geese
- Barnyard fowl: chickens and turkeys
- Birds that fish: pelicans, cormorants
- Birds that live on ice: penguins

One way to understand this is to imagine a frame for best examples of birds as having the roles as the properties listed above (small, songbirds, flying, etc.). The closer to the frame, the better the example. This is a principle for characterizing Rosch's "best examples."

COGNITIVE REFERENCE POINTS

Rosch (1975) observed that certain examples in categories were "cognitive reference points" characterizing goodness of example, which could be tested by reaction-time experiments. Take countries. A sentence such as *Mexico is like the United States* evokes a faster reaction time than *The United States is like*

Mexico. Or take numbers. The sentence *98 is close to 100* is recognized faster than *100 is close to 98*. The "United States" and "100" in these cases are cognitive reference points for Americans; "Mexico" and "98" are not.

The point of this phenomenon is that in Aristotelian categories, similarity is symmetrical: if A is similar to B, then B is just as similar to A since they share the same properties. But in categories with cognitive reference points, the symmetry disappears.

PARAGONS

Paragons are ideal members of a category, as opposed to ordinary members. There are also antiparagons, or disasters. Consider baseball players. Well-known Hall of Famers are better examples of baseball players than ordinary players or terrible baseball players. Among the best examples are legendary players such as Babe Ruth, Joe Dimaggio, Mickey Mantle, Jackie Robinson, Willie Mays, Roy Campanella, Hank Aaron, and Yogi Berra. Excellent contemporary or recent players such as Buster Posey, Marcus Semien, Francsico Lindor, Matt Olson, and Matt Chapman are next, and after excellent players are the ordinary good players (there are lots of them). Terrible players such as the legendary Bob Uecker and Marv Throneberry are antiparagons.

HEDGES

On any linear scale there are regions top to bottom that define "hedges," words that indicate general positions on the scale: terrific, fantastic, excellent, very good, good, pretty good, okay, somewhat good, not very good, bad, terrible, and a disaster (see Lakoff 1973). These are not clearly distinct regions but instead seem to be defined by somewhat overlapping bell curves along a linear scale, as suggested by Lotfi Zadeh (personal communication).

BASIC-LEVEL CATEGORIES

One of Rosch's (1973) most remarkable discoveries is that certain categories are basic level, while other are subordinate or superordinate (Table 4). Basic-level categories function best from certain cognitive perspectives.

Basic-level categories are defined by **mental imagery, motor programs, gestalt perception**, and **maximal knowledge**. Subordinate and superordinate categories deviate from these properties, with subordinate being special cases and superordinate being more general. Basic-level categories tend to have

TABLE 4. Basic-level scale

SUPERORDINATE	BASIC LEVEL	SUBORDINATE
mammal	dog	corgi
furniture	chair	rocking chair
fruit	apple	Pink Lady

shorter words, are easier to remember in psychological experiments, tend to be learned earlier by children, and tend to be used in neutral contexts; for example, *There's a dog on the porch* but less likely *There's a corgi on the porch*, not *There's a mammal on the porch.*

What we have here is a scale defined by basicness, with basic highest and the others less high.

OVERLAPPING CATEGORIES

There are well-known cases of neighboring categories that are mutually exclusive at their centers (membership value = 1) but also have membership values lower on a scale and may overlap with neighboring categories at the extreme ends. Examples that have been studied are cups versus bowls and armchairs versus couches. Cases where you can't distinguish between a cup and a bowl are not cases that are best examples of cups or of bowls. The same is true for armchairs and couches.

FRAME CLUSTERS

A category that is defined by multiple frames that have been integrated via neural bindings is commonly multidimensional, with each of its component frames defining a "dimension." Lakoff (1987b) cites the case of the category *mother*, which is defined by four frames with bindings across them: the birth frame, the genetic frame, the marital frame, and the nurturance frame. In the simple case, when one says *I visited my mother*, you would assume that she visited the woman who gave birth to her, from whom she got half of her genes, who is (or was) married to her father and/or who raised her. When one or more of these conditions are not met, there are special terms indicating a nonnormal case: *birth mother* (she gave birth to you but gave you up for adoption), *genetic mother* (she donated the egg), and *stepmother* (she didn't give birth to you but is married to your father). The result is a "radial category"

(Lakoff 1987b), with a central case (the norm) defined by the component frames of birth, sources of genes, marriage, etc. The central case is the best example, with lesser examples failing to fit one or more of the central cases.

RADIAL CATEGORIES

The title of the book *Women, Fire, and Dangerous Things* (Lakoff 1987b) comes from the Australian Aboriginal language Dyirbal, as described by R. M. W. Dixon (1972). Dyirbal has four classifiers, markers that divide all nouns into one of four categories. The first two, *bala* and *balan*, have men and women as central categories. The categories are then extended on the basis of cultural frames (e.g., men hunt, women cook). In each case, the extension of the category is apparently extended via a metonymic mapping from the central case (man or woman) to another frame element. For example, there is a significant Dyirbal myth that the sun is the wife of the moon, which extends the woman category to the sun and the man category to the moon. Since the sun is hot, the sun gets extended to fire. Since fire is dangerous, fire gets extended to dangerous things (fighting spears, the sharp-toothed stonefish and gar fish, and the Hairy Mary caterpillar, whose sting can burn painfully for a month or more). The book title itself, *Women, Fire, and Dangerous Things*, denotes a radial category. Other slightly different analyses have been proposed, but they are all radial.[13]

RADIAL CATEGORIES AND FRAME VARIATION

Radial categories come in three types: cluster-based, substitution-based, and mapping-based.

- **Cluster-based**: The classic example is "mother," which is defined by a cluster of four frames with bindings across them: birth, marriage, nurturance, and genetics. In the prototypical case, your mother gave birth to you, is the wife of your father, and raised you, and you inherited half your genetic properties from her. The radial variation occurs with the inhibition of one or more of these frames. Your birth mother gave birth to you, but that may be all. Your genetic mother donated the egg. Your stepmother is the wife of your father but didn't give birth to you.
- **Substitution-based**: A dim sum restaurant is a Chinese restaurant where you may be choosing food from carts that are wheeled, and the check may be calculated by adding up the number of dishes the food

comes on. Thus, compared to the prototypical restaurant, there are substitutions for food type, method of food choice, and method of tallying the bill.

- **Mapping-based**: These apply mostly to words, with conceptual metaphors and metonymies mapping the central meaning of a word onto noncentral meanings. For example, "up" has a vertical height as its central meaning but a quantitative increase in *Turn the radio up* and *My stocks went up* (via the more is up metaphor) and *I'm feeling up today* (via the happy is up metaphor). *She bicycles to work* extends the central sense of "bicycle" as an object to its use as an action verb (via the instrument for action metonymy). The same metonymy is at work in *They gunned him down*, where the gun is the instrument and the shooting with the gun is the action.

There are special cases that can stand metonymically for a whole category with respect to probability of occurrence. Daniel Kahneman (personal communication) called them "salient exemplars," that is, well-known and easily recalled cases. Kahneman frequently cited the example of a DC-10 crash at O'Hare International Airport in Chicago that was shown over and over on television. People got the idea that DC-10s were likely to crash when they actually had an excellent safety record. Many people immediately after that refused to fly on DC-10s.

Why? The TV viewer's neural circuitry for the crash was activated and reactivated over and over each time it was shown, each time making the circuitry stronger. The stronger the circuitry, the more likely the event seemed until the probability of a crash became high in people's minds.

In the 2016 election campaign, Donald Trump kept citing two well-known examples of crimes by Mexican immigrants: one a rape and one a murder. He then claimed that Mexican immigrants were rapists and murderers. Salient (well-known and easy-to-recognize) examples can stand metonymically for a whole category.

FROM WHAT TO WHY: RADIAL CATEGORIES

We are concerned not just with descriptions governing what happens. Science has the deeper job of explaining *why* the phenomena described exist.

The existence of radial categories has radically changed traditional ideas about categories. A radial category has the shape of a wheel with spokes. There is a central member with many kinds of variations on the radial category: metaphorical variations, frame-based variations, variations composed by bindings and integrations, and even shifts from one to two dimensions.

The classical example is "over," whose central sense combines "above" and "across," as in *The plane flew **over** the bay*. Radial examples include, among others, the following:

- *The tub overflowed.* A fluid went above and across the top of a container.
- *He overate.* The amount he ate went above the amount he should have eaten.
- *Do it over.* Action is metaphorical motion, and redoing an action is metaphorically motion going over a path previously taken.
- *The road goes over the mountain.* Roads are seen metaphorically as entities traveling along a path, and here the road is spoken of as moving above and across the mountain.
- *Put the board over the hole.* Whereas a line across is one-dimensional, a board is two-dimensional.

Adele Goldberg (1995) has taken a major step toward an explanation of radial category structure. She has observed that *in a radial category, the center acts like a general case and the radial variations act like specific variations on the general case.* Her observation, in effect, reduces the explanation for the existence of radical categories to the behavior of general and specific cases.

WHY CATEGORIES ARE MEANINGFUL

Categories of all these kinds are meaningful for a reason. Their structures make use of all of the mechanisms of embodied thought—embodied primitives, conceptual frames, and conceptual metaphors—and mechanisms of combination, namely bindings and conceptual integrations.

The meanings of categories of various types arise from the meanings of their parts and the way meaningful wholes are normally related to their meaningful parts. In short, the structure of categories arises from the basic mechanisms of thought.

The Takeaways

Now we have the fundamental mechanisms of thought:

- **Hierarchical network of simple mechanisms.** All of conceptual thought arises from a small number of basic conceptual mechanisms that combine to create more complex conceptual mechanisms: primary embodied schemas, frames that use primary embodied concepts, metonymies that use frames, conceptual metaphors that map schemas to schemas and frames to frames, bindings that identify two semantic roles across schemas or frames as being identical, and conceptual integration that forms integrated composites of all of the above as well as control and gating nodes that activate or inhibit such circuits, cascade circuits that extend across brain regions, and activating and inhibiting binding circuits and integration circuits to form complex ideas.
- **X-Nets.** X-Nets are networks that "execute," that is, perform actions. Motor control is a well-understood example. There are neural networks that control how we move our bodies. "X-Net" is a computer science term used in the present book for computational models of neural control systems.
- **Motor control and thought.** Motor control and thought involve aspect, causation, and purpose. Hierarchical motor control applies to the physical body. The fact that at the highest level the neural computational structure of motor control characterizes the concept of aspect (the structure of events and actions in general) in all natural languages brings embodiment front and center into conceptual thought. Motor control constitutes a claim that *we understand all events and actions in terms of what our bodies can do using our neural systems.* The hierarchical motor control system also characterizes the general structure of cause and effect and of purpose as they apply in the body.
- **Exaptation ("repurposing").** The human motor control system evolved from the motor control systems of animals. The use of this system in basic concepts of conceptual thought and in logic schemas is an example of the repurposing of motor circuitry over the course of evolution for use in thought in human beings.
- **Primary embodied concepts.** Primary embodied schemas provide structure to what we see and how we move. They provide the embodied content for primary embodied concepts. Primary embodied schemas have a structure in terms of wholes with related parts called

"roles." They are gestalt structures in that activating one or more parts activates the whole, and activating the whole activates all the parts. Embodied schemas come with logic schemas that carry out inferences.

- **Frames**. Frames are very general conceptual structures that allow us to conceptualize types of experience in terms of compositions of embodied schemas. Frames come in hierarchies: there are lower-level subframes that are instances of higher-level frames. At the highest level, frames are embodied by primary schemas that can form composites. Most frames are cultural in nature, characterizing institutions, cultural practices, and so on. Frames have logics, that is, frame-specific inference patterns that derive from the logics of the primary schemas that are combined to form a frame. Roles in frames can be filled by individuals or types or by individuals that fit types. Frames can combine to form more complex frames via bindings, in which roles in one frame are neurally bound to roles in other frames.

- **Frame circuits**. Every frame is characterized neurally by a frame circuit. Negating a frame activates that frame. For example, the sentence "Don't think of an elephant" leads you to think of an elephant. Every time a frame circuit is activated, it is strengthened, and all the frame circuits above it in its hierarchy are also activated and strengthened. Thus, negating a frame activates the frame in people's brains. Donald Trump is the classic example of someone who exploited this fact. It doesn't matter to him if he is praised or attacked. In either case, he becomes more prominent in people's brains. The only way to avoid making him more prominent is not to mention him, to ignore him.

- **Words defined in terms of frames**. The great linguist Charles Fillmore (1976) discovered that all words and morphemes are defined in terms of frames and embodied schemas. The use of a word strengthens its frame circuit and all the frame circuits in its hierarchy.

- **Conceptual metaphors as asymmetric frame-to-frame neural mappings**. Neural mappings are circuits extending from one brain region (a "source") to another (a "target"). These mappings are structured so that frame roles in the *source* brain region map to corresponding frame roles in the *target* brain region. As a consequence, the inference patterns of the source frames are imposed on the target frames. Metaphors can be neurally bound together to form more complex metaphors.

- **Primary metaphors**. From primary embodied metaphors, we learn that just functioning in the everyday world on this planet has profound effects on systems of human thought. Because it is common for pairs of embodied experiences to occur together (such as parental affection and parental warmth), a wide range of primary embodied metaphors are acquired just by normal neural learning processes such as STDP.

- **Primary metaphors as circuits linking embodied circuits.** There are hundreds of primary metaphors in the form of circuits crisscrossing the brain from one embodied circuit to another, imposing modes of inference from one brain region onto another brain region.

- **Emotion metaphors.** From emotion metaphors, we learn that a wide variety of physiological correlates of emotions can give rise to a correspondingly wide variety of embodied metaphors.

- **Sweet talk.** The sweet talk experiment shows that even the embodied neural circuitry for taste—the gustatory region—can play an active role in metaphorical thought. As Adele Goldberg showed, "She talked to him sweetly" activates the gustatory cortex even though the sweetness is metaphorical, not literal (Citron and Goldberg 2014).

- **Idioms, images, and knowledge.** From idioms based on images, we learn that we have in our memories thousands of cultural images and knowledge about the images and that general conceptual metaphors can apply to that image-based knowledge to provide the meaning of those idioms that have accompanying images. The words in those idioms often name what is in the image, although what is in the image may not be in the meaning of the idiom. A classic example is "spinning your wheels," which can mean that one is using effort toward achieving a purpose without making any progress in achieving that purpose even though one is not literally talking about wheels. This is based on the conceptual metaphor that action is motion and that achieving a purpose is reaching a physical goal. Cultural images such as spinning one's wheels can thus play a crucial role in linking language to embodied meaning via conceptual metaphors. The point is that there is no clear separation between cultural images and linguistic meaning. The meanings of words can make use of cultural images, such as the mental image of a car spinning its wheels.

- **Conceptual integration.** Most unconscious thought is fragmented. Conceptual integration forms an integrated whole from unconscious fragmented parts. Conscious thought tends to be integrated, although there is also integrated unconscious thought. In many cases, unconscious thought arises from diverse unintegrated circuits across brain regions. Conceptual integration, operating unconsciously, changes

unintegrated unconscious thought to make it integrated and a candidate for consciousness.
- **Categories**. Categories are complex and meaningful, since they make
  use of the full range of conceptual neural mechanisms listed above.

The sequence of basic thought mechanisms presented in this chapter shows how these mechanisms build on one another to produce increasing complexity of thought: from embodied primitives (X-Nets, primary image schemas, and force schemas) to complex combinations of those to frames, metonymies that are based on frames, primary conceptual metaphors (map-ping schemas to schemas), complex conceptual metaphors (mapping frames to frames), idioms using cultural images and knowledge to mediate between language and embodied meaning, and conceptual integrations that use all of the above mechanisms as well as to whole systems of thought such as philos-ophies and branches of mathematics and to imaginative literature, mythology, religion, and politics, human enterprises that let out all the stops and use a wide range of conceptual integrations of more basic ideas to create highly complex thought processes.

3

The Neural Mechanisms of Thought

Scope and Challenges

We began this book with Antonio Damasio's (1994) prescient observation in *Descartes' Error* that "the immune system, the hypothalamus, the ventromedial frontal cortex, and the Bill of Rights have the same root cause" (p. 262). What he meant and what we have seen is that contrary to Descartes, who saw ideas as abstract, nonphysical entities, ideas are located in the neural systems of our brains and bodies, interact with the physical and social world, and are subject to the laws of biology, physics, and chemistry.

As we noted in chapter 1, brain research in neuroscience spans multiple levels, from the internal dynamics of a single cell and its biochemical interactions to neural circuits and networks performing specific sensory or motor tasks and to entire pathways and brain systems integrating perception, intention, and behavior for a living and acting body. While all of these multiple levels contribute to thought and language, cognitive phenomena appear to emerge at the circuit and systems levels, whether they are individual ideas or systems of thought, individual words, or larger grammatical structures. Our computational models of how the brain functions are system-level models, although they are pieced together from smaller circuits that contribute to the overall functioning of the brain.

At the outset, we must acknowledge that little is known from large neuroscience efforts to map the brain (such as the connectome project) to show how specific neural networks create meaningful ideas and language. Our hypothesis is that the science of thought and language starts with neuroscience but, in addition, must include cognitive science, linguistics, and computational modeling. We have learned from cognitive science and linguistics that *the cognitive mechanisms for ideas includes frames, conceptual metaphors, conceptual integration, image schemas, and force-dynamic schemas.* We have sought to understand as best we can how these cognitive mechanisms for ideas can be constituted by neural circuitry that functions to characterize those cognitive mechanisms.

Our goal in this chapter is to provide a proposal for the general neural mechanisms capable of characterizing those cognitive mechanisms. The presentation is theoretical and computational and relies on a great deal of prior research in the neuroscience of perception and the control of behavior. However, our theoretical account is largely still to be validated with detailed experiments. We hope that this book spurs and accelerates such efforts.

As this book shows, it is not yet easy to come up with a unified, scientifically accurate and specific understanding of human thought. The current chapter takes us even deeper than we have gone so far. We use all the facts and evidence presented in the first two chapters to come up with our hypothesis of what it takes for our neural systems to create and constitute thought, together with details of what kinds of computation may be involved.

In the next section, we give a description of the detailed operation at the cell level. After that, we go through the basic firing sequence in detail. We then follow up with the basic mechanisms for learning based on synaptic changes and modulation. After that, we move from single cells to the formation and use of circuits and networks. The neural accounts motivate our hypothesis of the types of neural circuits and cascades that constitute the cognitive mechanisms discussed earlier. The next two sections outline our computational models of the basic neural components and circuits that are necessary for thought and language. Our account includes detailed *neural mechanisms such as gestalt circuits, gates, binding circuits, cascades, integration circuits, and convergence-divergence zones.* And we will show how *the neural computational theory of dual modeling links the computational modeling of neural mechanisms with the computational modeling of cognitive mechanisms.*

This is not simple. We have tried to present the essential components of our modeling and hypotheses in a fairly approachable manner while pointing the reader to further details with specific references in the relevant sections. One source of difficulty is that the neural models of how thought might arise are strikingly different from our everyday understanding of thought. When we consider all the neural details required for human thought, it is an amazing feat of human biology that any human being can think at all.

Neural Firing: The Basic Steps

The actual firing of a neuron is more complex than we stated in chapter 1. We are about to go through the twenty steps required for even one neuron to fire just once. Why does this level of detail matter?

Recall that the brain contains about 86 billion neurons and between 1,000 and 10,000 connections per neuron. That's hundreds of trillions of neural firings. A neural firing occurs in about five thousandths of a second (5 milliseconds). That's 200 times a second. And to have any experience of anything interesting at all, there must be a huge number of neurons firing that fast a huge number of times, not randomly but instead in a pattern that characterizes exactly the right experience.

But via our everyday conscious experience, we do not sense any of this. Given what little we can consciously sense, we just think and we just experience with no awareness of anything further, certainly no conscious awareness of what goes on in the vast neural universe in our brains. For even one neuron of the 86 billion in each of our brains to fire, it takes all of the following twenty steps.

THE FIRING PROCESS

Firing does not occur all at once. Before the firing process for a single neuron begins, there is a voltage difference between the inside and the outside of the neuron's cell body of about minus 70 millivolts (thousands of a volt). The "minus" indicates that there is more negative charge inside the cell body than outside.

The firing begins as neurotransmitters from the presynaptic (input) neuron begin to bind to the postsynaptic (receptor) neuron. When sodium neurotransmitters bind to receptors, inward-transporting sodium "channels" (which are complex proteins with a channel structure) begin to open in the cell membrane, letting in positive sodium ions. As the sodium ions accumulate inside the cell, the inside voltage rises gradually by 15 millivolts from minus 70 millivolts until it reaches minus 55 millivolts, the threshold for neural firing. At minus 55 millivolts, inward-transporting sodium channels along the axon open, letting in more sodium ions, which increases the voltage further, making the inside even more positive. As the sodium channels open step-by-step along the axon, progressively more sodium enters, creating a positive feedback loop, and the internal positive charge rises explosively along the axon. This change in charge moves along the axon and is called an "action potential."

This explosive increase in positive charge affects the sodium channels physically, and at around plus 40 millivolts the sodium channels reverse and start to let out sodium. Around then, outward-transporting potassium (positive-charge) channels and inward-transporting chloride (negative-charge) channels open, moving positively charged potassium out and negatively charged chloride in, thus reducing the positive charge of action potential until it drops to the original level of minus 70 millivolts. Because the outward-transporting potassium channels close more slowly than the sodium channels, the downward drop of the action potential continues for a short time past minus 70 millivolts, thus overshooting the original level. At that point a "refractory period" begins, during which the various channels are restored to their original state. Then the firing process can start again.

The action potential (the change in positive charge) produced in the firing process thus has the shape rise, peak and reversal, drop, overshoot, and refractory period.

This complex process can happen in *each one of the 86 billion neurons in our brains*. It is hard to conceive of all that neural activity going on in each thinking human brain containing nearly 100 billion neurons. We just think. We are not aware of the incredibly complex physical process needed for human thought. The physical process that allows each human being to think is that complicated: in 86 billion neurons! The Darwinian theory of human development is correct at Darwin's level of detail. But what lies behind it at the neural level is vastly more complex.

SPIKING AND TIME GAPS

Each commonplace neuron either does or does not produce an action potential during a short time period, and the action potential goes in one direction: down the axon. From the perspective of a somewhat longer time period, the action potential as measured by a voltmeter produces the picture that the voltmeter draws: a "spike," rapidly rising and falling (Figure 2). Most neurons generally produce a series of about six or seven spikes and then stop for a while during a recovery period. From this longer perspective, each spike is drawn as a vertical line of a fixed length, where the time between spikes is a gap. The resulting spike plot is a sequence of such lines separated by gaps. The

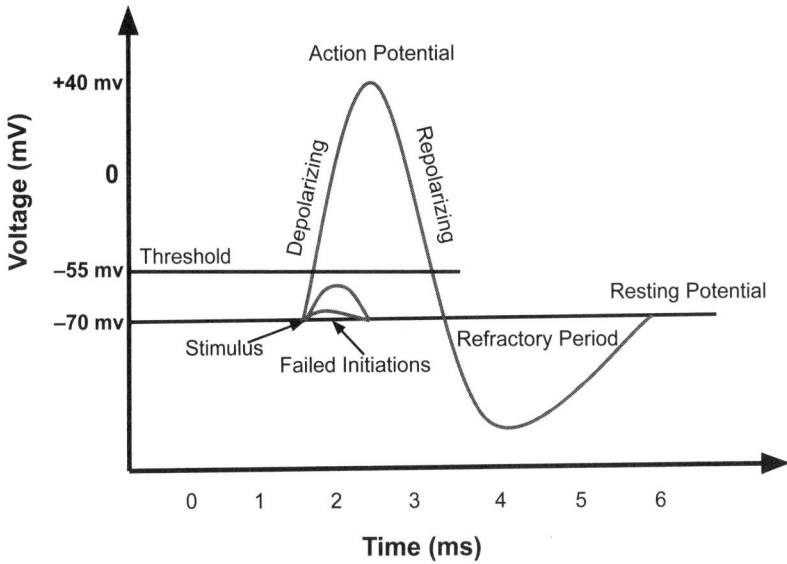

FIGURE 2. A simplified depiction of the main events that go into the generation of an action potential or spike. The figure shows the shape of the unfolding action potential (the change in positive charge) produced. The shape has a characteristic rise, peak and reversal, drop, overshoot, and refractory period. An incoming stimulus from a synapse changes the resting potential (where the inside of the membrane is more negative, −70 millivolts, than the outside). The change makes the inside more positive, and when a threshold is reached, rapid polarization results through an influx of positive sodium (Na⁺) ions, reaching a peak of +40 millivolts. This triggers a set of cascades (see the text for further details) where potassium (K⁺) ions are pumped out in an episode of repolarization, which results in the inside temporally becoming more negative than the resting potential (hyperpolarization) before reaching the steady state of the resting potential.

spiking models use both the time of each spike, time intervals between spikes, and the rate of the spikes (number of spikes per second) to encode information from a firing cell.

The neural "firing" process usually focuses on the "postsynaptic" neuron. The process occurs in stages. The following account is a simplified description of the different stages of the firing process:

1. **Neurotransmitter release**. The presynaptic neuron releases neurotransmitters into the synapse.

2. **Polarization**. The initial state is polarization, at which there is a "resting" potential, a difference in charge between the inside and outside of the cell.

3. **Neurotransmitter binding**. Neurotransmitters in the synapse chem-

ically bind to the sodium-gated transmembrane proteins, which then straighten out, opening the sodium ion "channels" (also called "gates" and "pores") in the dendrites and cell body.

4. **Influx.** Huge numbers of positive sodium (Na⁺) ions are "attracted" by the large negative internal charge of the cell (−70 millivolts) and flood in through the open channels.

5. **Depolarization.** As more positive ions enter the cell, the internal negative charge becomes more positive, that is, less negative. The internal charge goes from minus 70 millivolts to minus 55 millivolts, a positive increase of 15 millivolts, about a sixtieth of a volt, an extremely large and fast change in voltage for something as small as a neuron, and across something as narrow as a cell membrane.

6. **Threshold.** At minus 55 millivolts, a threshold is reached. The voltage gates for sodium ions in the axon hillock open. That is, bent sodium-gated transmembrane proteins in the axon hillock straighten out, allowing more sodium ions to rush in, increasing the internal positive charge both at each location and in the total positive charge inside the cell.

7. **Sequential openings.** The same types of transmembrane protein gates occur next to the axon hillock and all the way down the axon. As the positive charge increases at one location along the axon, the voltage-gated transmembrane proteins straighten out and open up at the next location, letting in more sodium ions there. This cycle occurs in sequence all the way down the axon.

8. **Action potential.** The result is called an "action potential," a wave of increasing positive charge starting at the axon hillock and moving all the way down the axon to the axon terminals. Those who think of neurons as metaphorically "communicating" tend to call this powerful increasing positive charge a "signal."

9. **Sodium channels sequentially closing.** As the interior charge becomes sufficiently positive overall, the sodium channels become inoperative, since there is no longer a sufficiently strong internal negative charge to attract the external positively charged sodium.

10. **Reaching a peak.** After a while, the accumulated moving charge reaches a peak, with the internal positive voltage increasing along the way. At the turning point, the resting potential is reached and then exceeded, turning the interior of the cell positive to about 40 millivolts from minus 55 millivolts, a powerful 95-millivolt change in a couple of milliseconds across an extremely small membrane!

11. **Calcium (Ca++) ion gates open.** There are thousands of axon terminals at the end of each axon. Each terminal has a membrane with positive voltage–gated transmembrane proteins structured to let in calcium ions when open. As the action potential wave reaches the terminals, the positive voltage causes the transmembrane proteins for calcium ions to open, letting in calcium ions at the terminal.

12. **Vesicle opening.** The axon terminals contain vesicles with neurotransmitters inside. When the calcium ions reach the vesicles, there is a chemical reaction in which the vesicles are fused to the cell membrane and open up.

13. **Neurotransmitter release.** As the vesicles open up, the neurotransmitters flow out into the synapse, starting the process all over again at the next neuron.

14. **Potassium pumps open.** In addition to the fast-opening sodium channels, the cell membrane along the axon also contains very complex transmembrane proteins that are straightened out and opened by *positive* voltage and that act as potassium (K^+) pumps. They open slowly, gradually pumping positive potassium ions (K^+) *out* of the axon, pumping two potassium ions out for every three sodium ions in.

15. **Potassium ions pumped out.** More slowly than the internal positive charge increased along the axon, the internal positive charge slowly decreases along the axon, lessening the internal charge as the pumping process proceeds.

16. **Resting potential reached.** As the positive potassium ions start to move out, pumping proceeds, and the interior of the cell becomes less positive, that is, more negative. At some point the original resting potential is reached, and then the negative voltage is exceeded as more positive potassium ions are pumped out.

17. **Potassium pumps slowly close.** The slow-closing potassium pumps close gradually until they all close all the way down the axon. At this time, the negative voltage has fallen below the original resting potential. This is called "hyperpolarization," and the time it lasts is called the "refractory period."

18. **Repolarization.** Because of the negative internal charge that is lower than the resting potential, sodium attracted by the negative charge begins flowing in until the resting potential is reached.

19. **Neurotransmitter release and reuptake.** After a while, the neurotransmitters that were chemically bound to transmembrane proteins

are released. The transmembrane proteins bend again, and the channels close. The neurotransmitters diffuse into the synaptic fluid. In "reuptake," protein molecules called "transporters" at the presynaptic terminals bind to the neurotransmitters and move them back in to the terminals, where chemical reactions locate them again in vesicles. Not all the neurotransmitters undergo reuptake after release from binding, and some of those subject to reuptake never binded to the receptors of the postsynaptic neuron and were "stranded" after being released by the presynaptic neuron.

20. **The initial state is restored.**

This whole process takes only about 5 milliseconds, five thousandths of a second. This is a short enough time to allow about 200 such processes per second.

SPIKING

Each depolarization process is called "spiking," and the action potential produced is called a "spike" because it appears as a sharp rise and fall on the apparatus for recording neural activity. Such recordings usually show six or seven spikes in a row. Seven spikes in each event of spiking takes $7 \times 20 = 140$ steps in about 35 one-thousandths of a second. Each such process is just one spiking event for just one neuron out of tens of billions that undergo this process.

Spiking neurons produce activation through their axons on connected synapses. This activation leads to biochemical changes at the synapse— strengthening or weakening—that may lead to the shaping and formation of new circuits, which is the only way you can have new experiences or learn anything.

SUMMARY

The sequence of electrical and biochemical events responsible for the transmission of information within a cell (neuron) and between cells through synapses is critical to understanding how the brain/mind works. The bridging models we will be presenting in this chapter are even further simplified than the basic account here. We model the action potential as a single spike generated at a certain time instant and at a specific rate (number of spikes per second). We do not computationally model the dynamics of neurotransmitter release or the biochemical reactions both at the synapse; in addition, cross-membrane ion transport within the cell is not computationally modeled. We conjecture that these simplifications preserve the critical aspects of neural firing that are relevant to the system-level questions regarding thought and language. But the selection of the appropriate level of detail for a bridging model is an open and ongoing empirical question. Our hope is that this book will provide a starting point for more detailed investigations with more faithful and granular models of information propagation and learning.

What we have presented is a simplified account of the processes involved in the generation of an action potential or a single spike emitted by a neuron. There are many fascinating and important details such as the biochemical models of neurotransmitters, the role of calcium channels, the complex processing in the dendrites of certain cells, and the biochemistry involved in learning and synaptic change (synaptic plasticity) that have been left out of this simplified account.

Think for a moment about how remarkable this is. No one ever just learns anything or has any experiences at all without all of this happening, a very complex process for each of a vast number of neurons over a vast number of extremely short time periods. This vast complexity is required for anyone to learn anything at all!

We are not and could not be consciously aware of all that is required for human beings to learn even the simplest thing.

What Is Structured Neural Computational Modeling?

THE REALITY AND THE FUNCTIONAL MODEL

Actual neural circuitry in human beings with real neurons is enormously complex, involving cells containing a universe of chemical and genomic complexity and an organization of nearly 100 billion cells, each with thousands of "connections": a total of hundreds of trillions of connections, a universe so large as to defy even supercomputers.

A *functional* model of such circuitry seeks to use the theory of computation, together with real computers, to provide a relatively simple yet biologically plausible model limited to the crucial aspects of how human brains *function* to give rise to conceptual thought and language.

We will call a *structured neural computation* model an SNC model, and we will call an SNC model that seeks to characterize the computational *functions* needed for human thought and language an fSNC model.

Throughout this book, it is crucial to distinguish what is real and what is modeled. Bear in mind as we proceed that one can *model* the flow of a river, but the model won't get you wet. All scientific research uses models, including neuroscience, cognitive linguistics, and experimental embodied cognition. But models of what happens in the brain are *not the real physical aspects of the brain* being modeled. What is assumed is that the models capture important features about the brain that is relevant to what is being studied. That is what scientific knowledge derived from modeling does, and it is assumed that it is accurate for relevant scientific results.

The goal of fSNC modeling is to explain cognitive phenomena with neurally plausible computational models. The result is a "bridging model," that is, a computational model that creates a "bridge" between what is known theoretically about neural circuitry and what is known theoretically about thought and language. But the theoretical models are not the thing itself. *The approach taken is to construct computational simulations of the cognitive functions needed for thought and language using simplified models of neural circuits.*

We believe that the fSNC model we are using makes valid, relevant, useful, and sufficient scientific contributions to the three sciences it links: neuroscience, cognitive linguistics, and experimental embodied cognition.

WHAT DOES IT MEAN FOR AN FSNC MODEL TO "WORK"?

An fSNC model cannot "work" in isolation. It depends on adequately capturing scientific results in three other fields. The fSNC model has to link the neuroscience and the cognitive linguistics (which studies what conceptual systems are from a neural perspective and how they fit language). There is a further scientifically constraining source of results. Conceptual thought and language make use of not just the brain alone but also the embodied brain, the brain neurally linked to the body, as the body-brain combination *functions* in the physical and social world. Such *functional* behavior is studied in the field of experimental embodied cognition.

The job of the fSNC model is to provide a precise computational model of what it means for the body-and-brain neural system to manifest the *functional* properties of thought and language on the basis of the *functional* properties of the brain-and-body neural system. Of course, our account is just a computational model and is thus theoretical and speculative. Indeed, a central motivation of such modeling is to suggest targeted experiments for cognitive and brain science. In chapter 2, we already saw an instance of this kind to collaboration in the modeling and experimental work on metaphor. In this chapter, we will see another very productive collaboration on the notion of simulation semantics, the use of imaginative simulation for understanding. But to get there, we will need to establish the foundational framework and models of the fSNC model.

One further caveat is in order here. The mechanisms of thought discussed in this book are to a large extent carried out unconsciously, and there is a huge question about a neural mechanism and explanation for conscious thought. There are many first-order problems in the study of consciousness and the mind-body problem that neither we nor anyone else (as far as we know) has any account of. We hope to spur research into this profound and important area by setting out the basic mechanisms of unconscious language and thought in computational terms.

In summary, the fSNC model is a crucial element of our theory, since it bridges the functional computational properties of the embodied neural system with the functional computational properties of the conceptual-linguistic system and shows how they match up. Let's dive in.

THE BASICS OF OUR FSNC MODEL

The basic ideas of the fSNC in the model we will be presenting step-by-step are rather simple, and they should be generally understandable one step at a time. Actual computer implementations of these ideas in programs are, of course, much more complex.[1] But understanding those implementation details is not necessary for understanding the theory presented here. We will begin in this section with only the most basic ideas behind fSNCs in general and elaborate on other basic details later in the next two sections. Note that for readers familiar with neural network models, this section is mostly a repeat of the computational details.

STRUCTURE AND ARITHMETIC

SNC modeling can most easily be understood in terms of two factors: structure and activation flow. It starts with the concept of a "node," which is taken to refer in the brain to an ensemble of neurons, either a group with an internal structure or just a group of isolated neurons. In some rare cases, "node" can refer to just one neuron.

This structure (Figure 3) is a simplification of the neurochemical and electrical events described in the previous section on the "firing sequence."

In the fSCN model, neurons contain (1) inputs (*dendrites*) that electrically transmit a signal from the incoming synapse to the cell body. The positively charged ions are sodium, magnesium, and calcium. The negatively charged ion is chloride. The charged ions flow from the dendrites into (2) the *cell body*. When the positive charge in the cell body reaches (3) the *threshold* charge, the cell "fires." In "firing," the cell sends a positive charge—(4) the *signal* or "action potential"—down the axon. Details of the action potential are not modeled; each firing event is treated as a single spike. At the end of the axon are thousands of pods containing (5) *neurotransmitters* and (6) *calcium channels* that open in the presence of firing and let in calcium. The fSCN model of

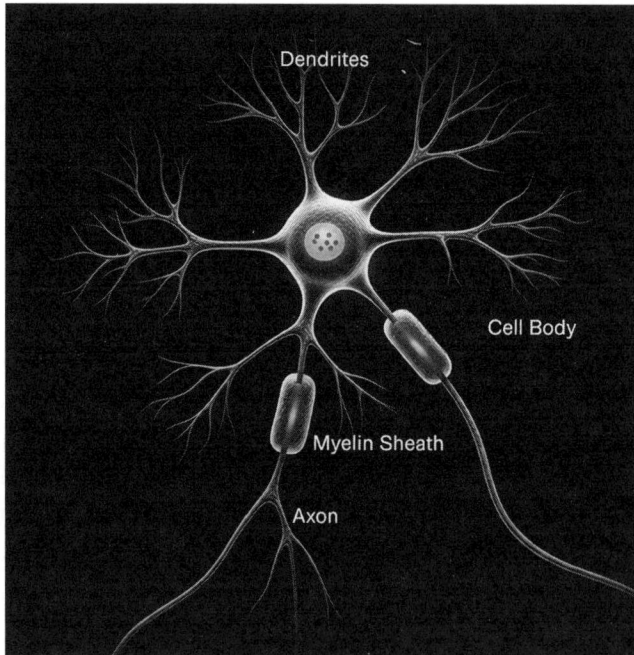

FIGURE 3. A simplified model of a neuron comprising a set of inputs (dendrites) that electrically transmit a signal from incoming synapses to the cell body or soma, which connect to axons that carry the signal to other downstream neurons through synapses. When the cell body reaches a threshold level, depolarization occurs, and the cell fires, sending an action potential down the axon, which then makes synaptic connections to the dendrites of downstream neurons, propagating the signal along the neurally connected pathway.

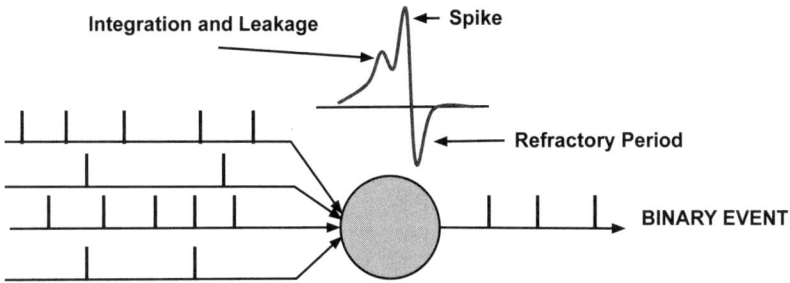

FIGURE 4. A simplified computational model of the neuron in Figure 3. Synaptic transmission is modeled by a set of real numbered weights. Action potentials are modeled by single spikes. Each incoming spike changes the membrane potential. When the combined integrated effect of all the incoming spike reaches threshold, the postsynaptic cell fires a spike, continuing the cascade to downstream cells through their synapses. This computational model is called the integrate and fire model.

the synapse is modeled as a one-dimensional scalar quantity called a "synaptic weight," where a higher weight implies more strongly connected pre- and postsynaptic cells. The synapse also contains additional neurotransmitters for the four types of ions. These are called (7) *modulators.* These modulators change the effect of the neurotransmitters released when the presynaptic neuron fires. They increase the effect in the case of positive ions (sodium, magnesium, calcium) and decrease the effect in the case of negative ions (chloride). Modulation in fSCN corresponds to "multiplicative" connections, where the synaptic activation is multiplied by a "gain" (which could be inhibitory or excitatory).

Another aspect of relevant neural structure is time (Figure 4). The times that are relevant are (1) the signal time, that is, the amount of time between the firing of the cell body and when the signal reaches the end of the axon; (2) the time it takes to cross the synapse and make the next neuron fire; (3) the "refractory" period, that is, the time between firings of given neuron; and (4) the minimum average time delay (positive or negative) between the firing of the given neuron and the firing of the neuron terminating at the same synapse firing most closely in time to the given neuron.

In summary, the computation in a neuron can be modeled in an fSCN by a set of incoming input neurons ($x1 \ldots x4$ in Figure 4), with spikes at various times. Each incoming spike changes the cross-membrane potential from minus 70 millivolts based on the synaptic efficiency (which is captured by a weight between 0 and 1). The overall effect of the change in potential is the integration of the effect of all the incoming spikes. When the integrated potential reaches the threshold of minus 55 millivolts, the postsynaptic neuron "fires" and emits an output spike. This simple model is called an "integrate-and-fire neuron."[2]

The model computes the accumulation of activation over time given the various input sources and times of firing. and fires a spike when the threshold is reached. The spike is treated as a single instantaneous event, which is binary in that the node emits a spike or not.

There are many variants modeling the exact nature of the functions used to compute the action potential and emit a spike. The more realistic models take into account many of the factors of the firing sequence, based on the potential (voltage) difference across the membrane followed by the electrical cable properties of the axon. The Hodgkin-Huxley model is a realistic model based on an electrical circuit model that takes into account the charge and the resistive properties of the signal and transmission through the axon. fSCN models such as ours do *not* take the cable properties of the axon into account.

We work often with an even more simplified model, which does not contain the actual numbers of amounts of millivolts or milliseconds. Instead, in the simplified model those amounts are modeled using simple numbers. For example, suppose the firing threshold of a neuron is modeled by the number 4 and that at a particular time a modeled neuron gets activation (that is, total positive charge) from two other modeled neurons that supply weighted activations of 2 and 3, and the weight at the synapses from the incoming neurons is 1. Then, since $2 \times 1 + 3 \times 1$ (weighted sum) adds up to 5, which is greater than 4, the threshold will be reached and exceeded, and the modeled neuron will "fire," passing activation to the next neuron it is connected to. The amount of the output of the firing neuron depends on the activation function. Activation functions could be linear (proportional to the weighted sum) or nonlinear (a commonly used function is the S-shaped sigmoid function).

A node that is made up of a collection of neurons can also be represented by an n-tuple of numbers, representing the thresholds of a group of neurons. A single neuron fires all at once. The neurons in a group can fire at different times, and subgroups of neurons in a larger group can have different connections to subgroups in other nodes. N-tuple representations can model such real complexities.

THE STRUCTURE MODELED AT THE NETWORK LEVEL

The first aspect of neural structure is at the network level. Nodes are represented as having "connections" to other nodes across chemically filled liquid synapses. The network of connections constitutes one kind of structure. The connections in the model have a direction: from the neurotransmitter-emitting axons of the neurons in each presynaptic node to the neurotransmitter-receptor dendrites of the neurons in each adjacent post-synaptic node. They are modeled as a graph, with directed arrows connecting the nodes in the graph.

THE WEIGHT OF CONNECTIONS

The connections in a network are represented mathematically by a directed graph, with the neural model mathematically represented by the nodes in the graph. Each arrow in the graph characterizes a connection between nodes. Associated with each such arrow linking two nodes is a number, the "connection strength," also known as the "synaptic weight" of the connection.

The synaptic weight is usually calculated by a rule using arithmetic. A common rule depends on the amount of modulation, that is, a positive or negative number representing the amount of positive or negative neurotransmitters in the synapse either coming from elsewhere or remaining in the synapse after previous firings. The idea is that modulating neurotransmitters can have an overall positive (increasing) or negative (decreasing) effect on the neurotransmitters coming from inputs to the node.

SYNAPTIC TUNING RESULTS IN "USAGE-BASED," OR "EMERGENT," LEARNING

The effect, over time, of the strengthening and weakening of synapses is metaphorically called "tuning," based on the model of stretching or loosening a violin string, changing its rate of vibration, and making its rate of vibration (its tone) go up with more stretching and go down with less.

Tuning occurs naturally as circuitry is used in real experience. Some experiences result in the strengthening of given synapses, and other experiences result in weakening. Over time, regularly used synapses become "tuned" automatically through use. In learning a language, such tuning is referred to as a "usage-based" or "emergent" model.

CONNECTIVITY AND FUNCTION

The function of a node in a model—its role in conceptual thought—depends on what it is connected to. The connections can branch *out from* the node or *in to* it or both. A node connects to another single node or can loop back, connecting to itself. But its function—if a single node can be said to have a function—depends on its *overall connectivity*.

Neural computational modeling requires the choice of a model of synaptic weight change. Different choices depend on what a computational model is trying to do.

Synaptic weights are "tuned"—that is, automatically adjusted to optimally fit their role in the network—in accordance to certain principles. There are three basic model types: the Hebbian Model, spatiotemporal extensions, and artificial neural networks.

The Hebbian Model

The Hebbian model, named for neuroscientist Donald Hebb (1972), increases weights for synapses between nodes in which the presynaptic node fires shortly before the postsynaptic node fires. This situation is seen as causal: presynaptic firing causes postsynaptic firing. When the presynaptic node fires *after* the postsynaptic node, there is no causation. The presynaptic node is firing to no purpose and dies off.

Spatiotemporal Extensions

More realistic rules use both spatial information (all incoming nodes to a synapse) and temporal information (when the incoming node spiked). Such models are called spiking models of neural networks. The spike-timing-dependent plasticity (STDP) theory makes use of the relative timing of spikes across different neurons entering a synapse.[3] These are models that we are primary concerned with in this book.

Other theories such as the BCM theory, developed by Bienenstock, Cooper, and Munro (1982), use spatial and temporal *averages* of firing (e.g., firing rates) plus a nonlinear (sigmoidal) activation function. It is common for neural modeling to be either rate-based or timing-based.

Artificial Neural Networks

Artificial intelligence researchers have developed what are called "artificial neural networks." The term "neural" is used, since they use the same types of computational structures that are used in modeling neural networks; that is, they make use of nodes that are connected through weighted connections forming a network. At the node (neuron) level, a node performs (computes) the spatial summation (weighted by the synaptic weight) of activation from the connected incoming nodes. The overall sum is then passed through a nonlinear (Sigmoid, ReLU, tanh) function to produce the output (a number).

At the node level, in most artificial neural networks the time of the incoming spikes is not taken into account. In more biologically realistic networks, which are also called "spiking networks," the time of the spike is also taken into account.

At the network level, these artificial neural networks often do not model the detailed connectivity of biological pathways. Instead they rely on simple patterns of locality (as in convolution networks) or start from a dense, fully connected network with random weights. This makes them typically require much more computational power than fSNC models, which attempt to directly model the functional circuits and connections whenever known. Since such massive computational power is now readily available, artificial neural network models have become an important addition to artificial intelligence.

Artificial neural networks have become useful for complex tasks where there are clear correct or incorrect answers. As early as in the Persian Gulf War (1991), they were developed to distinguish mines in the Persian Gulf from rocks, given complex sonar readings. The difference between a mine and a rock is clear. Artificial neural networks have been developed to distinguish real from forged signatures to produce a machine-generated pronunciation of a written English text and, more recently, to learn to translate between languages and to win complex games such as Go against human players. Many of these models extend simple networks with reward-based learning or reinforcement learning. Such models contain many interconnected layers of artificial "neurons." They "learn" over a massive number of trials by gradually increasing the synaptic weight of the neurons in proportion to how much that increase in weight decreases the error at the output, gradually decreasing the synaptic weights depending on how much the error increases.

The most common type of artificial neural network model uses the error in predictions (called the loss) to compute a direction and an amount to adjust the individual weights of the networks that will reduce the loss. This method is called gradient descent, and the most popular implementation is called the backpropagation model. Backpropagation tunes weights based on computing the gradient (how much is the error sensitive to the weight) and optimizing the weight to minimize this error. These changes use the mathematics of calculus (the mathematics of continuous change) and thus use functions such as the S-shaped sigmoid functions to guarantee continuity so that the calculus can work.

The backpropagation model is the standard model underlying commercial artificial neural networks. However, complex theories of optimizing change have been developed so as to minimize error. A whole industry has developed around such models. Google and Facebook have bought companies begun by great researchers in the field and now have major research efforts using this technology.[4]

Backpropagation is applied in a *supervised learning* setting where the system is trained on specific training input and training label pairs. In contrast, pure Hebbian learning is *unsupervised*, since there is no need for special training inputs. Reward-based learning, or *reinforcement learning*, is learning actions to maximize a discounted future reward that may be available after every input. Classic models of conditioning and reinforcement can be modeled using reinforcement learning.

STRUCTURED COMPUTATIONAL NEURAL MODELING

In the simplest neural network models, structure is solely emergent from the activation of the network and the tuning of the synaptic weights. However, in brains, much of neural activity comes from highly specific circuits and structures that are both

- a product of genetic coding, and
- tuned in an activation-dependent process of strengthening and weakening.

"Structured" neural computational models of the sort used in our approach are meant to model this dual fact of preexisting structure due to genomics plus activity-dependent tuning due to experience.

Our idea and SNC models go back to a research program initiated by Jerome Feldman (1982) and are inspired by various discoveries of highly specific neural structures within neuroscience. We saw an early example in chapter 1 of the gestalt node. We now give specific details of the basic types of computational circuits and functions that comprise our structured models of functional neural networks.

Basic Circuit Types

We are moving toward an account of the circuitry for ideas and the language used to express and communicate those ideas.

Ideas, we argue, arise ultimately from neural **exaptations**—that is, from the **repurposing** of neural circuitry originally used by animals in the course of evolution or the **repurposing** of neural circuitry that evolved in humans for needs other than ideas for perception, action, emotion, attention, and so on.

The *form* of language in both grammar and lexicon, we argue, also arises via such repurposing from while maintaining connections to still other parts of the sensorimotor and emotional system, those for sound, prosody, rhythm, gesture, and shapes (for letters in writing). Such repurposing results in specialized language circuitry in the brain using the same basic types useful for other social, motor, perceptual and cognitive behavior.

Those kinds of repurposing are part of the embodied circuitry in the brain. The big question is, of course, how this happens. Given our embodied circuitry, how have our brains formed, first, the complexities of ideas (including abstract ideas, reasoning, simulation, and imagination) and, second, all the complexities of linguistic form (e.g., words, morphemes, and grammatical constructions) as well as the complex relationships between ideas and the linguistic forms that express those ideas?

Given the embodiment structures of the kind discussed in chapter 1, we have identified a number of circuit types that seem to us to be important in characterizing such complexities. We are hypothesizing that those circuit types can be accurately modeled by functional structured connectionist models that capture the necessary generalizations.

The types of circuits that we are proposing are all learnable by normal neural learning mechanisms, electrochemical processes that modulate and change synaptic strengths of neural connections, that is, purely neural mechanisms that are not specific to any subject matter.[5] In addition, many of the circuits we propose are "functional" in that they have counterparts in mammalian sensorimotor systems as identified through single-cell recordings.[6] Our account, by necessity, uses the structured connectionist framework introduced earlier as a minimal computational abstraction from the details of these studies and recordings.

Our specific hypothesis is that there are integration circuits that have formed complex compositions of the basic circuit types discussed below and that they have resulted in structures useful for language and thought. This section is devoted to an understanding of those basic circuit types. How these circuits are integrated to form more complex circuits will be discussed in later sections.

We begin with neural recruitment, which, we hypothesize, is how such circuits arise.

We are born with about 86 billion neurons in our brains and between 1,000 and 10,000 connections each. That's on the order of magnitude of hundreds of trillions of connections. In childhood, the most-used connections are strengthened by use, resulting in the formation of fixed circuits that are formed when they are activated repeatedly because they serve important functions. Half of those connections, the least-used half, die off, leaving behind a brain highly structured by fixed, highly functional connections that are regularly used, together with a huge number (estimated to be in the hundreds of trillions) of connections that are neither fixed nor function regularly.

The probability would seem to be high that a great many (possibly billions or trillions) of connections form simple networks, available for use in the right situations or more than just "available."

A given connection, or small potential "circuits" made up of connections, might form and be used on random occasions when they happen to be in the right location in the brain to minimize a flow of activation that carries out some function that happens to arise in a given context. If that keeps happening, the connections can be strengthened and a circuit can be "recruited" because it happens to be useful for serving that function. In general, such a happenstance circuit would be strengthened by the reward systems of the brain. If a recruited circuit carries out a desirable function, that circuit can be strengthened by what we would call "reward circuitry." Circuits that serve important functions can be formed in this manner.

Consider an example. Suppose you are growing cherry tomatoes in a garden or a pot. When they ripen, many of them are entangled in a mess of branches with leaves and still-ripening tomatoes that you don't want to knock off. You want to get to the ripe ones, grasp them in your fingers, and pick them off without harming the branches and the tomatoes that are still ripening. To pick the ripe ones, you have to move your body into position, adjust your back, and coordinate the movements of your shoulder, elbow, wrist, and fingers in new ways appropriate to the particular tomatoes and differently for each tomato embedded in branches, each time coordinating all this with vision slightly differently each time. To do this in each case, you partly use fixed motor control networks you have used before, but you also have to recruit available unfixed networks of connections that can connect with the fixed networks and allow you to move in a new and somewhat different way each time. If you keep growing cherry tomatoes, those randomly used connections may be "recruited" to become cherry tomato–picking circuits.

We are constantly using available networks of local connections in such ways all day, every day, not using them enough for them to become fixed but enough for them to remain alive and available. Moreover, the same happens in thinking, as we form new combinations of existing ideas each day.

We hypothesize that such a large number of simple networks without a fixed function happen to be formed and regularly become available to be recruited for functional use in the right circumstances.

And we hypothesize that existing fixed circuits are formed via what might be called **neural Darwinism**, a kind of natural selection: the "available" simple networks become fixed when they happen to be used and rewarded—that is, strengthened via use—over and over in everyday experience, strengthened enough for their synapses to become permanent.

CASCADES

The central idea in the theory of integrative circuitry is that of the cascade. As we mentioned in chapter 1, we will be using the word "cascade" in two senses:

- **Cascade flow**: A flow of activation that goes across brain areas, typically across multiple brain areas. In common cases there is a bidirectional flow of activation, and it extends from one embodied brain region to another or many others. An active cascade will typically activate fixed circuits, circuits that characterize contexts important in everyday life, and may in addition activate nonfixed connections that happen to be "available" to carry a flow of activation most efficiently.
- **Cascade circuit**: A functional circuit that goes across areas in the brain, controlling and turning on and off fixed circuits. The fixed circuits that are components of a cascade, when turned on, extend and participate in active cascades, controlling where the flow of activation goes.

Our theory of cascades is set within the neural computational model for how neural systems function to produce effects. We hypothesize that cascade circuits are made up of basic circuit types.

EXAMPLES OF BASIC CIRCUIT TYPES

To comprehend how the neural system functions in thought and language, we turn from neural circuitry to the modeling of neural circuits and their effects using methods from computer science. In this section, we use the neural computation metaphor, which allows us to use precise forms of computation to model the functioning of the neural system. All of the circuits discussed in this section are part of our computational model of the neural system.

To emphasize the distinction between a model and the thing modeled, recall that a river can get you wet, but a computational model of the flow of the river cannot get you wet. The equivalent is true of the computational abstractions presented here. You can think using a real neural circuit but not using a computational model of one. They are functional models and computational hypotheses that capture the specific circuit structure and dynamics. Their main purpose is to guide further research on the actual neural implementation of these functions.

A central part of our neural embodiment hypothesis is that complex brain circuitry is made up of "small" simple types of basic circuits. The claim is that they form simply and naturally, given the neural tool kit and natural electrochemical mechanisms.

So far as we can tell, it appears that at least the following basic circuit types are prevalent in thought and language:

I. Circuits that provide basic structures
    1. Simple connection circuits
       a. Activation and inhibition
       b. Disinhibition
       c. Modulation
          ii. Divisive modulation
          ii. Subtractive modulation
    2. Gating circuits
    3. Gestalt circuits
II. Combinatorial circuits
    1. Binding circuits
    2. Mapping circuits
    3. Topographic maps and image schemas
    4. Comparison circuits and winner takes all
III. Control and coordination circuits
    1. Sequencing circuits
    2. Embedded circuits
    3. Concurrent circuits
    4. Conditional circuits

5. Shifter circuits
6. Substitution circuits
7. Priming circuits

Combinatorial circuits are the glue that combines simple circuits to form composite structures and cascades capable of carrying out various functions. Control and coordination circuits are circuits that provide basic operations of coordination and control useful for motor, movement, and behavior control. As we shall see, these circuit types are necessary for thought and language and structure our understanding of events, actions, and language.

These circuit types serve very different functions depending on what they are connected to. What they do depends on their role in larger circuits and their locations in the brain.

Simple Connection Circuits

Figure 5 shows simple circuit connections between two nodes.[7] Node A activates node B (shown by the directed arrow from A to B). Node C inhibits node D (shown traditionally by a line with a circle at the end). And given that node F, if activated, would inhibit node G, node E disinhibits node G by inhibiting node F, which keeps F from inhibiting G.

A basic circuit can be activated in one of two ways, depending on whether it is functioning in terms of the base rate of firing or the timing of spikes.

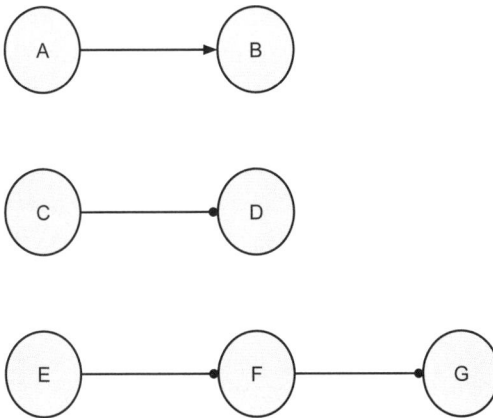

FIGURE 5. Simple connection circuits, showing a computational model of (a) excitatory (node A activates node B), (b) inhibitory (node C inhibits node D), and (c) disinhibitory (node E inhibits node F, which inhibits node G) circuits. In the disinhibitory case, if normally node G is inhibited with node F being active, activating E will inhibit F, thereby activating G. In this case, E disinhibits G.

- Base rate: A basic circuit is activated ("turned on") if it is functioning above the base rate of firing or is inhibited ("turned off") if it is firing below base rate. Here, "inhibited" ("turned off") does not mean not firing; it only means firing below base rate.
- Spike timing: A basic circuit is activated if presynaptic neurons are spiking at a time just earlier.

Gating Circuits

Circuit "gating" is a normal, natural phenomenon that allows modulation of functional circuits by controlling their activation or inhibition. Whatever can control that activation or inhibition is called a "gate." The brain has multiple gating systems. Some important gating systems are

- the loops between the cortex, the thalamus, and the basal ganglia, which includes projections from different parts of the basal ganglia including a disinhibiting signal from the basal ganglia to the association nuclei in the thalamus;
- the thalamic gating from projects of different association nuclei in the thalamus to different parts of the cortex; and
- interneuron gating within the cortex, such as gating withing the association cortices.

Figure 6 shows a simple gating circuit. This is an inhibitory circuit (one of several types of gating found in both cortical and subcortical structures in the brain). Here A provides a flow of activation to B. The gate G is inhibitory, shown by the shaded circle on the link from G to B. When the gate G is activated, it blocks the transfer of activation from A to B. But if G is not activated, activation can flow uninhibited from A to B.

Gates can be excitatory (enable flows of activation in specific circuits) or inhibitory (as in Figure 6, where a gate when active blocks the flow). As we saw in the neural tool kit section, neurotransmitters present in the synapse bind to transmembrane proteins that open up, letting in positive and/or negative ions.

Gates operate in a variety of mechanisms, including oscillations caused by the synchronized firing of quick-firing inhibitory neurons that change the relative phase of firing in the circuits they impact.[8]

Gating may allow the possible activation of a neuron if it results in the neuron being in the "up" state, in which it is "primed" to fire but just needs

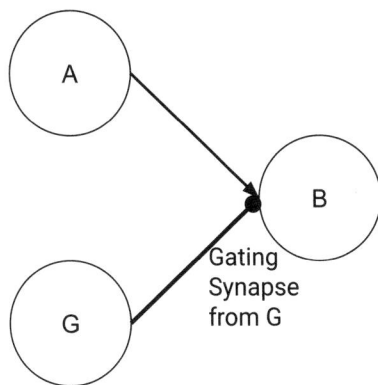

FIGURE 6. A simple gating circuit. This is an inhibitory circuit, one of several types of gating found in both cortical and subcortical structures in the brain. Here A provides a flow of activation to B. Gate G is inhibitory, shown by the shaded circle on the link from G to B. When gate G is activated, it blocks the transfer of activation from A to B. But if G is not activated, activation can flow uninhibited from A to B.

a bit more activation. Gating could also be inhibitory, leaving the neuron in a "down" state, making it less likely or unable to fire. Gating neurons that result in such "up" and "down" states result in what are called "bistable" neurons.

Gating neurons are found throughout the cortex as well as in important connecting subcortical structures such as the thalamus. The process of gating uses modulation—positive or negative input connections—to tip the scales quickly and readily for or against the firing of gating neurons.

All of the basic circuit types can be gated, that is, modulated so that they can be easily activated or inhibited by external connections.

In our computational model, a "gate" is metaphorically represented by a single node. When we talk of a circuit having a "gate," we are talking about the computational model with a single node serving a gating function in the computational model. Under the neural computation metaphor (which provides a computational model of a neural system), computational gates are mapped onto one or more real and complex neural processes that carry out a gating function.

In the computational model, if the gate is open (active), gating is modeled as atemporal (not time-dependent). That is, the gate either activates or inhibits, regardless of timing.

Gating and modulation occur everywhere in the brain.

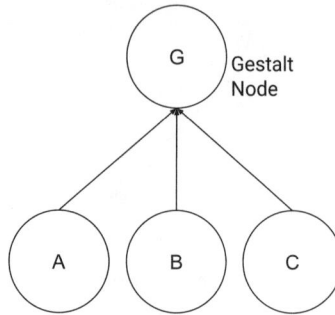

FIGURE 7. A simple gestalt node (G) with roles (A, B, C). The thresholds and synaptic weights have been left out of the figure. They meet the following constraints: If G is active, then A, B, and C are all active. If one or more of A, B, and C is active, then G becomes active, and as a result, all of A, B, and C become active.

Gestalt Circuits

A gestalt in general is a whole that consists of its parts. In many cases, the parts do not exist independently of the whole, although they can. The general idea is that a gestalt is more than the sum of its parts taken individually. A gestalt schema, in the notation of cognitive linguistics, consists of a gestalt symbol, two or more role symbols, and relationships among the gestalt and role symbols.

A gestalt circuit has parts called roles. The gestalt circuit shows how the parts are related to one another to form a gestalt. Figure 7 shows a simple gestalt node (G) with roles (A, B, C). The thresholds and synaptic weights have been left out of the figure. They meet the following constraints: If G is active, A, B, and C are all active. If one or more of A, B, and C is active, then G becomes active; as a result, all of A, B, and C become active.

Such simple circuits arise relatively readily via widespread neural recruitment. Simple circuits are made up of **circuit nodes**, **circuit connectivity**, and **activation thresholds**.

- **Nodes.** The gestalt circuit has three or more nodes: a gestalt node and at least two role nodes. For example, consider a movement. A movement has a source of movement (where the movement begins), a path of movement, and a goal (the endpoint of the movement). A gestalt circuit for a movement would have a gestalt node for the whole movement and three role nodes for the source, path, and goal.
- **Connectivity.** The gestalt node is connected to each role node. Each role node is connected to the gestalt node.

- **Thresholds and input activation**. Each node has a firing threshold. If the input activation is above the firing threshold, the node fires. Otherwise, it doesn't fire.
  - The activation input from each role node is above the gestalt node threshold. That is, the firing of any role node will make the gestalt node fire.
  - The activation from gestalt node to each role node is above the role node threshold. That is, the firing of the gestalt node will make *all* the role nodes fire.
  - The gestalt node is the control node of the gestalt circuit. If the gestalt node is inhibited, the gestalt circuit as a whole cannot fire. If the gestalt node is activated, the whole gestalt circuit fires.
  - Suppose there is a gating node that normally inhibits the gestalt node. The gestalt node and the whole gestalt circuit will then fire only if the gating node is inhibited. In such a situation, the gestalt circuit fires by disinhibition, that is, the inhibiting of the inhibiting gating node.

As an example, if the gestalt node characterizes surgery, the role nodes would include surgeon, patient prepared for operation, nurse assisting with the operation, anesthesiologist, scalpel, operating room, etc., all of the people and things that normally come to mind when you think of surgery. The idea of surgery activates *all* of them. *Each* of them activates the idea of surgery.

SUMMARY AND PREVIEW

Connection circuits, gating circuits, and gestalt circuits are among the most basic circuit types that, so far as we know, recur in the brain. But what about more complex combinatorial circuits such as mappings, bindings, and frames as well as complex sequencing and coordination required for all behavior? In the next section, we will review the current evidence from neuroscience on complex behavior coordination and control.

In digesting the amazing results on behavior from neuroscience and the biology of motor control, we arrive at a central hypothesis regarding the neural underpinnings of thought and language. Our results on embodiment suggest that the neural circuits for behavior *have been repurposed from animals during human evolution to provide the cognitive operations human beings use in thought and language*. At the end of the next section, we will discuss this hypothesis and thus be ready to move on to our computational models of circuits for behavior coordination that are composed of these basic circuits and that serve as the building blocks for quite complex cognitive structures and operations that occur naturally and often.

Coordination and Cascades: Evidence from Behavior Control

It is not easy for a child to learn how to tie her shoes. She has to keep her foot in a certain position and then coordinate a sequence of right-hand movements precisely with a sequence of left-hand movements. Coordinating complex movements requires synchronization and precise timing. Developmentally early examples of such coordination include pointing, where the movements of eyes, hands, and arms are synchronized in time and directed at a single motor target.

HOW IS COORDINATION ACCOMPLISHED?

Physical coordination appears to be carried out by cascades that control the timing and precise modulation of

- various executing networks (X-Nets) for sequential motor movements,
- attention, and
- feedback from the totality of perception.

Our hypothesis is that in thought and language, coordination works in a similar way: via cascades that control the timing and the turning on and off of circuits for thought and language. In addition, the same kind of circuitry would apply across different brain regions to perform very different tasks.

The neural embodiment of language hypothesis claims that the same types of coordination patterns found in motor actions are crucial for thought and language. Specifically, the construction and activation of cascades parallels the kind of coordination found in motor control. In this section, we discuss the basic kinds of coordination found in motor control and their connection to cascade formation and activation. The next section details our computational models of coordination circuitry and their application to language and thought.

There has been a lot of research on motor control emphasizing the role of motor synergies (Bernshteĭn 1967). **Synergies**, as we saw in chapter 1, *carry out specific routines (e.g., reaching, extending a finger, turning the palm upright) that are then composed (put together) into more complex behaviors.* Synergies typically have parameters that define their internal complexity. They form a set of building blocks that can be arranged in multiple circuits to produce complex behavior. Arbib (1998) proposed hierarchical sensorimotor couplings, what he called "schemas for complex tasks that are composed of multiple synergies."

Consider the motor schema for grasping as described by Itti and Arbib (2006) regarding schema theory:

> As the hand moves to grasp an object, it is *preshaped* so that when it has almost reached the object, it is of the right shape and orientation to enclose some part of the object prior to gripping it firmly. Moreover . . . the movement can be broken into a fast phase and a slow phase. The output of three perceptual schemas is available for the control of the hand movement by concurrent activation of two motor schemas: *Reaching* controls the arm to transport the hand towards the object and *Grasping* first preshapes the hand. Once the hand is preshaped, it is . . . only the completion of the fast phase of hand transport that "wakes up" the final stage of *Grasping* to shape the fingers under control of tactile feedback. (p. 297)

SCHEMAS GUIDE HOW THE BRAIN PERFORMS ACTIONS

Whether you are hitting a tennis ball, eating dinner, or hammering a nail, you are bringing together perception (e.g., seeing the ball, dinner, and the nail) and action (hitting, eating, and hammering). In each case you are following *coordinated schemas across different brain regions for seeing and acting that are tightly coupled.*

Perceptual schemas take perceptual inputs and link them to motor schemas that perform complex coordination and, via neural computation, produce motor output. Such schemas are active circuits distributed over different regions of the brain, coordinating perceptual, motor, and cognitive circuitry to yield circuits and systems of circuits that can carry out "functions" such as putting on your glasses, sawing a board in two, and cracking an egg into a frying pan.

A range of recent evidence (Graziano et al. 1997; Rizzolatti et al. 1998; Rizzolatti and Wolpert 2005; Gallese et al. 1996; Grafton et al. 2009) further suggests that such coordinating circuits are

- hierarchical in functionality,
- relational (applying to objects with multiple affordances, such as can openers),

- parameterized (for speed, trajectory, dynamics),
- event-driven (interruptible and adaptive to changing tasks and world states), and
- distributed across multiple cortical (premotor, primary motor) and subcortical (cerebellar and basal ganglia) regions of the brain.

Rizzolatti and his group in Parma, using single-cell recordings in awake monkeys, have mapped out several functional circuits for carrying out specific tasks. They found multiple interactive "loops" extending across prefrontal-to-parietal areas of cortex. Within each parietal-premotor-prefrontal circuit, *all forms of sensory information are integrated in carrying out motor actions* (Rizzolatti et al. 1996). Grafton, Aziz-Zadeh, and Ivry (2009) present further evidence revealing neural architectural constraints and flexibility in the motor system.

In addition to the basic findings on sensorimotor coordination circuits, work at the Graziano lab at Princeton University has led to a view of the motor cortex that is broadly consistent with the schema idea proposed earlier. In this view, the function of the motor cortex is not to decompose movement into constituent muscles and joints or into elemental movement parameters such as direction and speed. Instead, it appears that the motor cortex is based on the *foundational synergies tuned to produce some of the most complex components of the movement repertoire.*

The key experiments that led to this hypothesis came from microstimulation at the behavioral scale (half a second, or 500 milliseconds) instead of the usual shorter timescale (e.g., 50 milliseconds or less). The longer behavioral timescale roughly matches the time for the monkey to make arm movements to an object in the vicinity. Graziano (2018) and coworkers stimulated for half a second in different parts of the monkey cortex.

Stimulation in different regions of the cortical map evoked different movements that closely resembled common categories of actions from the monkey's normal repertoire. For example, when sites within one region of the map were stimulated, a hand-to-mouth movement was evoked. The movement included a closure of the hand into an apparent grip, a turning of the wrist and forearm to direct the hand toward the mouth, a rotation of the elbow and shoulder bringing the hand through space to the mouth, an opening of the mouth, and a turning of the head to align the front of the mouth to the hand. This complex coordinated movement occurred reliably on each stimulation trial and could be replicated even when the monkey was anesthetized. This clearly leads to the notion that *the common behaviors learned by the monkey and useful for everyday interaction were "packaged" into performable schemas and could be activated as a whole, integrated functional package.*

In a series of profound experiments, Graziano and Gross (1995) demonstrated neural networks that participated in multimodal tracking of the peripersonal space of a monkey.[9] Peripersonal space is the "bubble of space around the body" created by sensorimotor neurons and circuits in the cortex that control that part of the body (Graziano 2018). The tracking is done neurally through an internal approximate body simulation, or a body schema. The body schema simulation computes information about the location, shape, and state of the body; tracks the limbs; and enables accurate trajectory prediction and planning. Some beautiful work by Atsushi Iriki and Miki Taoka (2012) shows how tools can be incorporated into the body schema, which then enables the extension of the peripersonal space of the limb (in this case the hand) to be extended to the bubble around the hand together with the instrument held in the hand (e.g., a stick to reach farther). Space was tracked around the body, *space that could stretch and shrink* depending on tool usage. In addition, Graziano (2018) and colleagues, using the body schema, have shown how the peripersonal space is fundamentally responsible for maintaining a protected safety zone, the size of which may shrink based on familiarity and on conventionalized friendly or intimate social signals. Simulation of the body to create a body schema has thus been shown to have deep psychological, physical, cognitive, and social ramifications.

In summary, the evidence points to multiple circuits that carry out complex functions based on complex coordination circuits. Those circuits connect

- the premotor cortex, the parietal cortex, the primary motor cortex, and the supplementary motor area in the cortex with
- subcortical regions in the cerebellum and basal ganglia.

These circuits are formed initially through the use of multiple frontal regions. But when repeated and rewarded, they recruit more striatal and posterior regions in the cortex and become integrated packages. These circuits coordinate and orchestrate complex behavior, coupling perceptual and motor processing to perform specific functions and tasks.

How do these complex coordinated circuits get formed to produce complex behavior? For this we turn to the work of Ann Graybiel and her group at

MIT who have studied the interaction of rewards with the learning of complex coordination circuits.

FROM COORDINATION TO PACKAGES VIA REWARD

Graybiel and Smith (2014) have done extensive work showing how coordinated behaviors that start out new can become routinized and then packaged into chunks. This occurs when there is a complex interaction between the motor control and reward systems.

Behavior packages are formed by the interaction of cortical systems with the basal ganglia. There is a "reward prediction signal" based on the neuromodulator dopamine that "weights the value of our actions" by changing synaptic strengths, or "weights." "By monitoring our actions internally and adding a positive or negative weight to them, the brain reinforces specific behaviors, shifting actions from deliberate to integrated packages" (Graybiel and Smith 2014).

Reward mechanisms apply to coordinated actions and sequences of actions. When those actions are motivated (they are carrying out purposes that, when successful, generate rewards), they can be reinforced: the circuitry for performing purposeful actions is converted into integrated and routinized packages via neuromodulation within the norepinephrine system (motivational/attentional) and the dopaminergic systems (positive reinforcement based on reward prediction).

The mechanism for the "chunking" and packaging of actions involves both cortical and subcortical circuits, specifically the large loops that connect the basal ganglia, the thalamus, and cortical regions.

CORTICOSTRIATAL LOOPS AND BEHAVIOR CHUNKING

A large part of the frontal cortex receives inputs from the basal ganglia conveyed via the thalamus. These same cortical regions project to not only the basal ganglia (mainly to the striatum) but also other brain regions including the thalamus. The dopamine receptors and the dopamine-producing cells in the basal ganglia (substantia nigra) and in the brainstem (the ventral Tegmental area) are an integral part of the reward system in the brain. This circuitry "loops" between the cortex and the striatum and back to the cortex.

Upon repeated performance of a behavior that is based on being rewarded, the corticostriatal loop becomes strongly active, and the connections are strengthened. This leads to the coordinated action circuitry being chunked into a single atomic unit. The chunk relies on circuits in the striatum and depends on midbrain dopamine signals. When such chunks become integrated into packages, they become further imprinted, that is, made permanent and automatic. A specific region in the ventromedial prefrontal cortex appears to be able to control the activation and suppression of the acquired package.

LIGHT TRIGGERS AND THE ROLE OF THE VENTROMEDIAL PREFRONTAL CORTEX

Recently, using optogenetic techniques, researchers genetically engineered light-sensitive molecules in cells in the infralimbic (IL) cortex (a region in the ventromedial prefrontal cortex associated with primate emotion regulation).[10] Light-sensitive molecules, when placed in neurons using genetic engineering, discharge electrons that makes the neuron emit spikes (fire) or become inhibited (stop firing) when the light shines on them. Thus, in a remarkable way, researchers are able to target specific cells (those that have these protein molecules engineered into them) and turn them on or off using light.

In this case, the researchers were able to turn off the triggering of specific entrenched behaviors and routines (labeled "habits" by the researchers) by turning off this specific IL region in rats. Turning off the IL region for seconds appeared to completely block the routine. Indeed, if there was a competing routine (the rat ran on a different trajectory and to a different reward), that routine was acquired. Then, turning off a different ensemble of cells in the IL region (those controlling that new routine) allowed the previously learned habit to reemerge, since that habit was *competitively suppressed but not eliminated*. This research suggests that a circuit forms between a prefrontal region (the IL region) and a subcortical region (basal ganglia). This overall integrated circuit is responsible for the learning, activation, competitive suppression, and general control of routinized behavior.

LOOPS AND PACKAGES

Loops and packages are summarized as follows:

- Complex actions are acquired through the interaction of coordinated circuits in the prefrontal cortex and other cortical and striatal regions.

Initially, specific circuits in the striatum (part of the basal ganglia) monitor the new action closely by maintaining activation as long as the action is ongoing. Cells in the striatum are active throughout the performance of the newly learned action. Over time with repetition of the action, a control circuit learns key control transitions (the start and end) of the action and is active only for those phases of the action.

- The repeated performance of a successful (rewarded) action engages the reward circuitry in the brain, which creates integrated packagings (chunks) where circuits involving the basal ganglia are recruited to perform each whole complex action. This leads to decreased direct prefrontal involvement and more posterior (sensorimotor) subcortical region involvement, freeing up the prefrontal executive and planning system to learn novel behaviors.

- After an action has been packaged and its control circuitry is "moved" from the prefrontal regions to routines in posterior cortical and striatal regions, there is still a cortical control circuit connecting these regions to the ventromedial prefrontal cortex (in the IL system in the rat experiment) that can gate the switching on or off of the package. This IL-specific control node is then available for other cortical modulation for triggering or controlling the complex action and coordinating the learned package with other actions to create more complex packagings.

THOUGHT AND LANGUAGE PACKAGES

Our conjecture is that the process of behavior chunking and packaging underlies the development of complex thought and language.

We use the term "cascade" for these integrated and routinized packages that appear in thought and language. As we have outlined earlier, *cascade circuits are **thought packages*** that appear at multiple levels to structure our conceptual system. *Language involves the coordination and binding of conceptual circuits constituting the content* (what we want to say) *to the form* (how we say it) *of an utterance.* Language-mediated packages are learned in the same way as behavior routines. *Such thought-and-language packages are known as **constructions**.* Every construction is a chunked mapping between form and meaning. Constructional packages occur at all levels of language, from morphemes to grammatical constructions to narratives.

The connection between mechanisms for sensorimotor packaging and cognitive packaging thus applies to all the conceptual and linguistic integrated structures that we describe in this book: constructions, image schemas, frames, metonymies, metaphors, integration, scenarios, and narratives.

SUMMARY

Behavior control requires coordination of multiple circuits that are bound together into cascades of controlled activation and binding to accomplish complex motor behaviors. Our hypothesis is that these binding, mapping, and coordination circuits have been exapted (repurposed from animal behavior during evolution) for language and thought for human beings.

Meaning for neural circuitry is created by the brain performing actions, both internally triggered actions (as in imaginative simulation) and externally directed actions to achieve goals in the context of current and predicted future situations.

To make this hypothesis more precise, the next sections dive into the details of our models of these combinatorial and coordination circuits.

Combinatorial Circuits

Combinatorial circuits are the glue that enables multiple basic circuits to form composite structures and cascades. So far, we have seen examples in which there is a binding of actions to objects, mappings, frames, comparisons and scales, spatial locality, and topographic maps. In this section, we will illustrate computational models for these types of neural circuits composed from the basic circuits that will be described below.

There are a few types of these combinatorial circuits that appear to apply across different neural domains and functions:

- Binding circuits
- Mapping circuits
- Topographic maps and image schemas
- Comparison circuits and winner takes all

BINDING CIRCUITS

The notion of binding has been used to identify multiple issues related to information association and integration in the brain. Feldman (2013) has a useful characterization of the four different types of binding that are lumped together in the literature, and the details of how they work are referred to as the **neural binding** problem:

Type 1. One version of neural binding concerns the *overall integration of our multiple experiences and systems* into a unified whole that we perceive as our **conscious self**. Consciousness requires integration, and the process of integration takes much longer than simple neural firing.

Type 2. Another type of binding consists of the real-time emergence of *stable associations between structures* distributed in the brain (this is also called the "variable binding problem"). This can occur via the interaction of many circuits and is not a single straightforward process.

Type 3. A third kind of binding is when the stable associations between structures becomes materialized in a binding circuit. For instance, in the examples we will see below, the motor action of pushing, once learned, consists of multiple behaviors, such as moving your hand while keeping the palm open, making contact with the object (which may depend on the type of the object, such as a bottom or flat surface), and then applying force. Thus, the push behavior is composed of multiple actions bound together in a coordinated fashion. This type of binding circuit both creates the coordination structure and controls the activation to follow this structure so as to accomplish the overall behavior.

Type 4. A fourth kind of binding is the specialized usage of previously learned circuits to associate stimuli across multiple modalities. An example of the fourth kind of binding is the learned association between the sound of a car processed in the acoustic regions in the temporal lobe and its appearance processed in the visual pathway in the occipital lobe. This type of binding commonly occurs across modalities such as vision, touch, motion, hearing, and smell.

The kind of binding that leads to our unified perception of the world through consciousness (Type 1 above) requires more than just neural integration circuitry. What is requires in addition is an account of awareness and of the "feel" or "quality" of experience, such as the sound of a cello, the experience of the color red, or the experience of an orgasm (where a dictionary definition doesn't help). The technical term in philosophical discourse is "qualia." It is not addressed here, and very little is known about the mechanisms underlying such phenomenological and subjective experiences.[11]

The second kind of binding (dynamic binding) has been the subject of study in computational neuroscience for several decades, and there are many potential computational mechanisms that have been offered as solutions to this problem. Among the influential proposals are the use of temporal synchrony for real time binding of structured associations.[12]

More recently, there has been an attempt to solve the dynamic variable binding problem with a mathematically rigorous framework called assembly calculus (Papadimitriou et al. 2020). Assembly calculus models operate on assemblies of neurons, such as project, associate, and merge, which Papadimitriou et al. state "appear to be implicated in cognitive phenomena, and can be shown, analytically as well as through simulations, to be plausibly realizable at the level of neurons and synapses." The architecture and underlying operations have been used to model real-time emergence of syntactic structure in language processing. The approach and the results are a nice generalization of previous approaches. They implement the desirable property of the short-term plasticity and updating mechanism ins Hebbian learning and is very compatible with the circuits presented in this book. Whether this is the right approach is to be empirically validated. In short, the problem of dynamic binding appears to be solvable, and the best approaches to date are consistent with the techniques presented here.

What we *are* concerned with in this section is the *learned coordination* from experience of multiple regions, modalities, and circuits (associating the visual appearance and the sound of a waterfall or the learned movement of the right and left hands in tying a shoelace). Such bindings are often unconscious parts of the structuring of our everyday experience of the social and physical world we inhabit.

SIMPLE BINDING CIRCUITS

Imagine that you are using a wrench to tighten a nut that has become loose from a bolt. You put the wrench into position around the nut and tighten the wrench. At this point, the location and orientation of the nut and the wrench become the same; there is a tightness relation between them, and you can move the nut by moving the wrench, with the location and orientation of nut-and-wrench changing together. You see this, you understand this, and you act accordingly. Organizing what you understand, there is a "neural binding" between the part of the circuitry in your brain characterizing the nut position and orientation and the circuitry characterizing the wrench position and orientation. Since the wrench can be used for other things, the neural circuitry used to understand it must be independent of the neural circuitry for the nut, and that different neural circuitry must be in a different brain location. The "binding" must be accomplished by a circuit that identifies the two nut and wrench positions and orientations as being the same. This cannot be a fixed circuit; it must be an available circuit. And since you have just imagined all this in your mind without seeing it or moving anything, it must all be done via thought, without actual vision or movement.

Once recruited, such a binding circuit can become dedicated (no more available to do other bindings) if the activity is imagined or repeated often enough and/or is salient enough (in terms of emotional salience or external reward). Then the binding circuit functions like a gate that, when activated (open), connects the two nut and wrench positions and orientations, making them the same for the time when the gate is open.

Figure 8 shows a representation of a simple binding circuit, with A bound to C (gated by gate G1 represented by a triangle). Binding constitutes a basic mechanism of neural compositionality, the basic way in which different schemas are combined, or "put together," into larger wholes. Basically, there has to be a way to form larger circuits from smaller circuits when the two circuits are in different places. Binding circuits constitute one of these mechanisms.

Imagine two gestalt circuits. These could, of course, be in separate parts of the brain. Suppose you want to identify semantic role A with semantic role C, forming a new complex circuit in which A and C are identified as being the same entity. We saw a case like this in chapter 1. The restaurant frame uses both the business and food serving frames, with the customer in the business frame being the same as the eater in the food serving frame. Since the circuitry for these component frames would, presumably be located in different brain regions, there would need to be a neural circuit binding customer to eater, conceptualizing them as the same person.

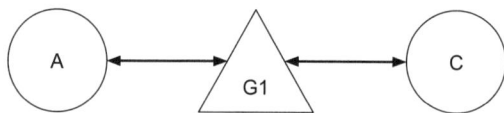

FIGURE 8.   A representation of a simple binding circuit, with A bound to C, gated by gate G1 (see Figure 6 on p. 197), represented by a triangle. Binding constitutes a basic mechanism of neural compositionality, the basic way in which different schemas are combined, or put together, into larger wholes. Basically, there has to be a way to form larger circuits from smaller circuits when the two circuits are in different places.

Here are the properties of the binding circuit for cases like this scenario:

- A gating node G that gates the bidirectional connection between A and C coactivates A and C, thus synchronizing the firing of A and C. When G is firing at or below base rate, the gate is closed (inhibited through lack of positive activation). Then A and B function independently, since they are unconnected.
- When G is active (which would happen if the gestalt node for the restaurant frame is active), the gate is open, and activation passes in both directions between A and C, with each activating the other and both coming to fire in synch. The rest of the brain then cannot tell the difference between them. The reason is that they are part of the same circuit and are directly linked together, and one fires exactly as the other fires. There is no way for outside circuitry to distinguish them based on when they fire. To outside circuitry, internal circuit location and firing rates make the two nodes look like one node.
- The time course of the binding connection is very small. Any difference in timing is minuscule and is within a window defining neural "simultaneity." Neural simultaneity is not absolute simultaneity, independent of all neural systems. Neural simultaneity between two neurons within a neural system means that the time difference between the firing of those neurons must be so small that the firing of any circuit connecting those neurons cannot depend on that time difference.

MULTIPLE BINDINGS ACROSS FRAMES

In Figure 9 we have two frames, F1 and F2. F1 has roles A and B, and F2 has roles C and D. And perhaps there are other roles not shown in the figure. Note that as described in chapter 2, frames are collections of gestalt circuits (the computational model of gestalt circuits was described earlier in this chapter in the section "Basic Circuit Types").

Bindings between frames enable roles of one frame to be bound to the roles of another frame. For example, F1 might be the business frame, with A being the salesperson and B the customer. F2 might be the food service frame, with C being the host and D the guest/eater. In the restaurant frame, A (the business salesperson) and C (the food service host) are the same person. And c (the customer in the business frame) and D the (guest/eater) are the same person.

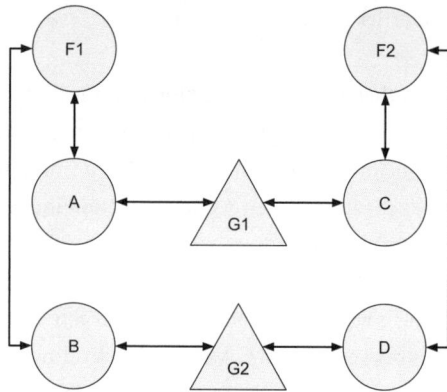

FIGURE 9. Multiple binding circuits. Here there are two frames, F1 and F2. F1 has roles A and B. Roles are represented by gestalt circuits (see Figure 7 on p. 198). F2 has roles C and D. And there are perhaps other roles not shown here. Note that as described in chapter 2, frames are collections of gestalt circuits (the computational model of gestalt circuits was described earlier in this chapter in the section titled "Basic Circuits"). Bindings between frames enable roles of one frame to be bound to the roles of another frame.

THE DISINHIBITION HYPOTHESIS FOR NEURAL BINDING

We hypothesize that many neural binding circuits are gated via disinhibition. This is in analogy to the motor control system, where disinhibition appears to be a central mode of behavior coordination through loops connecting cortical regions to the basal ganglia and the thalamus (Narayanan 2003a). The idea is simple. A binding circuit consists of two activation circuits firing in opposite directions. When the activation is flowing, the connected nodes are firing synchronously and function with the rest of the brain as if they were a single node. That is, the rest of the brain cannot tell the difference between them by attempting to determine which unique single node is firing at a given time. At a given time, two different nodes can be firing synchronously.

- The gate controlling the flow is inhibitory. The gating node inhibits the bidirectional flow of activation. The binding can be turned on only by the disinhibition of the gating node. That is, the gating node is inhibited from firing, which keeps it from inhibiting the flow in the binding circuit.
- More recently, findings from animal behavior in vivo demonstrate that salient events (those that are readily noticed and more easily remembered) often elicit disinhibition of projection neurons that then become active to form memories. Multiple behavioral functions such

as auditory fear learning and spatial navigation arise from disinhibition at timescales ranging from milliseconds to days. The data to date suggests that disinhibition is a common circuit mechanism contributing to learning and memory.

- Disinhibition is fast. The flow of activation is ready to go as soon as inhibition is turned off.

SINGLE MAPPING CIRCUITS

Imagine that you are hanging a picture on a nail already hammered into a wall. There is a wire behind the picture from one side of the picture frame to the other side. You need to get the wire onto the nail in just the right place so that the picture will be upright and the sides of the frame will be vertical. You hold the frame by the sides with one hand on each side, loop the wire over the nail, and start sliding the picture back and forth to find the right spot for the wire to rest on the nail. To find that spot you must establish a correlation between the right orientation of the picture and the right place for the wire to rest on the nail. Mentally, you are establishing a correspondence between the picture's orientation and the spot on the wire where you feel the picture can maintain the right orientation; in other words, you have to establish a mapping from the picture's orientation to spots on the wire. Neurally, you have to have in your brain a circuit in which both the picture and the wire are characterized and their relative locations are fixed. To establish the mental conceptualization between the orientation of the picture and the location of the wire on the nail, there must be a neural mapping that carries it out.

The picture's orientation is the *source* of the mapping, and the location of the wire on the nail is the *target*. The structure of picture, frame, and wire happens, cognitively, to be characterized as a conceptual *frame*, which is in turn characterized in the neural computational model as a gestalt circuit.

Such mapping circuits have the properties

- A (the correct orientation of the picture) and B (the correct location of the wire on the nail), which are in the same frame circuit. By fixing the correct orientation of the picture, you are fixing the correct location of the wire on the nail and vice versa. A activates B and B activates A, with G as a gate. The gate G only allows A and B to activate each other.

In the neural embodiment hypothesis, such mapping circuitry is determined by physical actions whereby a single mapping is needed from a source element of a frame to a target element of the *same* frame. When used in thought, the mapping from one part of a frame to another part of the same frame is called a **metonymy**, or a **metonymic mapping**.

Since ideas are expressed in language, metonymic mappings in the conceptual system commonly appear in language. The classic example is from the restaurant frame, where one waitperson says to another *The hamburger wants his check*. In the restaurant frame, there is a customer, who is the eater, and the product consumed, which is a dish, in this special case a hamburger. There is also a pairing of particular dishes ordered with the customers ordering them. This conceptual pairing occurs within the restaurant frame, allowing the particular dish (the hamburger) to stand metonymically for the customer who ordered it and will therefore get a check, an invoice for payment. All of this is part of the restaurant frame and the metonymic mapping within that frame, and thus *The hamburger wants his check* can be processed unconsciously in a few hundred milliseconds.

Multiple Mapping Circuits

METAPHOR MAPPING

How do we understand what it means to be a member of a club? We understand a club as a metaphorical container, with an interior and an exterior. The club members are in the interior. Those who are not members are in the exterior.

This requires a mapping with multiple parts:

- The container interior is mapped to the club.
- The people in the container are mapped to the members.
- The people outside the container are mapped to nonmembers.

There are, of course, lots of things that we metaphorically understand as containers, and there are lots of such mappings. To control which mapping is operative in the brain, there have to be gatings between the container schema with its parts and what they map into. Learning a specific metaphor mapping corresponds to learning the gating circuits (nodes in the computational model). In the above example, each role in the container schema has to map, via a neural connection, to the right corresponding role in the club schema; that is, the people inside the club as a container map to club members in the membership frame. For those neural connections to be active, the right gating node must be activated so that activation can flow from the container schema circuitry to the club schema circuitry. This is necessary to, for example, understand Groucho Marx's famous quote, "I wouldn't want to belong to any club that would have me as a member."

When these conditions hold, there is strong activation from the container roles to the club roles, and there is weak activation (or none) in the opposite direction from the club roles to the corresponding container roles. Thus, mapping circuits such as metaphor are projections from the source to the target, while binding circuits are bidirectional.

Figure 10 shows a metaphor mapping circuit with a source frame and a target frame. Role A of the source is projected (notice the unidirectional arrow) to C in the target and B in the source to D in the target. In the example in the text, the source would be the container frame, and the target would be the

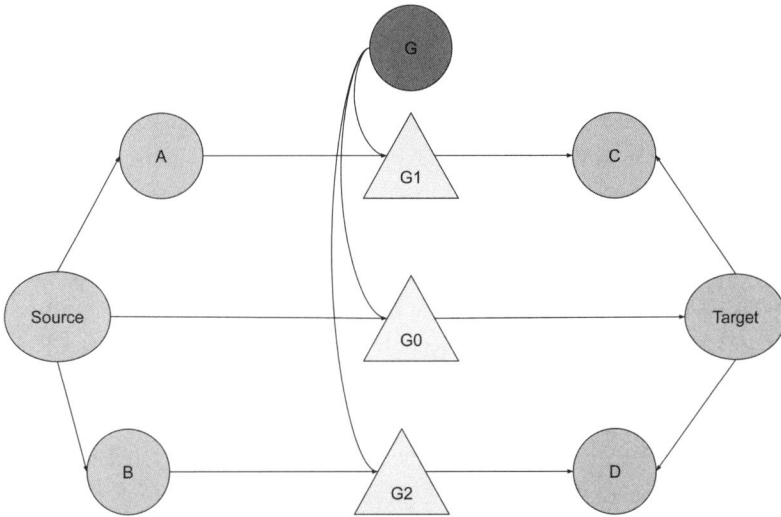

FIGURE 10.   A metaphor mapping circuit with a source frame and a target frame. Metaphor circuits combine frames and binding circuits (see Figure 9 on p. 212) with gating circuits (see Figure 6 on p. 197). In Figure 10, role A of the source frame is projected (notice the unidirectional arrow) to C in the target frame and B in the source frame to D in the target frame. The projections (gates G0, G1, G2 in the figure) are activated with the metaphor that is active (gate G). When active, the connected roles are active. Note that this is a projection, not a bidirectional binding. In the example in the text, the source would be the container frame, and the target would be the club frame. The role A is the interior of the container frame mapped to role C in the club frame. Individuals binding to role A (entities in the interior of the container) would be conceptualized as members of the club.

club frame. The role A is the interior of the container frame mapped to role C in the club frame. Individuals binding to role A (entities in the interior of the container) would be conceptualized as members of the club.

A metaphor mapping circuit includes the following aspects:

- Circuits characterize two schemas or frames and their respective roles. Call one the source and the other the target.
- There are gated strong activating connections from source roles to corresponding target roles, and there are very weak or nonexistent activating connections in the opposite direction.
- The gates, G_0, G_1, and G_2, are all gated by the node G.
- If the source frame is firing and gate G is firing, then gates G_0, G_1, and G_2 will all be firing, so the target frame and all its roles will be firing.
- In thought, such a multiple mapping is called a **conceptual metaphor**, or a **metaphor mapping**.

INTEGRATIVE MAPPING

Consider the number line in mathematics. Numbers are seen metaphorically as points on a line. The points on the line are not just used metaphorically to understand what numbers are; the relationship is more powerful than that. On the number line, the points are numbers, and the numbers are points on a line. Numbers and points on a line are integrated and are taken to be the same things. There is not just a conceptual metaphor linking points and numbers. There is also a *binding* between the points and the corresponding numbers, which makes the points into numbers.

Neural binding is the mechanism for understanding two entities characterized by different circuits as being identical. Each binding is gated; when the gate isn't activated, the entities are conceptualized separately. Only the active gating of the binding circuit creates the neural "identity."

The result of activating an integrative mapping circuit is to conceptualize a single integrated entity simultaneously, which is structured by all of the schemas, frames, and/or metaphors. Here is the integrative mapping for the number line integration, where the equals sign (=) indicates a binding:

- Integrated structures: The A line are the real numbers.
- The metaphor: Real numbers are points on a line.
- The binding: Points on the line = real numbers.

- Larger numbers are to the right of smaller numbers on the line.
- The difference between two numbers = the distance between their corresponding points.
- The inhibiting gates for the metaphor and the binding keep the source and target nodes of the metaphor as different entities, allowing us to understand the difference.
- An integration node I that, when active, disinhibits the two gates, allowing the corresponding source and target nodes of the metaphor to be understood as the same entities, that is, allowing the points on the line to be understood as being numbers.
- This permits us to understand the numbers and points on the line literally as different entities and metaphorically as the same entities.

What results is an integrated whole concept: the number line, a central idea in mathematics. The number line is an internally consistent single idea even though it is composed of parts that are conceptually distinct: a line, which is continuous and spatial, and numbers, which are separate entities.

It should be born in mind that neurally if the integrative node I for the mapping is not active, then the real numbers and the line would be thought of as separate mathematical entities. In the version of mathematics in which numbers are *defined as being sets*, there is a different integrative mapping between numbers and sets, where zero is the empty set, 1 is the set containing the empty set, 2 is the set containing zero and 1, and so on. These two versions of the foundations of mathematics are defined by different metaphorical mappings: from points on a line to numbers and by sets to numbers.

The general concept: In chapter 2 in the section titled "Conceptual Integration and Consciousness," we discussed the Grim Reaper example and the general concept of an integration circuit. A single integration circuit contains an integration node and connections that activate or inhibit those control nodes and gates that allow for the creation of an integrated whole. That is what is done in the Grim Reaper example, and it is also done in the number line example.

MAP NETWORKS FOR IMAGE SCHEMAS: A HYPOTHESIS

We do not know what kind of neural circuitry carries out image schemas. However, in the 1990s as Terry Regier was mulling over his dissertation (Regier 1996), he came up with an idea: a network of connected topographic maps, which we will call **map-nets**, might work for understanding image schemas in terms of neural circuitry. Although Regier ultimately rejected the idea, it is interesting enough to be worth considering.

Regier's conjecture was based on facts about maps. Relatively small two-dimensional topographic maps exist throughout the cortex. The cortex has six layers, with connections within each layer and across layers. Maps can occur in each layer with connections across layers. In addition, map layers in the brain can take neural inputs from elsewhere in the brain.

WHAT WOULD A MAP-NET BE LIKE?

A map-net, if such existed, would be a collection of layers. It would be cross-modal—that is, usable in vision, motor control, and touch—and would have the following properties:

- There would be an input layer connected to another brain region or regions.
- Each map would be a collection of neurons that preserves closeness, given its input. What "closeness" would be closeness of would depend on its input.
- Each map-net would have layers.
- Each layer in the map-net would be a two-dimensional structure of neurons (or nodes).
- Each layer would have specific internal connections among its neurons (or nodes).
- Each layer would take input from and/or provide output to at least one other layer in the network.
- The connections may be either activating, self-activating, or inhibiting.
- A specific map-net is defined by the layer-internal and cross-layer connections and the connections to the input map.

THE CONTAINMENT MAP-NET

Boundaries of objects and regions are in general assumed to be computed in early vision. The parietal lobe contains maps in the where pathway that take input ultimately from the early visual cortex. Suppose one of these is the boundary layer.

- The boundary layer: Layer 1 takes input from early vision. A "boundary" in a spatial map is a sequence of self-activating neurons constituting a closed curve. The neurons need have no connections between them.

When Layer 1 ceases to take boundary input, the neurons in Layer 1 all return to base rate firing, as do the neurons in Layer 2.

- The exterior layer: Layer 2 consists of self-activating neurons that are arranged with connections that directly flow from the outside edges of the map to the inside of the map. When Layer 1 gets boundary activation, neurons at the periphery of Layer 2 begin firing, and the activation moves from the outside periphery of the map to neurons on the inside of the map.

The boundary neurons of Layer 1 are connected by strong inhibition to Layer 2. As activation in Layer 2 spreads and stays active from outside to inside, the spreading activation encounters the strong inhibition from the boundary layer in Layer 1, and the spreading activation stops where there is inhibition. The result is that the neurons in Layer 2 characterize the exterior; they are active and stay active.

As a consequence of the exterior-to-interior spreading activation stopped by the inhibition from the boundary map, the container can be any size and any shape and in any location. In short, it is topological in character.

- The interior layer: Layer 3 consists of self-activating neurons that take inhibition from the active neurons in Layers 1 and 2 that characterize the boundary and exterior. This inhibition from the exterior and boundary layers—Layers 1 and 2—will turn off the corresponding Layer 3 neurons, leaving active those neurons characterizing the interior.

Portals are discontinuous regions along the boundary that allow for motion in and out of the interior.

- The portal layer: The neurons in the portal layer—Layer 4—are self-activating. Layer 4 takes strong inhibition from the interior and exterior layers and also takes inhibition from the boundary layer. The portals are extremely active regions of neural activity along the boundary layer, which are strong enough to override the inhibition coming from boundary layer (Layer 1). Those active neurons along the portal map characterize the portal.

The entire map network characterizes the container schema. Each map in the network characterizes a distinct semantic role in the schema: boundary, exterior, interior, and portal.

Such a structure will capture the logic of the container schema. This is far from trivial. It is remarkable that a map layering can characterize the logic of an image schema!

Of course, this is hardly a complete account. Neural maps are two-dimensional. Containment can be three-dimensional. Three-dimensional containment can be accomplished by two or more uses of a two-dimensional map in different orientations.

But that raises further issues. What is an instance of an image-schema? Two instances could be two activations over a short time. But then the instances of three-dimensional containment would have to occur in two different orientations. How would that be done?

Moreover, most real-world containers are solid, and their solidity exerts force to keep the contents in the container and other things out. The dimensionality, solidity, and reproducibility issues would have to be dealt with in a full account.

Do we believe this two-decade-old conjecture of Regier's? There is no evidence for it. Regier has not accepted it himself. How would it arise? Perhaps it would arise as a generalization over experiences with many kinds of containers. But there is no theory as to how this would work. Could it be innate, since human beings everywhere have such a general container schema, and would this map-net characterize it?

At this point, Regier's map-net conjecture remains interesting but unsubstantiated. And what it characterizes is only a two-dimensional container without force dynamics. Why mention Regier's conjecture? Because it at least allows us to imagine how a map-net theory might work, given that there are maps throughout the brain (and the body).

Let us return to cases that are widely assumed to be real.

LATERAL INHIBITION AND WINNER-TAKES-ALL CIRCUITS

Figure 11 shows a simple winner-takes-all circuit. X is active if A and B are active, and Y is active if B and C are active. Either X or Y, whichever is more active by mutual inhibition between X and Y, drives the other to be inactive and thus is the "winner." Comparison circuits compare two nodes, X and Y, which are mutually inhibiting each other. The "winner" can be defined as either (a) the node with more activation or (b) the node that is activated first. The node that "wins" stays on; the one that "loses" turns off.

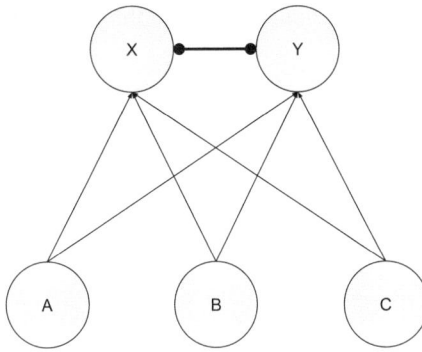

FIGURE 11.   A simple winner-takes-all circuit. X is active if A and B are active, and Y is active if B and C are active. Either X or Y, whichever is more active first, by mutual inhibition between X and Y, drives the other to be inactive and thus is the "winner." If X is more active first, it will be the winner, and then Y will be inhibited, and if Y is more active, it will inhibit and thus deactivate X.

The mechanism for this is mutual inhibition, where Node 1 and Node 2 inhibit each other. Thus, only one can "win" by staying on. Suppose they are both activated at the same time and Node 1 gets more activation than Node 2. Node 1 will overwhelmingly inhibit Node 2, while Node 2 will not be activated strongly enough to inhibit Node 1.

In the case of timing, if Node 1 is activated first, it will inhibit Node 2, and Node 2 will not be able to be activated at all.

A multiple comparison circuit is called a winner-takes-all circuit, where all nodes are mutually inhibitory. Winner-takes-all networks are commonly used in computational models of the brain, particularly for distributed decision-making or action selection in the cortex. Winner takes all is a general computational primitive that can be implemented using different types of neural network models, including both continuous-time and spiking networks. Important examples include hierarchical models of vision (Riesenhuber and Poggio 1999) and models of selective attention and recognition (Carpenter and Grossberg 1988; Itti et al. 1998). Winner-takes-all circuits are known to be very powerful computationally (Maass 2000) and also have very efficient realizations in artificial neural hardware (Lazzarro et al. 1988).

Feldman and Ballard (1982) proposed winner-takes-all circuits for computational models of action selection. The architecture used a simple rule: in an interconnected network of action modules, any module that receives a higher input from a competing module than its own will set its output to zero. This turns all those modules that receive more activation than they send into losers, and the single module that sends the highest activation to all the others is the winner.

One question this line of thinking leads to is the following: Given the kinds of initial connectivity patterns and biological learning algorithms such as STDP, what kinds of available circuits are more likely to be formed, and do they consist of the basic circuits needed for language and thought? There is some work on this that suggests that the kinds of gestalt circuits we proposed are fairly natural products of learning algorithms. In chapter 2 we presented the formation of mapping circuits for metaphor, which is a direct result of prior connectivity tuned by STDP. This provides a reason to take STDP seriously.

Control and Coordination Circuits

We are now in a position to develop models of the structured connections that control the coordination behaviors outlined earlier. More generally, all behaviors, from *Drosophila* (flies) and worms such as *Caenorhabditis elegans* (Bargmann 2012) to primates and humans make extensive use of neuromodulated activation in a coordinated and integrated fashion, linking perceptual stimuli, memory, and motor action to produce seamless behavioral responses.

Using an analogy to sensorimotor control (described in results in chapter 1 in the section titled "General Features of Cascades"), we hypothesize that there are cascades of neurons that extend within and across brain regions and contain coordinating circuitry: cascade circuits that monitor, coordinate, integrate, and control other neural cascades. Coordinating circuits can directly coordinate and integrate the flow of activation in other typically complex circuits, occur as part of hierarchical cascades, and are learned when complex cascades are put together. Because coordinating circuits act directly in that they are learned as part of the cascade, their effect is quick and takes little energy. They often occur as part of larger fixed circuits (such as frames) but sometimes may be temporarily recruited in a given context. In a cascade, both activation and inhibition flow, which is structured by the circuits in the cascade.

There are several basic subtypes of coordination circuits that appear to play a large role in language and cognition:

- Sequencing circuits
- Control circuits
  - Iteration
  - Suspension/resumption/cessation
- Concurrent circuits

- Conditional circuits
- Substitution circuits
- Priming circuits

SEQUENCING CIRCUITS

Performing a complex action requires sequencing and coordination. X-Nets in computer science were designed to model such sequential motor control. X-Nets have key states and actions (transitions) such as a (possibly complex) precondition, a start, a center, a test for whether purposes have been achieved, a finish, a final state, and a consequence. Actual X-Nets that have been constructed by computer scientists have additional complications (loops, pauses, returns, abandonment), but the sequencing circuits for the stages of an action are simple.

If you are at node n and about to shift to node $n + 1$, then in the overall circuit, node n is active, and node $n + 1$ is not active. The sequencing circuit with three nodes activated sequentially looks like Figure 12 at three stages, with arrows between shaded and blank nodes showing activation from a prior stage to a subsequent stage. As Figure 12 depicts, in a sequential activation of three nodes, the active node is represented as dark at any given time. There is an activating circuit connecting from node n to node $n + 1$, and there is an inhibiting gating connection from node $n + 1$ to node n. The circuit works as follows:

- Node n is active.
- The connection to node $n + 1$ is activated.
- Node $n + 1$ is active (for a single neuron, we would say it is spiking, but a node is an ensemble, so we use the word "active," which implies that the ensemble is firing as a circuit with a sequence [spatial and temporal] of individual spikes).
- Node n is inactive (in the "down" state) and node $n + 1$ is active.

FIGURE 12. Sequential activation of three nodes. Sequential activation controls activation flow in a chain of nodes. Shown in the figure is activation propagating across three nodes (from left to right); the active node is represented as dark at any given time. Notice how the first (leftmost) node is active initially, which activates the synaptic connection (modeled as the leftmost arrow in the chain) and then activates the next node in the sequence until the activation travels to the final node in the chain.

Iteration

Actions typically occur in sequential stages. At each stage, S_n, there is an embedded circuit that carries out the action at that stage. When stage S_n is activated, the embedded circuit is started. When the embedded circuit is completed, S_n is inhibited, and S_{n+1} is activated.

When actions are iterative, there may be "loops" in which a sequence is repeated until a condition is met. If the condition is not met, the loop's output is inputted to the loop. If the condition is met, the loop's output activates the next stage.

Suspension/Resumption/Aborting of Actions

Actions are extended in time, and unexpected events could temporarily suspend an action, which may be resumed, restarted, or aborted. An ongoing action could therefore be interrupted and its performance temporarily stalled. Language and discourse routinely express interrupted actions and their consequences (consider the words "slip" or "stumble").

Concurrent Actions

We often have to or want to do things that require coordinating *both* feet or *both* hands or singing and dancing *together*. It is not easy to do something complicated for the first time, such as learning a dance routine or a tai chi set, learning to pivot on a double play in baseball, or learning how to do a triple lutz in figure skating. The first time requires a lot of effort and energy: paying close attention at all times, keeping track of where you are, anticipating the next movements as you're doing the earlier ones, and trying to consciously control your body movements and consciously coordinate them in just the right way (Figure 13).

As you practice dozens, hundreds, or even thousands of times, you unconsciously *recruit unconsciously coordinated circuitry* that directly bypasses the bumbling consciously controlled attempts and allows you to do the pirouette; in tai chi coordinate cloud hands with your feet, head, breathing, and eye movements, precisely timing the shift of your yin hand and yin foot to your yang hand and yang foot and vice versa; or, as a second baseman in a double play, approach second base with the right-length steps and speed to snatch the shortstop's toss with your glove hand, hit the bag with your left foot and with your knee bent, switch the ball to your throwing hand, and leap into the air, turning and throwing with one motion while rolling over out of the way of the

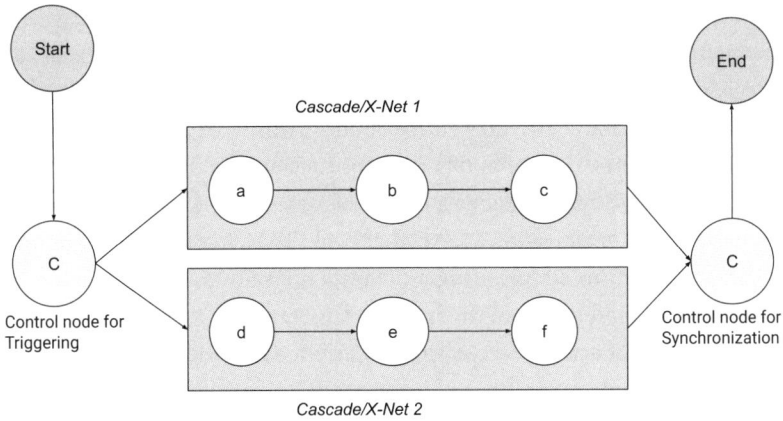

Cascade/X-Net 1

Start

End

a → b → c

C
Control node for
Triggering

C
Control node for
Synchronization

d → e → f

Cascade/X-Net 2

FIGURE 13.   Concurrent actions allow multiple actions or action sequences to be performed at the same time. In the figure, X-Nets are used to model the concurrently active
action sequences. These X-Nets are represented by nodes a, b, and c, which are individual
actions in X-Net 1 and nodes d, e, and f in X-Net 2. There is also a circuit consisting of two
control nodes (labeled C) that allow for the controlled performance of the overall cascade.
One of the control node triggers all the action sequences, which then can be performed
in parallel. The second control node, titled "synchronization," waits for all the X-Nets to
complete before finishing the overall controlled cascade. In addition, there is a start node
that activates the controlled concurrent cascade and an end node that gets activated at the
end of the controlled concurrent cascade.

runner's slide. (Don't ask us about the triple lutz!) It is the recruitment of coordinating nodes that allows us (at least some of us) to acquire shoe-tying as children and other feats of coordinated mastery. As we shall see, the same kinds
of coordinating circuitry may be used in not only doing these physical actions
but also *imagining* doing them and, as we shall see, thinking and speaking.

Imagine reaching out and grasping a glass of water. You have to reach out
toward the glass but not the center of the glass if the glass is wide. You have
direct your reach to the outside of the glass. As you see your hand get close,
you have to trigger the grasping X-Net as your hand continues to move. Your
fingers open first. As the opening between the thumb and the forefinger get
close to the glass, you activate a new movement directing the opening to the
center of the glass. As your fingers go around the glass, the finger-closing circuit is activated. When the finger touches the glass, the force-dynamic finger-
tightening circuit is activated long enough to get touch feedback indicating
a tight-enough grip. And if you want to drink the water, a lot more detailed
muscle-control circuitry for hand, elbow, shoulder, lips, mouth, tongue, etc.
will have to be coordinated.

At most stages there is a "controlled concurrency." While staying activated, one circuit signals the activation of another: while reaching for the glass as you get close enough, your hand opens up, and so on. This requires multiple X-Nets and rhythmic circuitry (these circuits, which are known as **central pattern generators**, control bodily rhythms such as heart rate and breathing rate). The rhythm circuits define a notion of same time. Each rhythm circuit reaches maximum activation at a rhythm peak. Two circuits neurally bound to the same rhythm circuit firing at its peak are said to be firing "concurrently." Neural bindings from rhythmic circuits to two or more X-Nets define what "concurrent actions" are. Such concurrent actions that are stable over time are commonly referred to as "homeostasis."

The computational model has the following aspects:

- There are two (or more) X-Nets simulating action sequences with stages of each action. In the Figure 13, these X-Nets are represented by nodes a, b, c in X-Net 1 and nodes d, e, f in X-Net 2.
- There is also a circuit consisting of control nodes for triggering the two X-Nets (node C).
- There is a control node that requires both X-Nets to complete before further processing can take place. This is the Control node synchronization node in Figure 13.
- There is a start node that activates the controlled concurrent cascade and an end node that gets activated at the end of the controlled concurrent networks.
- In addition (not shown in Figure 13), there may be binding circuits tying actions in X-Net 1 with the actions in X-Net 2, the concurrent actions in different X-Nets. To carry out such a conjoining circuit, one must perform both X-Nets, with the actions bound across them occurring at the same times.

Conditional Circuits

A conditional circuit controlled by node B is a circuit that is activated only if a given condition is met (Figure 14). This is a simple disinhibition-based control circuit for conditional activation. A circuit contains two nodes, A (the condition) and B (the controlled node), and a gating node G:

1. Gating node G gates node B.
2. G is active and inhibits B.
3. At steady state in the absence of any input, B is inhibited.

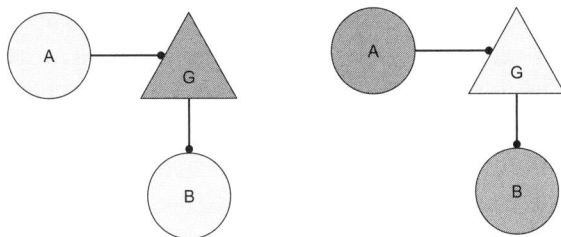

FIGURE 14.   A simple disinhibition-based control circuit for conditional activation. The circuit contains two nodes—A (the condition) and B (the controlled node)—and a gating node G. Gating node G gates node B. If G is active, it inhibits B. If A is active, G is inhibited, and activation can flow freely between A and B. Further details are provided in text.

4. When the condition A is met, node A is active.
5. When active, node A inhibits gate G.
6. Gate G, when inhibited, disinhibits B.
7. Activation then passes from node A to node B, making B active.
8. Thus, when condition A is met, B becomes active via disinhibition.
9. When condition A is not met, B remains inhibited because G is active and is inhibiting B.

Substitution Circuits

We often have to make minor adjustments in routines, such as plugging in a lamp where there is a dresser between the plug and the lamp. Under certain conditions (e.g., there is a space heater in the way), you have to *substitute* a different hand and arm action for the one you would normally use if the plug were unobstructed. We call the X-Net for the normal case the default X-Net.

A substitution circuit is a conditional circuit in which the following occurs:

- B controls the new X-Net to be substituted in.
- C controls the default X-Net to be substituted for.
- A is the condition and is not active.
- Gate G is active and inhibits B.
- C is active, controlling the default X-Net.
- When the condition is met, A becomes active.
- A inhibits C and G, at once inhibiting the default X-Net and activating the new X-Net via disinhibition of B.

Often, not all possible substitutions are or even can be enumerated. One hallmark of human reasoning and planning is the ability for on-the-fly opportunistic substitution. We believe that a best-fit neural optimization mechanism should provide a framework for investigating the nature and mechanism of this process, which people are engaged in almost all the time. This problem is related to the real-time (in actual performance) binding between structures (the variable binding problem referred to earlier in the section "Binding Circuits"). We don't have a single solution to this problem, but there are promising neurally plausible approaches based on Hebbian short-term plasticity such as assembly calculus (Papadimitriou et al. 2020) that can be applied toward a solution.

Priming occurs when exposure to a stimulus causes a system to be more likely to become active in subsequent presentations of the same or similar type of stimulus. Priming is a widely studied phenomena in psychology and neuroscience and appears to be evolutionarily conserved. For instance, in his recent dissertation work on *Drosophila melanogaster*, Yang (2020) found that a prior encounter with an aversive stimulus produces long-lasting changes in the animal's neural and behavioral response to a novel odor. While some extreme and surprising claims of priming in language and thought have been controversial—such as the replication of results of Bargh et al. (1996)—it is clear that priming is a fundamental evolutionary strategy for associational influences in memory and learning.

A minimal priming circuit has three nodes, N_1, N_2, and N_3 (Figure 15). We will assume for this simple circuit that N_2 can is bistable and has a low and high state (unprimed and primed state).

- N_1 has an activating connection to N_2.
- N_1 has an activating output just less than the firing threshold for N_2.
- N_3 has an activating connection to N_2. When N_3 fires, it adds just enough activation to the activation from N_1 to make N_2 fire.

N_1 is said to "prime" N_2.

Effectively in this simple circuit, N_2 must have at least two stable states: unprimed and primed, in addition to the firing state.

CIRCUITS STRUCTURE SIMULATIONS

We are always simulating. Our present simulation is what we are keeping track of at present, and our preexisting circuitry defines how we understand

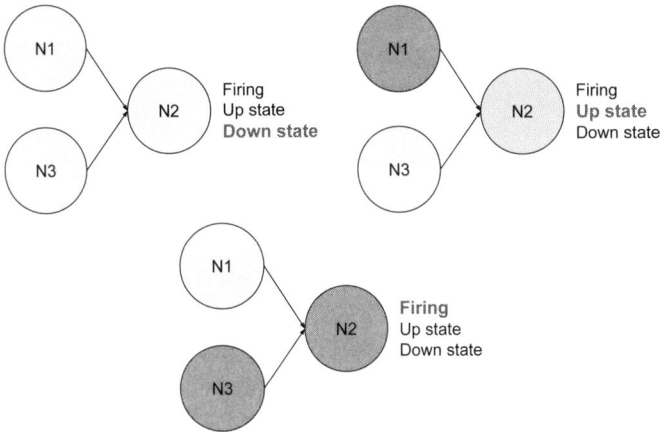

FIGURE 15. Shown here is a minimal priming circuit. We assume here that the primed node N2 is bistable and has a low (unprimed) and a high (primed) state. The minimal circuit has three nodes: N1, N2, and N3. The basic operation of the circuit is that when N2 is primed (in the high state, as is the case after N1 fires), minimal additional activation (such as N3 being active) is sufficient to push N2 to fire. N1 thus primes N2 in this circuit.

the present. In addition, we are performing many other simulations at once: a relevant past, immediate possible alternatives to the present, and desired, needed, or externally mandated alternatives to the present as well as imagined situations of all sorts: goals, possible difficulties, expectations, what you or others ought to do, consequences, and so on. In cognitive linguistics, such simulations are called **mental spaces**.[13]

MEMORIES AND SIMULATION

Active cascades, made of circuits that structure the cascade and direct the activation flow, result in simulations. As others have pointed out,[14] what is called a "memory" is not a "stored" past event that is retrieved from a static store of memories. A "memory" is a partial simulation activated by a neuronal group, which is then "filled in" by existing fixed circuitry that happens to be activated at that time. Which circuits are active can be internally and externally triggered (based on attention, tasks, moods), modulated (based on rewards or intrinsic alertness or arousal), and manipulated (by providing strong contextual cues). Thus, as Elisabeth Loftus (2003) has famously shown, what is experienced as a "memory" may be only partially remembered, with the rest filled in by current understandings, which may have changed significantly since the "remembered" event. Thus, what we understand as the "remembered past" is at best partially remembered, with details filled in by current, unremembered understandings that have arisen since the "remembered" event.

The role of simulation in dreaming and in memory consolidation is now robustly established in multiple studies. György Buzsáki (2019) in his important and wonderful book *The Brain from Inside Out* describes some amazing experiments with mouse head-direction neurons that fire when the mouse's head is facing a particular direction. In this case, the researchers recorded from a region in the thalamus (head direction is encoded in multiple regions) and were able to reconstruct the head direction from a neural network based on the spiking patterns of a population of neurons in the awake mouse. The surprising finding was what happened during the period after when the mouse fell asleep:

> The exciting part of the experiment began when the mouse fell asleep. Because during sleep all peripheral inputs are either absent or constant, one might expect that head direction neurons will be silent or just fire randomly. Yet, to our surprise, not only did head direction neurons continue to be active, but they replicated the sequential activity patterns seen in the waking mouse. Neurons coding for nearby orientation continued to fire together, while neurons with opposite head direction tuning remained anti-correlated. During rapid eye movement (REM) sleep, when brain electrophysiology is strikingly similar to its awake state, the neural patterns moved at the same speed as in the waking animal, recapitulating the activity seen when the mouse moved its head in different directions during waking. In essence, from the temporally shifting activities of cell assemblies, we could determine the mouse's virtual gaze of direction during REM sleep. During non-REM sleep, the activity was still perfectly organized, but now the virtual head direction drifted ten times faster than it did in waking and in REM sleep. Because no head movement accompanied these changes in the sleeping animal, the temporal organization had to rely on internal mechanisms rather than external stimuli.[15]

READY FOR INTEGRATION

Circuits that are ready for integration are the basic circuit types used to model actions, events, and many forms of imagination, thought, and language. They are ubiquitous, both in embodied circuitry and in circuitry exapted for thought and language. They perform vital functions, combining to create complex cascade structures and to do some of the work of integration. But there is an even more fundamental form of integration that applies throughout the brain and is responsible for the bulk of complex thought and language. We turn to it next.

Integrative Circuits and Simulation

In chapter 2, we introduced conceptual integration. Integration is a very general process. Here we introduce the computational properties of integration circuits and their connection to simulation and best fit.

Integrative circuits are one of the fundamental circuit types in the brain. What integrative circuits do is *coordinate nodes that control functions in various parts of the brain and put them under the control of one node* in such a way as to integrate their diverse functions. We have already seen examples of integrative circuits as in Figure 8, which coordinated the controlled parallel (concurrent) activation of two network cascades.

An important way to think about integrative circuits is to consider how they would be recruited, or "learned." Let us begin with integrated physical activities. Consider an infant learning to eat. She has to not just get food into her mouth; once the food is there, she has to integrate many activities at once. First she has to keep it in her mouth, keep it from slipping out and running down her chin. To keep it in her mouth she has to coordinate certain muscles. At the same time she has to start chewing—or mashing if she is without teeth—and then slowly start swallowing. All this requires coordinated muscular action in the mouth. Because this is done at the beginning, the circuitry used must be otherwise unemployed but present in just the right place to carry out the neural requirements. When it is done repeatedly, an integrative circuit is recruited.

Consider learning to walk up steps. The infant has to not just raise and lower her leg but also has to push hard enough to lift her body, lean forward just enough, and maintain her balance.

Imagine a terrain with streams and logs that have fallen across them. To cross a stream on a log, you have to learn to integrate balancing and walking: walking with one foot in front of the other while slowly trying to maintain your balance, using neural circuits that were never used for that before. If you do this repeatedly over a short time, the ad hoc circuits used become a recruited integrative circuit.

Consider trying to determine whether a fruit is ripe, say, a lemon on a tree in your California yard. You pick it up in your hand, test how heavy with liquid it is, squeeze it to see how hard or soft it is, and look at its color. All of these are integrated to determine ripeness, and as you do this regularly, an integrated circuit for ripeness determination is recruited. In each case, you have recruited one node that activates an integrative circuit.

INTEGRATIVE INSTRUMENTS

To understand a screwdriver, you have to know that it is a single thing structured so that its parts map onto parts of an integrated activity, that is, screwing a screw into a board so that the parts of the screwdriver each fit the parts of the integrative activity. The activity has the following parts: holding the screw upright in one hand and keeping it upright while the other hand does the following using different parts of the screwdriver:

- Match tip to slot: Keeping the tip of the screwdriver in the slot at the top of the screw so it does not slip or pop out.
- Match pushing on top to pressure, overcoming resistance of the board: Pressing down on the top of the screwdriver from the top sufficiently to get the screw to go into the board.
- Match turning the handle to the spiraling of the screw into the wood: Turning the screwdriver while pushing downward hard, keeping the tip in the slot.

Each "matching" is carried out by a neural binding of instrument part to integrated activity. Cognitively, the screwdriver is a single instrument whose parts match up with integrated actions. Such a collection of matches is carried out by an integrated mapping. The integrated mapping is a simultaneous activation of gates for neural binding circuits that bind parts of the screwdriver to the matching actions.

Again, such a circuit is recruited because learning to use a screwdriver requires using otherwise unemployed neural circuits over and over until their synapses get strengthened and they form a permanent fixed circuit.

GENERAL INTEGRATIVE CIRCUITS

We believe that most of the really complex feats of motor control and event recognition as well as really complex thought and language is carried out by cascades made up of general integrative circuits bound together and extending across much greater distances in the brain. Some integration circuits described in the earlier sections include the mapping and binding circuits as well as the complex control and coordination circuits described in the "Control and Coordination Circuits" section.

Generalizing over these circuits, we hypothesize the following simple structure for the integration circuits. Their power comes from their ability to coordinate complex cascades, which gives them enormous integrative capacity.

An integrative circuit consists of

- a control node, and
- two or more gating nodes connecting to other circuits, controlling their activation or inhibition.

There are activating connections from the control node to the gating nodes and from the gating nodes to the control node. The activation of the control node activates all gating nodes and, conversely, the activation of each gating node activates the control node.

That seems simple enough, and in structure it is very simple. Yet as simple as they are, integrative circuits of this kind perform remarkable functions for two reasons:

- First, they coordinate content from multiple areas of the brain.
- Second, they are compositional. That is, integration circuits can be embedded inside other integration circuits.

Each integration circuit can therefore create highly complex integrated "packages" of diverse content, which can be quickly activated and integrated within a short window of time. What such circuits do is to localize the control of the integration of content that may be far-flung.

Since activation can pass in both directions, such circuits are both divergent and convergent. They can both disperse information coming from the control node and collect information coming from all the gating nodes. And the bidirectional connections circulate activation, keeping that part of the larger cascade flowing until it is turned off by a shift to the activation of another cascade circuit.

Integrative circuits are thus a simple mechanism for forming complex circuits and localizing their control, for example, complex image schemas such as *into*, *across*, and *along*; complex frames such as the restaurant frame consisting of business, feeding, and hosting frames; and complex metaphors (see chapter 4).

BUILDING COMPLEXITY

One of the most profound properties of such integrative circuits is their ability, when used recursively, to build structured complexity by forming successive layers of embedded circuits. For example, we saw in chapter 2 that image schemas can be bound to other image schemas to form complex image schemas, such as *into*. A motion schema can be bound to the *into schema* to form a motion such as *He ran into the room*. A force schema can be bound to the integrated motion *into schema* to provide a complex *forced motion + into schema* as in *They pushed him into the room*. The conceptual metaphor that starting an activity is entering a bounded region can apply to the *forced motion + into schema* to yield metaphorical cases such as *They pushed him into running for president*. Then, the about-to role in an X-Net can be further added to yield *They are about to push him into running for president*. Successive applications of basic circuits and integration circuits can build up such complex ideas and combine them into fixed cascade circuits that are integrated unconsciously, automatically, effortlessly, and almost instantly prior to consciousness.

EMBODIMENT AND MEANINGFULNESS

When the conceptual content of a sentence is embodied, it becomes meaningful. In a sentence such as *They are about to push him into running for president*, the embodiment consists of X-Nets, a force schema, a motion schema, and a container schema as well as the conceptual metaphors whose sources are embodied (indicated in italics): action is *motion*, an activity is an *action sequence*, activities are *containers* for *their actions*, and causes are *forces*.

Here's how the words in the sentence above work via an integration of all these elements:

- *About to*: The stage just before the start of the X-Net sequence.
- *Push*: A direct cause understood (via the conceptual metaphor "causes are forces") in terms of a pushing force.
- *Running for president*: An activity conceptualized metaphorically as a

container. The entry into the container is the declaration of running for president; being inside the container implies running for president.

- *Into*: A neural binding of the meaning of *to* (*motion*) and *in* (*containment*) to characterize motion beginning outside of and ending inside of a container. Metaphorically, the container is the activity of running for president.

The sentence gets its meaning via the activation, bindings, and integration of all of these embodied circuits, which occur in different and sometimes distant brain regions. Integration circuits not only allow for the building of integrated content but also can combine to form cascades that reach across the brain and integrate the various embodied contents that the meaning of the whole depends on.

INTEGRATION AND SIMULATION

Integrative circuits are useful for doing things in the world with the least effort. We hypothesize that they also serve an imaginative function. Imagination is crucial to understanding. As we function in the world, our brains automatically produce an understanding of what we are experiencing and doing. That understanding is conscious in part but is mostly unconscious. We call such understandings "simulations."

A simulation can be triggered by what we do at a given moment, by memory, or by various forms of imagination: dreaming, reading fiction, watching a movie, and so on. Unconscious simulation is, of course, far more extensive and provides the understood background for conscious awareness of whatever we are doing, such as working, planning, daydreaming, remembering, and dreaming.

The language we use and perceive is available to conscious awareness. But what is used to understand that language is overwhelmingly unconscious, and it has to be. Unconscious understanding uses the massively "parallel" and branching structures of our neural systems, structures that link up language to all the embodied brain circuitry in the many diverse and dispersed regions of the brain used in understanding.

Conscious thought is sequential; unconscious thought is parallel and branched. The flow of conscious awareness may shift with attention, changing direction as it flows. But we are not consciously aware simultaneously of dozens or hundreds of things at every instant. Rather, the unconsciously thinking neural system can be and often is cascading in many branching directions at once. Unconscious neural thought cannot be linearized, so we cannot be consciously aware of most of what we are unconsciously thinking via our neural system.

Although the brain circuits used in simulation are dispersed throughout many physically separated brain regions, simulation is "integrated"; that is, even our unconscious thoughts fit together optimally in coherent "packages" that define "common sense."

Once you learn an integrated package, it is used automatically, easily, and quickly. But learning such a package is a more complex process. We have fixed coordinated circuitry for tying our shoes. We didn't have it the first time we tried.

SIMULATION AND DISPLACEMENT

Simulation is real and important for our mental functioning and appears to be central to the way we understand anything at all. But if you study what people simulate all the time, much of it is not about reality but instead is "displaced" to future possibilities, memories, narratives, regrets, fears, and things to be relieved about. These are all forms of imagination.

Simulation involves the ability to *displace* oneself from attending only to the direct sensory input and to enter the realm of the imagination. A lot of our everyday activity starting as children and continuing into adult life involves *displaced simulations*, where we imagine things beyond the here and now (such as those stories that begin "long ago and far away"). Displaced simulations can be reconstructions of situations from our past, hypothetical imaginations of some future situation, autobiographical narratives about oneself created to make sense of past experience, and so on.

HOW MIGHT DISPLACED SIMULATION WORK IN THE BRAIN?

Over the last two decades, there has been an accumulating body of evidence regarding such simulations and their recruitment of specific networks in the brain. The networks recruited in simulation include the brain's *default network* (DN), a set of functionally connected brain regions including ventral medial prefrontal cortex, the posterior cingulate cortex, the inferior parietal lobule, the lateral temporal cortex, the dorsal medial prefrontal cortex, and the hippocampal formation (Buckner et al. 2008).

The discovery of the DN could answer one central mystery about the brain that has been asked repeatedly since the 1920s: the brain consumes an almost constant amount (less than 5 percent variation) of energy independent of how much attentional effort or focus is involved (although certain phases of sleep use less energy). In other words, "doing nothing" (the brain at rest) consumes as much mental energy as hard mental effort. This is not true of physical effort, obviously (to the benefit of health clubs everywhere). In the early 2000s, neuroscientists at the Raichle lab at Washington University in St. Louis found specific pathways and systems of the brain (the DN) that were *more* active at rest than while working on a focused task. In other words, these areas consumed more energy (because of greater activity) when the brain was *not* engaged in a task than during the performance of an attentionally demanding task. Other regions in the brain consumed more energy (were more active) during a focused task than at rest. Since some regions are more active at work and others are more active at rest, overall the brain consumes a roughly constant amount of energy.

In recent years, there has been a flurry of work in neuroscience on establishing and questioning the functional significance of the DN. There are multiple conflicting theories about the exact role of the DN. These include mind-wandering and daydreaming, directed simulations, imaginative projections of the self onto others, and consolidating experience and memory in making sense of one's past. One productive line of inquiry suggests that the DN has an imaginative function. Buckner et al. (2008) and Schacter et al. (2009) propose that the primary function of the DN is to perform a mental simulation imagining oneself in the past or in the future. This projection and simulation of the self in different situations may be the exploratory substrate for planning, memory, and inference.

Importantly, mental simulations of hypothetical scenarios have been shown to engage core regions of the DN. For instance, perspectival simulation, in which we are able to simulate another person's perspective or point of view, activates core regions of the DN. Various forms of future envisioning, in which people project themselves into future situations, also activate the DN, as does the reconstruction/remembrance of past episodes in one's life (Buckner et al. 2008; Schacter et al. 2009; Spreng et al. 2013; Benoit et al. 2014; De Brigard et al. 2015). Activation of the DN can be modulated (made stronger) by attentional control and practices such as meditation. In addition, various psychological and memory disorders including ADHD, schizophrenia, autism, and Alzheimer's disease exhibit different types of abnormal functioning in the DN.

Further evidence of the link between mental simulation and memory comes from work on a key component of the DN: the hippocampus. Some of the early evidence on the role of the hippocampus in forming new memories of situations came from the patient H.M., whose hippocampus was surgically removed. This resulted in him being unable to form new memories of events or situations, a condition called amnesia. H.M. was unable to remember specific facts about persons (such as their name) or situations a few minutes after experiencing the event. He could meet the same person every day and would have to be continually reintroduced to that person, since he couldn't remember anything about the previous day's meeting. The disorienting and terrifying ramifications of such a disorder are captured in the fictional 2000 movie *Memento*.

Detailed brain studies (based on cellular recordings in the rat hippocampus) also identified the DN's role in forming spatial and cognitive maps of the environment. Specifically, excitatory neurons of the rat hippocampus show spatially localized place responses during exploration; hence, they have been called *place cells*. The discovery of place cells in the hippocampus and the related *grid cells* in the adjacent region called the entorhinal cortex (which appear to code metric information such as positional information and distance) forever changed our understanding of animal navigation and resulted in the Nobel Prize for Physiology or Medicine in 2014. John O'Keefe discovered place cells, and May-Britt Och and Edvard I. Moser discovered grid cells. Place cells fire when the rat is in the corresponding location. Nearby place cells fire for nearby locations. The place-cell sequence traces the path that the rat followed. Clearly, once you have such place cells, the current location might be informative about other locations that the animal cares about, such as the remembered goal or the set of locations defining a route. Some years ago, neuroscientists at Johns Hopkins (Pfeiffer and Foster 2013) showed that *before* goal-directed navigation, the rat hippocampus generates brief sequences of place-cell firings that progress from the subject's current location to a known goal location. This happens even if the specific start and end states are novel and the rat has never seen that particular combination. It has been shown more recently that even abstract memory requires spatial simulation based on hippocampal maps. This is directly predicted by our hypothesis about the metaphoric embodiment of abstract concepts in direct social, physical, and emotional interaction and experience.

THE ROLE OF SIMULATION IN OUR MENTAL LIFE

Pfeiffer and Foster (2013) found a high correlation between which simulated sequences were highly active and the immediate future behavior of the rat.

The researchers proposed that the rat is using mental simulation for navigational planning. We conjecture that this basic scheme of forward simulation for planning has been repurposed by human beings to provide imaginative simulation for understanding, as when you hear the beginning of an argument and you can simulate the rest. The repurposed integrated circuit recruits the entire DN to coordinate this simulation. Indeed, recent evidence suggests that there are functional connections (through a technique called *functional connectivity analysis*) from the DN to other areas that are responsible for cognitive control and might underlie our creative abilities. This research suggests that simulation may underlie creative thought.

Thus, the existing evidence suggests that imaginative simulation plays a central role in our mental life. Moreover, the DN may *coordinate* imaginative simulations via interactions with other circuits in the brain. All this evidence is suggestive, and further experimental investigation will need to test this hypothesis.

Simulation in Language and Thought

The most common use of the term "simulation" comes from computer simulations in the sciences, where certain aspects of a physical system that are of interest and that change over time are simulated. For example, in weather forecasting for a given general area, what is of interest is temperature, degree of sunshine or cloudiness, amount of precipitation (rain, snow, hail), wind and wind direction, barometric pressure, and approximate times of all of those.

In this book, we are doing two scientific simulations at once using the same neural computation for both: (1) a simulation of neural circuitry and its functioning over time and (2) a simulation of the mechanisms of embodied thought and language.

But we are also making a different use of the term "simulation," one introduced by Jerome Feldman (2006) in his classic book *From Molecule to Metaphor*. Feldman argues that as human beings, we understand via a human form of simulation: simulation using our embodied neural system to simulate via imagination. Feldman takes as his major example our understanding of natural language and characterizes what he calls "simulation semantics."

EVERYDAY SIMULATION

You are walking down a moderately crowded street. You are moving your legs and looking ahead. You twist your body to avoid bumping into someone who has just jutted out of a doorway. You see someone coming right at you. Do you move to your right or your left? Which is better? Which will avoid hitting the person while minimizing your change in route?

You are engaging in simulation. You are mentally simulating the reality of your situation, making decisions, performing actions, and updating as you go along. You are attending to what you are actively doing, factoring out what is irrelevant, and imagining what comes next. Do you turn left at the corner? Will that get you to where you want to go? Monitoring reality and imagining go hand in hand. Remembering is part of simulation. You remember that if you turn the corner, you will pass a great cheese store and you can get some cheese. You decide to turn at the corner.

DIMENSIONS OF SIMULATION

At the outset, we distinguish three different but intimately related ideas:

- Active simulation, which is both monitored and imagined and is carried out over time. This could include a simulation of bodily actions, thought, or both.
- Conceptual system, which is needed to perform the simulation.
- Simulation space, which is a real or imagined situation being simulated.

Imagined simulations are always partial and use the full conceptual system, including embodied schemas, frames, conceptual metaphors, bindings, integrations, and neural cascades.

SIMULATION SEMANTICS

Take a sentence such as *Sara walked into the café*. Feldman (2006) observed that you must at least be able to imagine what it is to walk into a café to understand the sentence. The semantics of understanding, in short, requires simulation. Here is what is required by such a simulation.

First, you need to imaginatively simulate walking: propelling yourself forward by moving your legs in the appropriate manner, pushing forward using alternating legs. Walking contrasts in manner of motion with skipping, dancing, running, limping, shuffling, stumbling, moonwalking backward, and so on. To imaginatively simulate walking, you have to know what "forward" means and imagine a normal human body with a face normally oriented in the same direction as normal motion. And there is more:

- In *Sara walked to the café*, the "to" indicates the goal of the walking. The sentence indicates that she reached the café, but the sentence does not indicate that she went inside.
- In *Sara is in the café*, there is containment within the café but no motion to it.
- In *Sara walked into the café*, there is both motion and containment. The café is a bounded region. If she walked into the café, she started outside, crossed the boundary through a doorway, and wound up inside. Motion and containment are integrated.

In short, the mental simulation of the meaning of the sentence requires a complex neural network to simulate the walking; embodied connections to the legs; a motion schema with a source, a path, and a goal; a container schema with an interior, an exterior, a boundary, and a portal; and bindings of a source of motion to the exterior, a goal of motion to the interior, and a path of motion through the portal.

And then there is the café frame, in the prototypical case serving coffee and accompanying food in a location that has chairs and tables. The café frame combines the business frame and the food service frame, with the customer of the business bound to the consumer of food in the food service frame. All of that is understood almost instantaneously (within about half a second) of the completion of the sentence. Yet, the simulation is dynamic, making use of a neural network for walking and moving from outside the café to inside it. The imaginative simulation may preserve the temporal order of the situation that is being simulated, but what is simulated may in reality take a lot longer (as in *Sara walked across the Golden Gate Bridge*) or shorter (as in *The sodium ion entered the cell body of the neuron*).

In the theory we are proposing, all of this is part of human imaginative simulation. Conceptual metaphor is part of human simulation as well. In understanding a sentence such as *Sara told me a story*, the conduit metaphor is used. We understand the story as originating with Sara and as passing from Sara to me by virtue of Sara's speech, and we know that it is inferred that I heard the story from Sara.

Moreover, in a sentence such as *Cancer took Chuck from us*, we are integrating three metaphors at once:

- Life is being here, and death is departure, no longer being here.
- A living person who is valued can be metaphorically conceptualized as a valuable possession. When he dies, we can speak of "losing him" or having him "taken from us."
- A phenomenon causing death is often conceptualized metaphorically as a person causing death. Here the phenomenon causing death is cancer, and it is metaphorically personified as the subject of "took."

In the theory we are proposing, this is a conceptual integration accomplished by an integration circuit that neurally integrates these three metaphors, creating an imaginative simulation structured by all three metaphors at once and understood instantly and effortlessly.

BEYOND LANGUAGE

The theory of human understanding via simulation goes beyond the understanding of language. This theory applies to the understanding of anything one is attending to, either consciously or unconsciously. Any aspect of a situation, real or imagined, present or remembered, awake or dreamed, will require much or all of the of apparatus of thought that we have been discussing in this chapter: conceptual primitives, X-Nets, frames, metonymies, conceptual metaphors, and conceptual integrations.

All human conceptual thought requires embodied human simulation using the neural system. We don't *just* perceive and act. We don't *just* imagine or think. Simulation is always taking place using particular neural circuitry that characterizes our conceptual systems, allowing us to think and imagine via simulation.

INTUITIONS

What happens when you have an intuition but don't know how to consciously spell it out?

Intuition is an unconscious thought that is partly conscious. You know what the intuition is about. You have to know some things about it. And you have to know that what it is about is relevant in your life. Spelling the intuition out involves a conscious simulation of its logic. In the process of making the intuition conscious, the unconscious content we attempt to imaginatively simulate may well be changed by the brain prior to consciousness to better fit what is already present in the brain; as Elisabeth Loftus (2003) demonstrated, memory of an event is structured by prior knowledge. Just as our perceptions can be unconsciously changed within 100 milliseconds, what we try to imagine can also be changed unconsciously and automatically before we become conscious of it.

We regularly simulate unconsciously in, for example, understanding rapid speech via simulation semantics. However, we may be neurally unable to consciously make an unconscious simulation—an intuition—conscious.

HUMAN VERSUS COMPUTER SIMULATION

At the beginning of this section, "Simulation in Language and Thought," we made the distinction between scientific simulation using computers and human simulation using the neural system. The theory we are presenting is about human simulation using the neural system, and the science we use to model human simulation is scientific simulation using computation and forms of computer science to do modeling.

It is important not to mistake one for the other. We issue this warning because such a mistake is commonplace. It is common to find people—even scientists—believing that human simulation *is* the computer simulation used to model the human simulation. This is the mistake of thinking that embodied human thought *is* disembodied computation as characterized by computer programs simulating what is coded by a human programmer. It is like the mistake of thinking that the map of a territory *is* the territory being mapped or the mistake of thinking that the computer model of the weather *is* the weather itself. A storm may get you wet, but a computer model of a storm will not get you wet.

The best book-length treatment of experimental research on human simulations that we know of is Benjamin Bergen's (2012) *Louder Than Words*. We strongly recommend it, just as we strongly recommend Jerome Feldman's (2006) discussion of simulation in *From Molecule to Metaphor*.

SIMULATIONS AS CASCADES AND MENTAL SPACES AS SIMULATIONS

Simulating a situation requires a cascade that activates the circuitry characterizing the situation. If the circuitry contains an X-Net, then the simulation can evolve over time.

Gilles Fauconnier's (1985) theory of mental spaces has been about situations as characterized in cognitive semantics. Mental spaces are ubiquitous in language and imagination. Whether we are thinking about the past or the future, a hypothetical situation, or a counterfactual analysis or are merely remembering an experience, we *run* the content of mental spaces using a simulation. In the account of the neural mind that we are proposing, a mental space is a neural simulation, which in turn is structured by a neural cascade of the sort we have been describing.

For example, Fauconnier (1985) spoke of someone's "reality space." We see this as a simulation (partial, of course) of the relevant aspects of the situation you are in. If you are simulating a sentence such as *Harry believes that Obama is a Muslim*, here is what is involved. First you need a cascade for your understanding of *Obama is a Muslim*, with all the implications of what you think would follow in such a situation. Second, using the frame semantics of believing and a cascade characterizing your assumptions about Harry, you need to simulate what it means for Harry to believe something. In general, having a belief about something important usually involves having other related beliefs, and you would be simulating what follows from those related beliefs, which would be necessary to understand the implications of those related beliefs. And if you are a person who acts on your important beliefs, then that simulation would play a part in your decision-making.

In computer technology, simulation has become a major industry in three enterprises: virtual reality, robotics, and driverless cars.

The Takeaways

What we are providing here is a general approach to how brains think. Our approach integrates research over four different fields: neuroscience, cognitive science, linguistics, and computer science. These fields are each developing, so we expect details of our approach to develop as well over time. That is part of the excitement of the endeavor. Indeed, neuroscience is developing so fast that new neuroscience will likely lead to new accounts of the neural mind in a number of respects.

Despite ongoing developments, we feel sure that certain aspects of our current approach are deeply correct:

- Thought is physical and is constituted by neural circuitry.
- There is no meaningful thought without embodiment.
- Complex thought is integrated, not just some collection of features.

- There is a big difference between an integrative *mapping* and a general integrative *circuit*. A simple integrated mapping occurs between two schemas (or frames) and is limited in that way. A general integrated circuit can integrate many different neural structures.
- Thought is not localized in brain regions that happen to "light up" in functional magnetic resonance imaging. Instead thought "adds up" via integration over cascade circuits.
- Neural circuitry used in the sensorimotor system that evolved via animals has been exapted (repurposed) for use in human thought.
- X-Nets, image schemas, force schemas, frames, and conceptual metaphors are real embodied modes of thought, central to all human conceptual systems.
- Bear in mind that there are not many basic circuit types in our neural computational model. We believe that the ones we cite are all there are (or at least come close) to what is needed neurally for thought.
- Fixed thought circuitry is "packaged" integrated circuitry that is constantly used to structure thought.
- Optimization in the neural systems occurs via least energy use for a given purpose, the shortest connections needed, and convergence zones, where neural circuits converge in a way that optimizes energy use for given needs of the organism.
- A crucial way to understand how brains think is the use of structured neural computation to model both the functional effects of the neural system and the details of thought and language. The bridging computational model uses two scientific metaphors—the neural computation metaphor and the cognitive computation metaphor—to carry out the computational modeling and, by this means, to pair neural circuitry with thought and language.
- Computational modeling is useful and often necessary when researchers are attempting to precisely characterize the neural system and what it does in thought and language. When such researchers are dealing with thought and language, any neural computational system must have the right kind of structure so that it can map onto the structure of thought and language and also onto the neural system.
- The simulation of experienced reality is constantly going on in the human brain, and imagining beyond what is experienced goes hand in hand with it. What are called "mental spaces" by cognitive scientists are simulations carried out neurally.

Maps are tools for understanding the territory mapped. We choose among routes to get to places. When we actually drive or walk following a map, we are blending the map with the territory, understanding the length and time of the trip in terms of the map, and asking which routes show least congestion, that is, the shortest total of red lines on our modern-day computer maps resulting in the shortest hypothesized travel time.

Applying a scientific theory is like following a map, blending the map and the territory while distinguishing between them. A GPS system depends on signals traveling at the speed of light. The calculations in the design of the system have to use the mathematics of Albert Einstein's theory of relativity to get the GPS system to work. We simply use the GPS-based maps as tools without going through the relativistic mathematics needed to construct them. GPS-based maps are very handy.

Neuroscientists use their tools to construct brain maps using brain scans. The scanners used in functional magnetic resonance imaging research are based on a theory connecting the ratio of oxygenated to deoxygenated blood to the amount of neural activity in a given region of the brain. Computer programs written by specialists and used by scientists are necessary to do the computations. A high ratio of oxygenated to deoxygenated blood while carrying out a task might show up on a brain map as red. The color red is in the brain map, not in the brain. The precise blood ratios to be colored red is determined by the neuroscientists. Because of the expertise of the neuroscientists, such maps can reveal much that is significant.

From the outset we have distinguished between the map and the territory: between our neural computational bridging theory (the "map") and the actual use of thought and language by human beings (the "territory"). Our theory is the result of more than four decades of research in four fields: neuroscience, computer science, cognitive science, and linguistics. We have relied on the metaphors of "connections" and "circuits" ubiquitous in neuroscience. Our neurocomputational models of the functional structure of brain "circuitry" are based on generalizations over what is necessary to model the circuitry needed for ideas, which has been studied in cognitive linguistics and experimental embodied cognition.

The circuitry types discussed throughout this chapter—gestalt circuits, conditional circuits, coordinating circuits, mutual inhibition, disinhibition, gates, binding circuits, integration circuits, and cascades—are modeling structures used in our theory of the neural mind. They are the most general modeling structures required by our bridging theory, which contains a single computational dual model, "dual" because it simultaneously models both thought and language on the one hand and the corresponding neural circuitry on the other hand.

In chapter 4 we will be applying our theory of the neural mind to language. We will be asking whether the circuitry types used in modeling ideas and thinking—gestalt circuits, conditional circuits, coordinating circuits, mutual inhibition, disinhibition, gates, bindings, integration circuits, and cascades—can also adequately model language, which is used to express and communicate such ideas. As in the study of ideas, we will consider general neural mechanisms: Hebbian learning, STDP, best fit (understood as a least-energy condition), and so on.

As in normal scientific practice, we will be using a blend of map and territory, that is, a blend of neural modeling and the linguistic phenomena modeled. On one level, this enterprise is descriptive: we ask whether our tool kit of neural circuitry types is sufficient to adequately describe the general circuitry needed for neural models of language. But because the tool kit of circuitry types is derived from research on ideas, not on language, *adequate description* rises to the level of *explanation*. A priori, it is conceivable that the repertoire of circuitry types needed for language could be entirely or significantly different from those needed for thinking with ideas, for example, before language is learned. If on the other hand the circuitry types needed for thinking are also adequate for describing language, then we have an *explanation*: preexisting circuitry types for ideas are all that is needed for a neural account of language.

What we need to do now is test this explanatory hypothesis. In this book, we will limit ourselves to English and to a nontrivial collection of English grammatical phenomena.

4

Neural Language

Thinking Neurally about Language

AN OPPORTUNITY

This chapter presents a remarkable opportunity to begin thinking seriously about language from a neural perspective. Language, of course, includes the ideas that each language expresses. Those ideas, as we have seen, are constituted by neural circuitry at the system level, that is, across brain regions and throughout the body. We expect no less of language, for a simple reason. In addition to the ideas that language expresses, language includes the grammatical structures and lexical items in which the ideas are expressed plus the neural linkages between the ideas and the linguistic forms—in grammar, morphology, and phonology—used to express those ideas.

Since we have some sense of the circuitry needed to constitute those ideas, we have a place to start: the circuit types needed for ideas. These are the types of circuits we discussed in chapter 3. We can then ask whether those circuit types for ideas are also adequate for grammar and morphology, both for the grammatical structures and lexical forms and for the form-meaning links used to express ideas.

THOUGHT AND LANGUAGE: SPECIFIC CIRCUITS
AND COMMON EMBODIMENT

Children learn to do a lot of thinking before they learn language. All the embodied neural circuitry for controlling the body and for the basics of human thought is present in the brains of children before they learn language. For a young child to function physically and socially in a child's environment, much of the circuitry for thinking and understanding basic experiences must already be present. In short, a significant amount of the human conceptual system is already in place *before* we learn language.

Since language expresses and communicates ideas, much of the idea circuitry must be present *before* language is learned. This fact about language learning therefore makes empirical sense of the question we are starting with: *Which of the circuit types, together with the neural tool kit discussed in chapter 3, that are present **before** language are also used **in** language, and is the circuitry needed for ideas also sufficient for the language used to express those ideas?*

Language is so complex, with so many kinds of problems, that answering this question is an extraordinary difficult task. Here is the strategy we will follow:

1. Start with the strongest hypothesis: that the circuitry types and neural mechanisms needed for ideas will suffice for grammar and morphology.
2. Try testing the hypothesis on a wide range of difficult cases to see if it works and, if not, where it fails.
3. We will start by limiting ourselves to grammar and morphology, leaving aside phonetics and phonology at first.

Note that the hypothesis here is that the same kinds of computational structures that we have seen in chapter 3 will suffice to characterize language use. *Of course, as we will see, these structures will be used in specific circuits* to capture the morphology, phonology, and grammar of a language. We expect the language phenomena to be distributed in multiple circuits of the brain rather than located in a specific region. Much of the recent work in the neuroscience of language is consistent with this view. With increasing evidence from the Human Connectome Project,[1] research from Eddie Chang's lab at the University of California–San Francisco (Chang et al. 2015), Eva Federenko's lab at MIT (Federenko 2016), and others, neuroscience has shifted from localized theory (such as the Broca's area of the brain) to a network model. We hypothesize that the neural circuits for gating, binding, and coordination described in chapter 3 are functionally connected to form a large left-lateralized frontotemporal language network composing and connecting linguistic form to ideas and meaning. Specifically, we will show how these structures are used in a language-specific way through constructions that link linguistic forms and descriptions to ideas and meanings.

Language links ideas and modes of thinking to linguistic forms in appropriate contexts. Language allows ideas to be expressed, understood, and communicated. Neural circuits that link form and meaning in this way constitute what in cognitive linguistics are called "constructions." *A language consists of constructions that characterize the general principles linking meaning and form in that language.*

THE PRAGMATIC NETWORK OF CONSTRUCTIONS

Linguists have known for decades that there is a relational network in which one construction is a variation based on the form of another. There are somewhat different constructions for making statements, asking questions, giving orders, etc. For example, the form of questions is a variation based on the form of corresponding statements: *Is Harry cooking dinner tonight?* is a variation on *Harry is cooking dinner tonight,* in which the auxiliary verb "is" appears before the subject "Harry" and the rest of the question is the same as in the statement. The form of the imperative construction is also based on a statement form, namely you—verbal auxiliary—verb phrase. The form of the imperative construction is just the verb phrase with an understood "you" as subject: *Wash the dishes! Take care of yourself!*

The relations between constructions are asymmetric—unidirectional—and based on *pragmatics,* that is, on the *ideas that function actively in carrying out the use of language in context.*

In our account of executing networks (X-Nets) for aspectual ideas, we have already seen how circuit types for purposeful physical motions have been exapted (repurposed via evolution) for carrying out and understanding abstract activities (e.g., planning) and the action of thinking itself, including reasoning from premises to conclusions.

Pragmatics is about using language to carry out communicative purposes. The most obvious cases are, of course, the speech acts: making statements (getting others to believe), asking questions (to get information), giving orders and making requests (to get others to perform actions), making declarations (imposing a situation on the world when one has the authority to do so), making promises (committing oneself to future action), communicating emotion (surprise, annoyance, anger, delight, sadness, and so on), and establishing rapport. And beyond speech acts, there are other major pragmatic goals of communication: to frame a situation as you want it framed, to direct or to divert attention, to express emotions such as anger or spite, to impugn the motives of others, to try out a suggestion, to shift blame, and to shift viewpoint, that is, doing any or all the above from someone else's point of view. The pragmatics of language is the study of all of these things and more.

The neural circuitry used in pragmatics is active: X-Nets—that is, neural networks that "execute" (do things)—carry out purposes (i.e., desired end states) and provide the means for achieving those desired states, including gathering resources, engaging in courses of actions, dealing with difficulties, and so on. Pragmatic meaning is meaning writ large, perhaps the most important aspect of meaning but often the least obvious. Why? Partially because context is most often understood unconsciously and taken for granted and because there are many situations in which it is common to hide one's purposes.

INDIRECT PRAGMATICS AND HIDDEN PURPOSES

Constructions in language often function to express implicit but not explicit purposes; that is, constructions often serve to convey one meaning under the cover of another. Thus, there are rhetorical questions, that is, making statements or imposing frames by asking questions (*What would it cost for us to buy an electric car?*). There are suggestive questions, making a suggestion by asking a question (*Why not open the window and let in a little more air?*) There are critical questions, making a criticism by asking a question (*Why don't you ever make the bed?*). There are statements that make a commitment by saying something about the future (*I'm going to take out the garbage after dinner*). And there are forceful imperatives made via a statement about the future (*You are damn well going to take out the garbage today!*). Indirect pragmatics is everywhere in language.

Pragmatic meaning is **conveyed meaning**. Technically, circuits with the ability to carry out an inference by moving from a premise to a conclusion can also be used to convey a pragmatic meaning indirectly by moving from the meaning of one pragmatic construction (e.g., a speech act construction such as a question) to another (e.g., a statement, a suggestion, or a criticism).

The pragmatics of constructions is central to everyday life and reveals a lot of what everyday life is about. But perhaps the arena where it is most central is politics, where the pragmatics of communication can be a matter of help or harm, of life or death, or even of war.

The collection of X-Nets for conventional, normally used ideas for actions is learned and remains relatively fixed. It is there in your brain, and when activated, an X-Net can carry out an action either physically, in imaginative simulation, or via language; an example is a rhetorical question, making a statement by asking a question: for example, in a home setting *Do you have the energy to take out the garbage?* and in a political setting *Should Trump be prosecuted for tax fraud?* These work because it usually does not take all that much energy to take out the garbage, and if it is assumed that Trump committed tax fraud, it could be a crime worthy of prosecution.

IDEAS BEFORE LANGUAGE LEARNING

Just as children learn to think before they learn language, they also learn ideas and corresponding actions before they learn language. Consider a parent's action of trying to get an infant to eat by presenting a spoonful of food to the child's mouth. The action is successful if the child eats. But suppose the child doesn't want to eat. She can refuse by turning her head aside as the spoon approaches or by crying, pushing the spoon aside, or spitting out the food. These are learned purposeful actions in response to perceived purposeful actions by the parent. Once learned, the action can be carried out via existing neural circuitry.

Children learn early to perform actions with hidden purposes, such as refusing to get to get dressed for bed by starting to play intently and persistently with a toy. This is an action whose goal is not overt. Ideas for action are learned very early in life, as is their use in hiding covert purposes.

GRAMMATICAL COMPOSITIONALITY: PRAGMATICS AND MAPPING CIRCUITS

The general principles governing the neural circuitry for grammatical constructions characterize a relational network, that is, a network characterizing relations between constructions. For example, consider circuitry showing how question constructions are based on statement constructions and how passive constructions are based on active constructions. This naturally results in grammatical compositionality. Consider a composition of a passive construction and a question construction. Since passive sentences (such as *Harry has been arrested*) are statements, they can form the basis for passive questions (such as *Has Harry been arrested?*).[2] The active-to-passive mapping circuit automatically forms a composite with the statement-to-question mapping circuit, resulting in passive questions.

GRAMMATICAL MAPPING CIRCUITS

The grammar of a language is a system of related constructions in which one construction is "based on" one or more others. The "based on" relation can be defined by asymmetric correspondences between the constructions, that is, correspondences that can be carried out by neural circuitry that "maps" one construction to another.

The difference between two directly related constructions is a composition of two differences: a difference in pragmatic meaning and the difference in form corresponding to the pragmatic difference. Another way to think about this is that the grammatical form "marks" (i.e., indicates) the nature of the pragmatic difference. That is, a grammatical form can have a purely pragmatic meaning.

At the center of the pragmatic system is the statement speech act expressed by a declarative speech act.[3] In a statement speech act in English, the subject typically appears before the verb. In the corresponding question speech act, an auxiliary verb appears before the subject (*Did you take out the garbage?*). In the directive speech act (e.g., *Take out the garbage*), the person directed and the auxiliary verb are omitted.

Knowing a language means that the circuits mapping one construction to another are fixed in the brain. Each such construction-to-construction mapping circuit has two parts:

- One part, the speech act subject matter, is fixed and is the same across the constructions.
- The other part states the pragmatic difference and the difference in form expressing (or "marking") the pragmatic difference.

Included in what is the same are the speaker and the hearer of the two differing speech acts. The shared elements of form (e.g., subject, verb) are part of what is the same, but the grammatical ordering of the elements in the sentence may be different.

Much is the same from construction to construction. Note that bindings as proposed here are transitive. For instance, if (1) node A is bound to node B (A and B are neurally indistinguishable in their firing patterns) and (2) node B is bound to node C (B and C are neurally indistinguishable in their firing patterns), then (3) node A is bound to node C (A and C are neurally indistinguishable). That is, a binding circuit linking A and C can be readily recruited. A simple example is a passive question:

A: Bill saw the accident.
B: The accident was seen by Bill.
C: Was the accident seen by Bill?

On the whole, the differences across related constructions are minimal with considerable shared circuitry. In short, the system shows **neural optimization**, that is, a minimum energy configuration. Compare such a system with an alternative in which each construction is completely separate neurally; for example, a passive questions would look entirely different from either a passive construction or a question construction. Natural languages just don't work that way, because such a language would violate the neural optimization that arises when the constructions are separate but where one construction can apply to the result of another construction, as when the passive construction applies to the result of the question construction to yield a passive question.

NEW MEANINGS FROM LANGUAGE

From a neural perspective, each construction consists of complex neural circuitry: circuitry for meaning, circuitry for form, and circuitry linking meaning and form. Since constructions are general and each linguistic utterance is specific, specific utterances tend to be made up of a structured combination of a number of general constructions. The neural mechanisms for putting together constructions to form more complex constructions are bindings, integrations, and cascades. These circuits constitute the ways that complex constructions are structured.

For example, the question form (e.g., *Has he been here?*) has the first auxiliary (*has*) of the corresponding statement form (*He has been here*) before, not after, the subject, with the rest of the statement (*been here*) kept the same. That sameness is achieved by a generalized neural binding that applies in all such cases.

Neural integrations occur when the cross-construction relations are not so simple, and then a choice must made to value one change over another. Consider an exclamation such as *How big your garden is!* How phrases are grammatical + pragmatic variants versions of what, when, where, who, whom, which, whose, why, and how questions, such as *How big is your garden?* In *How big is your garden?*, you are requesting information about the garden. But in the exclamation, you are in the process of experiencing the garden, and you know how big the garden is. You are using a statement (not a question) form, with the subject (*your garden*) before "is." But you are using the how phrase placement first. "How" by itself in a question is focused on the means by which something is accomplished (as in *How did you find all those native plants?*). In *How big your garden is!* you are expressing (1) that it is very big, (2) that a lot of effort and/or resources beyond the norm went into it, and (3) praise, a compliment. Compare this with the title *How Big Is Your Garden?* of a gardening book. Here the title might suggest that the book offers advice for creating gardens of varying sizes.

Neural cascades are needed when there are chains of bindings and integrations that produce constructions that put simpler cases together in complex contexts to form resulting complex constructions.

In short, language uses neural mechanisms to produce an unlimited number of new complex modes of thought: new complex meanings from simpler ones. This requires not just ideas alone but also ideas paired with form, that is, constructions. In the section later in this chapter titled "The Two-Type Language Contrast," we will review results by Len Talmy and Dan Slobin on how the details of particular languages create different forms of "thinking for speaking," generalizations over language types for special modes of linking thought and language that vary from language to language.

Language thus can serve the function of extending our existing conceptual systems in regular ways. Language creates new meanings from older ones when we form new *grammatical combinations of existing linguistic elements* to create new *conceptual combinations of existing meanings* that are appropriate to new contexts of use. Furthermore, this works differently for different languages, since the relevant constructions are often very different across languages.

SOUND SYMBOLISM

It should not be surprising that the forms of language should have a causal effect on what those forms mean. We have already seen this in the case of sound symbolism.

Sound symbolism arises because we use embodied image schemas, which are part of thought, to conceptualize the structure of our mouths as we speak. As a result, certain types of sound combinations can have image schematic meanings. For example, *-ip* words such as "drip," "clip," "slip," "dip," and "nip" have, as part of their meanings, a short path to a sudden stop, which is what happens to the airstream in your mouth when you pronounce *-ip* words. In other words, a central aspect of meaning—image schemas—is used to structure form—phonological form—when applied to our mouths! Sound symbolism uses a meaning-to-mouth correlation, with image schematic meaning fitting the mouth! In sound symbolism, meaning is embodied in a direct manner.

FROM THOUGHT TO LANGUAGE

The remainder of this section is dedicated to providing examples of how the neural mechanisms for ideas *might* also apply to language. We begin with Table 5, which reviews the neural mechanisms for thought that we have already discussed.

The left-hand column in Table 5 contains the types of neural circuitry we have hypothesized for ideas. We will discuss how those mechanisms *could* account for neural generalizations governing constructions as well. These are correlations that we believe are causal, that is, when the neural mechanisms used in thought apply to language expressing such thoughts.

NEURAL CIRCUITRY IN LANGUAGE

We will use the neural circuitry types for ideas in the left-hand column of Table 5 to show how to construct a corresponding table indicating how the same circuitry types for ideas apply to constructions in language. As an example, consider relations between constructions based on their pragmatics, as in the relation between statement constructions and corresponding question constructions. The mechanism for this is like that for primary conceptual metaphors but with an important difference. In the case of a primary metaphor, there is a neural circuit carrying out an asymmetric neural mapping from one circuit in one brain region to a different circuit in another brain region. In the case of a relation between one construction and another, there is an asymmetric neural mapping from one brain circuit to another related brain region.

TABLE 5.   How neural mechanisms for ideas are used in thought

NEURAL MECHANISM	USE IN THOUGHT
X-Nets used for sequence circuitry for bodily motions	In aspects of meanings, events and actions are understood metaphorically as motions
X-Nets used for monitoring and control of complex sensorimotor actions	Carry out the monitoring, control, and sequencing of complex thoughts
Gestalt circuits, in which the activation of the whole circuit has a greater effect than the activation of its individual subcircuits	In gestalt schemas and frames, the meaning of the whole is greater than the meanings of its individual parts
Binding Circuit, which makes the bound circuits neurally indistinguishable, thereby linking the circuits and creating an integrated circuit	Creates identity between conceptual elements in different thoughts, thereby linking the thoughts and creating an integrated thought
Activation circuitry, connecting activated neural ensembles and neural circuits	Activates connected ideas
Inhibition circuitry, which turns off neural circuits	Keeps existing ideas from being used at this point in thought
Disinhibition circuitry, which turns off inhibition circuitry, thus allowing activation to occur	Allows unused existing ideas to be used and selectively activates ideas in a cascade
Neural Integration Circuit, a hierarchical circuit that combines multiple cascades through gating and modulation	A conceptual integration network framing that makes otherwise inconsistent ideas into consistent wholes
Neural mapping, which involves asymmetric neural mapping from circuit to circuit and from one brain region to another	In a conceptual metaphor, asymmetric conceptual mapping occurs from frame to frame and from one subject matter to another

TABLE 6.   Construction-to-construction mapping

Asymmetric neural mapping from circuit 1 to circuit 2, substituting subcircuit 2 for subcircuit 1 while keeping the remainder of circuit 1	Asymmetric mapping from construction 1 to construction 2, substituting the subconstruction for pragmatics 2 for that of pragmatics 1 while preserving the remainder of construction 1

We can represent a construction-to-construction mapping in Table 6. Note the similarity to last row (a conceptual metaphor) in Table 5.[4]

The remainder of this chapter is dedicated to going into the details of difficult examples of linguistic complexity to see how well this hypothesis holds up.

Basic Circuit Types in Grammar

Science looks for explanations, beyond mere descriptions, although descriptions are hard enough to achieve. Throughout this book we have been seeking neural explanations. A particularly elegant example is our account of how spike-timing-dependent plasticity at the neuron level can provide an explanation for the asymmetry of conceptual metaphor as well as for the neural "choice" of source and target domains for primary conceptual metaphors. We consider this a truly stunning example of a neural explanation, and we adopt it because we take explanation to be more important than mere description.

We will ask throughout this chapter whether there are neural explanations for grammatical phenomena. In chapter 3, we noted that children engage in thought—even complex active pragmatic thought—before they learn language. We then asked the following questions:

- Which of the circuit types present *before* language are also used *in* language?
- In other words, does language require any types of circuitry in addition to the types needed for the ideas expressed in language?

A priori, one might suppose that the neural circuitry types used for language are completely different from those used for thought. If we were to find that the neural circuitry types used for thinking before language are also used for language as well, we would have a powerful explanation of *why* those circuit types are used for language: those circuit types were already there, and the brain tends to use the architecture it already has rather than starting anew, forming fully new and different types of networks and circuits.

In the search for explanation, we begin by asking how circuitry used for *ideas* also applies in *language* and what other additional circuitry types are needed in language.

Consider these basic circuit types for ideas: gestalt circuits, X-Nets, binding circuits, mapping circuits, integration circuits (including coordination circuits), and cascades. How do these basic circuit types show up in grammatical constructions, which characterize the relationship between form and meaning?

PHRASAL AND INFLECTIONAL STRUCTURES

Many languages, but by no means all, have a phrasal grammatical structure and morphological inflections. Phrasal structures include examples such as the noun phrase [The [cold peaches]], the verb phrase [enjoyed [the [cold peaches]]], and the clause [[William Carlos Williams] [enjoyed [the [cold peaches]].

What grammarians call "inflections" include pronoun variations such as *I, you, she/he/it, we, you,* and *they.* These are the nominative case versions of the inflectional categories person, number, gender with the values person: 1, 2, 3; number: singular, plural; gender: female, male, neuter. Thus, "I" is first-person, nominative, singular, while "she" is third person, nominative, female, singular.

MODELING PHRASE STRUCTURES

The theory we are presenting uses the form of neural computation presented in chapter 3 to concurrently model thought and language on the one hand and the neural circuitry that we take to be used for thought and language on the other. That is why we will be speaking of "modeling" in this section.

Phrase structure is hierarchical and linear. The hierarchical structure can be modeled using gestalt circuits. The gestalt node models the phrase as a whole, and the role nodes on the gestalt model the parts of the phrase. Phrases within phrases (grammatical "recursion") can be modeled with the gestalt node of an embedded gestalt circuit neurally bound to a role node of the "higher" gestalt circuit.

For example, in the sentence *I know Bill left,* there is a knowledge frame with a knower (*I*), the knowing (*know*), and the content. There is also a leaving frame with a leaver (*Bill*) and the leaving action (*left*). The leaving frame is neurally bound to the content of the knowledge frame. Thus, one gestalt circuit (e.g., leaving) can be bound to a role (e.g., *content*) in another gestalt circuit (e.g., knowing).

Phrase structure also involves linear order. The linear order can be modeled by X-Nets, which are sequential, linking role nodes asymmetrically to subsequent role nodes in the gestalt circuit that characterizes the neural structure of the phrase.

A phrase structure cascade neurally binds (1) the roles of the hierarchical

gestalt circuits to (2) the corresponding elements in the X-Net that characterizes the linear order of the grammatical elements in the phrase.

Consider a gestalt circuit with two roles. We will refer to it as a **gestalt doublet**.
Suppose that one role models a form, that the other role models a meaning,
and that the gestalt doublet as a whole models the form-meaning relation.

In a gestalt doublet circuit, the activation of the form node will activate the
gestalt node, which will then activate the meaning node. Similarly, activating the meaning node will activate the gestalt node, which will then activate
the form node. Any direct form-meaning relation can thus be modeled by a
gestalt doublet circuit.

In simple cases, each simple lexical item is a construction, a gestalt doublet
with a form role and a meaning role.

Polysemy, where a single lexical item has multiple related meanings, is very
common. Such cases are modeled by a doublet, whose meaning role is a central meaning, that has connections to members of a network of related meanings. Lakoff's (1987b, Case Study 2), covers the preposition "over" and provides a conceptual network with dozens of variations on the central meaning
of the word. Examples include *The deer jumped **over** the fence, There are **over**
a million cases of covid in the United States, The gas tank **over**flowed, Please
move **over** a little, Pour chocolate **over** each cookie, He lives **over** the hill from
here, The concert is **over**, The soldiers are spread out **over** the hill, Do it **over**,
but don't **over**do it.* Showing the systematic relationships among the dozens of
meanings of "over" was a real intellectual achievement by Claudia Brugman
(1981) her master's thesis "The Story of *Over*."

What about grammatical categories such as noun and verb? They are categories of forms with meanings of a certain kind. Nouns characterize entities (concrete or abstract). Verbs characterize happenings: ongoing actions,
events, and activities. Adjectives characterize properties of entities. Prepositions characterize image schemas, such as part-whole (the partitive "of," as
in *the handle of the knife*), containment (*in* and *out*), motion (*to, from, along,*

across), relative location (*behind, after, before, in front of*), and so on. Verbs, adjectives, and prepositions are predicates. Adverbs characterize properties of predicates (as in run **slowly**, **extremely** intelligent, **partially** in).

Grammatical categories are categories of phonological (or written) forms that are paired with their meanings via gestalt doublets. For example, the noun category consists of (1) forms such as *car* whose meaning is (2) the subcategory of entities defined by the car frame.

MODELING INFLECTIONAL STRUCTURES

There are certain universal concerns that are expressed grammatically in languages as what linguists call "inflections": relative time of events (past, present, future), number (singular, plural, sometimes dual), gender (masculine, feminine, neuter), and speech roles (speaker, hearer, subject matter).

A form of inflection can be simple, as in *cats* (a single morpheme *s* for a single inflectional meaning), or double (one morpheme for two inflectional meanings, as in *we*: first-person plural, including speaker). Double inflections are modeled by integrations of the various categories expressed (e.g., speaker included/plural, as in *we*).

Those are central meanings of inflectional morphemes. There are also variants, such as the royal *we* (said by a monarch in his or her official capacity) and the number of group members (as in the British *The committee **are** meeting*). Some inflections have a zero form, such as the singular in English (e.g., *cat*). Some inflections are shown as a vowel change for plurals in special cases (e.g., foot/feet, goose/geese).

DEFAULTS AND SPECIAL CASES

It is common for inflectional paradigms to have default cases and special cases.[5] For example,

- *Present* is a default tense, with *are* as the default value and *am* and *is* as special cases.
  - *Am*: Number: 1; tense: present; number: singular
  - *Is*: Number: 3; tense: present; number: singular
  - *Are*: Default: everything else
- *Past* has a default with *were* as the default value and *was* as a special case.
  - *Was*: Speaker 1, 3; tense: past; number: singular (as in *I was* and *she was*)

These obviously can be modeled using if-then/otherwise circuitry. If there is a special case meaning, then use a special case form; otherwise, use the default form.

Agreement makes use of neural binding circuits. For example, verbs agree with their subjects in person and number. This is done by binding the value of number for subjects to the same value for number in the predicate. For example, in a sentence such as "Worms are slimy," "worms" is plural, and "are" agrees with "worms" in number. *Agreement can be carried out by binding circuits identifying the value of the verbal category number—in this case plural—with the value for the subject category number.*

There are complications, of course.

- A conjunction of two singulars acts as a plural, as in *A dog and a cat were fighting*. This would require an integration circuit marking the conjunct noun phrase node for multiple singulars—such as noun phrase and noun phrase—into a plural with an integrated meaning.

Existential *there* takes a neurally bound copy of the number of subject following *be*. This shows up in *there* sentences:

- There *is* a present in the closet.
- There *are* presents in the closet.

This binding is preserved even in so-called raising constructions:

- There *is* believed to be *a present* hidden in the closet.
- There *are* believed to be *presents* hidden in the closet.

INTEGRATION CIRCUITS

Integration circuits can form integrated whole linguistic forms by piecing them together from parts and variations on parts. Obvious cases are *gotta, wanna, oughta, wontcha, dontcha, cantcha*, and so on. For example, the form sequence *will not* is integrated to *won't*, and *won't you* is integrated to *wontcha*, combining two existing reduction rules turning *ty* to *ch* and *you* to *ya*:

- [[*will not*] *you*]
- [*won't you*] by integration of [*will not*]
- *wontcha* by integration of [*won't you*]

Consider the grammatical framing of *Sara touched Jim*:

- *Sara* is subject, *touch* is predicate, and Jim is direct object.

Now consider the corresponding meaning frame:

- *Sara* is agent, *touch* is action; and *Jim* is patient.

The gestalt doublets (GDs) are

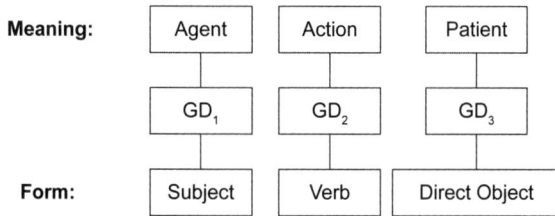

GD_1, GD_2, and GD_3 are the gestalt nodes forming each of the doublets linking meaning to form.

- **Integration:** A single node, I, integrating the control nodes—GD_1, GD_2, and GD_3—of all the doublets, to yield "I" $\leftrightarrow GD_1, GD_2, GD_3$.

Here "$\leftrightarrow$" indicates activation from "I" to each of GD_1, GD_2, and GD_3 and activation from all of GD_1, GD_2, and GD_3 to "I."

"I" is normally inhibited, as are GD_1, GD_2, and GD_3. This allows the meaning and the form to function freely in other constructions. When "I" is disinhibited, GD_1, GD_2, and GD_3 are activated, and the construction as a whole is activated. The result is one of the simplest constructions, as we saw in the figure above.

GD_1, GD_2, and GD_3 are the gestalt nodes forming each of the doublets linking meaning and form and jointly forming the roles of the gestalt of the whole construction.

There are, of course, other simple constructions: intransitive constructions, passive constructions, predicate nominal constructions, predicate adjective constructions, and so on. They are formed in basically the same way.

Note that gestalt doublet circuits create neural connections between forms and meanings. The gestalt nodes of gestalt doublet circuits act as gates. When the gestalt doublet nodes are activated, the meaning nodes and form nodes that they link are activated together. And when the gestalt doublet nodes are inhibited, there is no neural connection linking form and meaning.

The Scope and Power of Ideas in Grammar Due to Circuits for Ideas

We have hypothesized that the neural circuitry for ideas is sufficient to characterize language. This section explores the power that idea circuitry has to create a wide and complex range of linguistic phenomena. We begin with names.

NAMES

In the naming construction, there is a unique person, thing, or institution that occupies the meaning role of a lexical item. We will speak of the corresponding form role as the "name" of that person, thing, or institution. People have names. Frame roles that confer authority on people in institutions have names called "titles" such as president, senator, admiral, and so on. Each title is a doublet (with a gestalt node), and each person-name is a doublet (with a gestalt node). When a person fills a role in an institutional frame, the "filling" of the role by the person is carried out neurally by a circuit binding person to role.

THE TITLE-NAME CONSTRUCTION

An example is *President Joe Biden*, with the title *President* and the person-name *Joe Biden* and with the person Joe Biden filling the president role. Thus, there are three control nodes: one for the title, one for the person-name, and one for the gating node of the neural binding circuit that carries out the filling of the frame role. The construction is formed by an integration node that integrates all three nodes.

GRAMMATICAL CATEGORIES

Gestalt doublets open up another possibility for meaning-based language analysis defining grammatical categories in semantic terms. As we discussed in the previous section "Integration Circuits," nouns, verbs, prepositions, and adjectives are defined by what they mean, and their meaning is created via gestalt doublet circuits.

SENTENCE ADVERBS

What are called "sentence adverbs" also modify predicates but in aspects of meaning that are not expressed in surface form. A well-known example is "frankly" in sentences such as *Frankly, he's an idiot.* The meaning of this sentence contains a semantic characterization of the active speech act state that the sentence expresses, with the meaning of "frankly" modifying the speech act predicate state, as in *She frankly stated that he was an idiot.* The speech act predicate is part of the meaning but is not expressed in form. This leaves sentence adverbs such as "frankly" modifying a meaning (the speech act of stating) that is understood but not expressed in sentences such as *Frankly, he's an idiot.*

Another common example is "unfortunately," in sentences such as *Unfortunately, he failed the exam.* "Unfortunately" occurs in sentences expressing the speaker's emotion about the event expressed by *he failed the exam*, namely to be unfortunate.

DERIVATIONAL MORPHOLOGY

Derivational morphology also works in constructions that are carried out by integration circuits. For example, consider cases such as "laziness" and "sincerity." It is important to distinguish between them. Derivational morphology is

- a means of linking one *concept*—a property such as "lazy"—to another, an abstract quality such as "laziness," which is understood as a kind of entity, and
- involves the operations on *linguistic form* that, for example, adds -*ness* to the adjective "lazy" to derive "laziness," as opposed to a different syntactic operation with the same meaning by adding -*ity* to "sincere" to derive "sincerity."

THE GRAMMATICAL DESCRIPTION

There are two constructions for expressing a quality in terms of a corresponding adjective: one construction for adjectives that take -*ness* and one for adjectives that take -*ity*. Each adjective that can express a quality will take one or the other. This may be arbitrary for each adjective. But more likely, there seems to be a categorization of adjectives into Germanic and Latinate roots, with Germanic adjectives taking -ness and Latinate adjectives taking -ity. In that case, the constructions would be conditional on using that distinction. This can be implemented via gates from our neural tool kit.

THE MECHANISM

The -ness and -ity constructions would apply only if a quality were being expressed. Since adjectives can occur independently, their occurrence in -ness and -ity constructions would have to be gated. Only if the gates were active could the adjective occur as expressing a quality and be marked by -ness or -ity. In each quality construction, each of the gates controlling the -ness and -ity constructions could be activated via an integration node. Learning the construction would entail connecting the right integration circuits to the appropriate adjectives in the lexicon.

The classic example of complex derivational morphology is the word *anti-disestablishmentarianism*, which is formed by a number of different constructions: *anti-, dis-, -ment, -ary, -an,* and *-ism*. A construction with an integration circuit adding *-ment* to "establish" with the sense of the established Church of England would yield "establishment." Another construction adding the prefix *dis-* with the meaning of reversal would yield "disestablishment," the undoing of the established church. The prefix *anti-* means "against." A construction adding *anti-* to "disestablishment" yields "antidisestablishment" (*against the disestablishment of the church*).

A construction that combines *-ary* (characteristic of, as in "parliamentary") with *-an* is added to get *-arian* (a person functioning in that characteristic role, as in "parliamentarian"). A construction that adds *-arian* to "antidisestablishment" produces a the term that means a person in a characteristic antidisestablishmentary role, namely a supporter of antidisestablishment. The suffix *-ism* denotes an ideology. Thus, "antidisestablishmentarianism" is the ideology of people against the disestablishment of the church.

The point of this example is to display the combinatorial power of integration constructions in derivational morphology.

A slight complication arises when the derivational morphology takes on a special related meaning. Consider "refrigerator," analyzed as *re-frig-er-at-or*, as in an active physical object in the form of a container that causes things inside of it to be cold again. That's close. It actually means a machine that either causes things (mainly food) to be cold again or just cool or cold enough to preserve their natural state. The *frig-* (as in "frigid") means "cold," the *-at* means "causes," and the *-or* means an "agent."

It should be pointed out that "refrigerator" is not an arbitrary name for an appliance that does what it does. You would not call a tree a refrigerator, nor would you call a chair, a racetrack, or a dog a refrigerator. The meanings of the parts of "refrigerator" play crucial roles in the meaning of the whole, although the process is somewhat complex: a cascade of integrations is needed to put the meanings of the parts together in the right way.

HOW SEMANTIC ROLE MODIFIERS WORK

What is the relationship between the verb "run" and the adverb "slowly"? "Run" is an instance of an action. The action frame specifies properties that actions can have. Speed is one such property and is called a "parameter." Each parameter has possible values. The adverb "slowly" in *run slowly* specifies the value of the semantic role speed as slow. Thus, "run" is an action, with the value speed unspecified. *Run slowly* is an action with the value of speed specified as slow. This involves the neural binding of slow to speed in the running frame. The modeling of the neural structure is carried out with a gestalt circuit for the activity frame and binding circuits that bind "running" to the activity and "slow" to the speed role in the activity frame.

Adverbs perform the corresponding modification operation on adjectives. For instance, in *extremely slow*, "extremely" picks out the high value for the degree of speed. There are more complicated cases such as *devastatingly beautiful*. This case uses the general metaphor that sexuality is a physical force and the beauty frame in which being beautiful has sexual effects. This allows "devastatingly," via conceptual metonymy and metaphor, to fill the effects role in the frame for being beautiful.

There are complexities in how prepositions designate image-schema relations. The preposition "in" has a meaning that binds a trajector to the interior of a container, which functions as a landmark for the trajector. Degree adverbs such as "partially" and "completely" characterize how much of an overlap there is between the trajector and the container interior, such as partial or complete, which are roles within a linear scale frame.

WHAT IS A "HEAD"?

Run slowly, in its form, rather than meaning, is a verb followed by an adverb constituting a phrase, which means that it is structured neurally by a gestalt node with two roles: verb and adverb. In meaning, each verb is a predicate, and each adverb is the modifier of a predicate. As a whole phrase, *run slowly* has the meaning of a predicate, just like the verb "jog," which has a similar meaning. At the general level of predicates and modifiers, *run slowly* has the following semantic gestalt structure: predicate 1 is a general (unmodified) predicate, and predicate 2, having been modified, is a special case.

Correspondingly, when we do the same with the meanings run and slowly, we get the gestalt structure: "run" 1 is a general (unmodified) action predicate, and "run" 2, having been modified to have a slow speed, is a special case of running, that is, slow, not fast.

Both of these are instances of the same general phenomenon: a general concept plus added limiting information yields a special case of the general concept.

In traditional grammar, the general concept is called a "head," and the added limiting information is called a "modifier" of the head, since it modifies the meaning of the head. Thus, "run" in *run slowly* is the head of the construction, and *slowly* is the modifier.

Extremely slow works in a similar way. Slow is an adjective that specifies a value of a speed parameter. But "slow" itself introduces a degree parameter for a degree of slowness that is unspecified. "Extremely" specifies the value of that degree parameter. "Slow" is a modifier, and "extremely slow" is also a modifier. Thus, "slow" is the head of the phrase *extremely slow*.

Although these are simple cases of grammatical heads, the more complex cases work in basically the same way. Specific lexical items—nouns, predicates, modifiers, and so on—are constructions with forms whose meanings are roles in conceptual frames. Those frames contain other roles. For example, the verb "drink" is given meaning by a conceptual frame in which "drink" is an action of taking a liquid into the body through the mouth. In the lexical entry for "drink," the liquid is unspecified. There is a grammatical construction consisting of a verb and a direct object, for example, *drink beer*. The semantics of the construction binds beer to the liquid role in the frame. The result is that *drink beer* is a verb phrase, with "drink" as the head verb in the verb phrase.

This is the traditional notion of a "head" in grammar and works when there is an overt modifier and a modified entity in a simple phrase structure construction. Examples include *liquid nitrogen* and *delicious cake*, where nitrogen and cake are the heads and liquid and delicious are the modifiers.

But not surprisingly, there are complexities. Consider the sentence *I poured and enjoyed a delicious cup of coffee*. Coffee has two relevant properties: it is a liquid that is drunk and has a taste. The word "coffee" has those properties as word-internal components of its meaning. "Cup" is both a container-object and a measure; there are drinking cups and measuring cups. Thus, you drink a cup of coffee from a cup-object, and you measure a cup amount (8 ounces) of flour for a cake recipe. All of this is straightforward in a sentence such as *I poured and drank a cup of coffee and the coffee was delicious.* The first half of the sentence picks out the liquid-amount property of the coffee, and the second half picks out the taste property.

But in a sentence such as *I poured and enjoyed a delicious cup of coffee,* the phrase "a delicious cup of coffee" is a single phrase that is grammatically a direct object of both "poured" and "enjoy." You might expect a single phrase in a sentence to have a single meaning in that sentence. But this is not the case here. "Poured" applies to the container-object and the liquid property of the coffee, ignoring the deliciousness. "Enjoyed" applies to the cup amount and the taste property of the coffee. In short, the single phrase "a delicious cup of coffee" simultaneously picks out two different meanings—for measurement and for taste—at once. How is this possible?

This is a job for a neural integration.

In form, there is a conjunction of two verbs ("poured" and "drank") and in form a single noun phase ("cup of coffee") that is the grammatical direct object of both verbs. But the single expression "cup of coffee" has two meanings. The two meanings are related to the single form by an integration circuit consisting of two gestalt doublets, with two distinct meanings (one for amount and one for taste), that are integrated, resulting in a single expression. The coffee frame contains both a liquid property and a taste property. The meaning of "poured" applies to the liquid aspect of coffee, while "enjoyed" applies to the taste aspect of coffee, with "delicious" as a modifier of the taste of coffee.

INTEGRATION DETAILS

A cup is understood via the integration of two frames: the container-object frame and the measure frame:

- "Cup" constituted by a gestalt circuit. The gestalt has two roles: container-object and measure.
- The container-object frame is constituted by a gestalt circuit whose roles include content and size.
- The measure frame is constituted by a gestalt circuit whose roles are amount and substance measured.

- The integration is created by two neural bindings:
  - ◦ Amount is neurally bound to size.
  - ◦ Content is neurally bound to substance measured.

In *a cup of coffee*, "cup" refers to both the container-object and the measure, while "coffee" refers to both the content in the container-object frame and the substance measured in the measure frame.

Delicious coffee is constituted by frames for "coffee" and for "delicious." The coffee frame has the roles liquid and taste. Taste is bound to a linear scale for tastes, with the roles high and low on the scale. "Delicious" is a value bound to high on the taste scale.

As a noun phrase, *delicious coffee* is a head-modifier construction, with "coffee" as the head and "delicious" as the modifier and with "coffee" referring to a general category and "delicious coffee" referring to a subcategory of the coffee category.

In *a cup of delicious coffee*, the liquid in the coffee frame is bound to the content of the container-object frame for "cup."

All of these frames and the bindings across them constitute an integration, the integrated meaning of "a cup of delicious coffee." The important thing to notice is *the bindings between roles in different frames. In each case, the bindings integrate each of the given frames with their corresponding values into a single coherent meaning.*

If we add "large" to form *a large, delicious cup of coffee*, there is further integration. "Large" modifies "cup of coffee," with "delicious cup of coffee" as the head and general category and "large delicious cup of coffee" as the subcategory.

The meaning of "large" is defined by a linear scale, with "large" bound to the high point on the scale. The scale can be bound to size, defining a large size of the cup, and the scale can also be bound to the amount role of the measure frame.

"Large" also modifies the container-object and the amount in the measure frame, resulting in both a large container-object and a large amount of coffee.

In short, adding "large" extends the integration in two of ways, affecting both container size and content amount.

Now consider the sentence *I poured and enjoyed a large, delicious cup of coffee*. "Pour" takes as its patient and direct object the liquid of the coffee frame. "Enjoy" takes as the direct object and thing enjoyed in the enjoy frame the taste in the coffee frame.

These relationships pick out the integrated parts of the larger integrated whole: *a large, delicious cup of coffee.*

The meaning of the sentence is a very complex integrated whole, and yet we understand it instantly.

Integrations of this sort are commonplace. They are a normal part of how the neural system creates unconscious understandings that are integrated so quickly and naturally that we don't notice them consciously.

WHAT IS A MINOR RULE?

Lakoff (1965) is a lengthy study of types of exceptions to rules of grammar. An important type of exception is the **minor rule**, a rule that normally does *not* apply but can apply in exceptional cases. A standard example is the *-ee* suffix. It works like a case marker called "absolutive" in ergative languages, where the same case marker is used for direct objects of transitive verbs and subjects of intransitive verbs. Compare "employee" (the direct object of "employ") with "escapee" (the subject of "escape"). This case-marking pattern is rare in English and applies only in exceptional cases. From a neural point of view, it is introduced via a construction that adds the *-ee* ending to verbs, but the construction has a gating node that inhibits the construction with normal verbs, while it applies with exceptional verbs with connections that disinhibit the inhibiting gating node for the corresponding construction for normal verbs.

We have already seen another case, adjectival nouns with *-ity* versus adjectival nouns with *-ness*. Neither *-ity* nor *-ness* can apply to adjectives generally. You have to know which adjectives take which suffix. To keep the suffixes from applying to all adjectives, the general category adjective has to inhibit both circuits adding the suffixes. But then in order to get the right suffix on the right adjectives, each adjective has to disinhibit (turn off) the gating node that inhibits the correct suffix (either *-ness* or *-ity*) for that adjective.

The general point here is that all this complexity arises because of the history of English, which has incorporated words of both Germanic and Latinate origin. Because we think with the neural system in our brains, which works by activation and inhibition, that neural system has to incorporate words of both Germanic and Latinate origin, which work by different and mutually exclusive rules. Gating is needed to accomplish this neurally.

ABSOLUTE EXCEPTIONS

Consider the word "alacrity." The Merriam-Webster dictionary points that the term "denotes physical quickness coupled with eagerness or enthusiasm." The etymology of "alacrity" comes from the Latin adjective *alacer*, often translated as "lively," that is, physically quick coupled with eagerness or enthusiasm. There is no adjective "alacr" in English. One way to make it fit with other *-ity* nouns is to postulate a hypothetical adjective "alacr-" and have it, when activated, activate the *-ity* construction directly. Thus, "alacr-" will never apply alone but only as part of "alacrity" and with the right meaning. Such cases are discussed in *Irregularity in Syntax* (1970), the published version of George Lakoff's 1965 PhD dissertation.

SUMMARY

This is a remarkable miscellany of grammatical phenomena. The question to be asked is why such a range of phenomena should exist. It is because *the types of neural circuitry used for ideas also apply to the language that expresses those ideas*, in many ways to create a wide variety of linguistic phenomena.

Embedding and Composition

We think via conceptual composition, which is a fancy way to say that we think by putting ideas together. All the circuitry for conceptual mechanisms are forms of composition, whether using frames, coordination, bindings, metonymies, metaphors, and/or integrations. A common way that we put ideas together is what linguists call "embedding," that is, filling a role in a conceptual frame with another frame.

Conceptual embedding can be accomplished by binding the gestalt node for a frame to a role node in a higher frame. For example, in *I know that Sam ate lunch*, the knowledge frame contains the role of content. The eating frame is filled in with Sam as eater and with lunch as eaten and is bound to the content role of the knowledge frame.

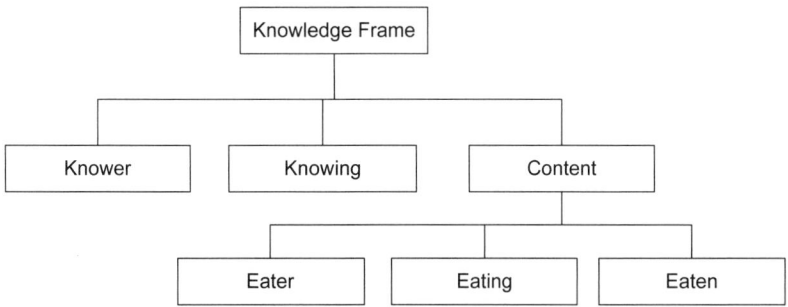

A special case of such an "embedding" occurs with frames concerned with either satisfying a desire one has or expressing an emotional reaction to a situation one is affected by. An example is *Sara wants to leave.* English has a grammatical construction for such cases. It is called the **equi construction**, since the person with the desire or emotion (Sara) is identical to (that is, "equal" to) the person in the situation being described as "leaving."

In the equi construction, a neurally bound role in the meaning frame for the sentence is omitted in the sentence. Since the content of that role is stated earlier, it does not have to be overtly expressed, and considerations of neural optimization (energy minimization) apply to leave it *known conceptually* but not *expressed linguistically* (which would take extra energy). Two things are left out: (1) the person who is the same as the *experiencer* of desire or emotion and (2) the tense, which is predictable from context.

In the case of desire (*Sara wants to leave*), the tense of "leave" is in the future, which follows from the meaning of "desire": a purpose that arises from the desire and fulfills the desire. With the emotional reaction to an event that has already occurred, the event is either (1) in the past (*I'm angry about Sara leaving early*) or (2) where the tense on the main verb predicts the embedded tense, that is, in the present (*It's fun swimming; It was fun swimming; It will be fun going to the beach*) or a hypothetical case (*It would be great to spend a month in Italy; It would have been wonderful to win the lottery*).

Here are some simple examples, with "__" indicating the omitted subject and tense marking of the embedded clause. In the meaning of the sentence, there is a neural binding between the semantic subject role and the meaning of the boldface noun (**Harry, Rhiannon**, etc.):

- **Harry** loves [__ swatting flies].
- **Rhiannon** wants [__ to leave].
- [__ Getting a parking ticket] bothered **me**.
- [__ To take a vacation] would be delightful to **me**.
- It amused **Natalia** [__ to see the clown do a somersault].
- **Oliver** avoids [__ skiing on dangerous mountains].
- **Paloma** doesn't like [__ to eat meat].
- Harry made **Rhiannon** [__ leave early].

The following is the equi construction, including the meaning, the form, and how the meaning is paired with the form.

The equi construction is very general. It applies to a huge range of cases. It can do this because it is defined only by constraints. Any pairing of form and meaning fitting these constraints, given the meaning-form pairings in the rest of the grammar, is permissible in English.

Here is the equi construction, with both its meaning constraints and its form constraints:

- Meaning:
    ○ There are two frames, frames A and frame B.
    ○ Frame A has at least two roles, A1 and A2.
    ○ Frame B is embedded as a filler of role A2 in frame A.
    ○ Role B1 is the protagonist in frame B.
    ○ Role A1 is neurally bound to role B1.
    ○ The tense/aspect of frame A is neurally bound to the tense/aspect of frame B.
- Form:
    ○ The form of role B1 is null.
    ○ The tense/aspect of the form of frame B is null.

The neural bindings characterize the referent of B1 as being the same as the referent of A1, and they characterize the tense and aspect of B as being the same as the tense and aspect of A. No forms for B1 or the tense and aspect of B are needed to characterize their meanings, since the conceptual neural bindings do that job by binding values for B to the corresponding values for A.

Neural optimization thus allows the elimination of those "unnecessary" forms as part of the construction.

What does it mean for neural optimization to "shape" such a construction? The construction is learned by every speaker of the language, *and in order for it to be learned, it has to fit the requirements of neural optimization, just as learning in general does.*

This construction has wide application but does not apply in all cases. Compare the following statements:

- *I think that I have a cold.*
- But not **I think to have a cold.*[6]

- *He feels that he should leave.*
- But not **He feels to leave.*

In some cases, neural optimization applies when the embedded clause is marked with a gerund (*leaving*), not an infinitive (*to leave*).

- *I considered leaving.*
- But not **I considered to leave.*

The equi construction seems to apply generally when frame A expresses emotion about the content of frame B, as in *I was surprised to see him so drunk* and also with "amazed," "alarmed," "amused," "disgusted," and so on. With the same class of predicates there is a gerund version in a preceding clause—*Seeing him so drunk surprised me*—and also with "amaze," "alarmed," "amused," "disgusted," and so on.

The construction applies to gerunds with verbs of perceiving: *I saw him running away, I heard him raiding the refrigerator, I caught him stealing,* and so on. The construction also applies with verbs of causing to infinitives and bare stem verbs, as in *I **forced** him **to apologize** to his boss, I **caused** him **to be fired**, I **made** him **leave** at midnight,* and so on. But the choice of infinitive or bare stem varies with the main verb; for example, "cause" requires the infinitive, while "make" requires a bare stem. Hence, we do not get **I caused him be fired* and **I made him to leave.*

The point is that there is a basic construction that is simple, there are some generalizations as to when it applies, and there may be some isolated exceptions. The generalizations can be stated with an if-then circuit: If the verb has a given meaning, then the construction applies. If not, the construction is inhibited. Another possibility is that a construction can apply only with certain exceptional verbs. This can be done with individual connections from the exceptional verbs activating a gating node for the exceptional construction.

The Coordination of Multiple Circuits in Grammar

MULTIPLE INDEPENDENT CIRCUITS

Consider the sentence *I figured it out that Sally had left.* "Figure out" is a composite verb that integrates the meanings of "figure" and "out" in a complex way. The verb "figure" refers to numbers and makes use of the metaphor that reasoning is arithmetic calculation, which occurs in sentences such as *I put two and two together, It just doesn't add up, That's the bottom line, That fact doesn't count,* and *It figures.* The "out" in "figure out" makes use of the conceptual metaphor that "hidden" is "in" and "known" is "out," as in sentences such as *She keeps her feelings in, He let it all hang out,* and *His secret came out.* "Figure out" therefore means to reason about a problem, with the result that the answer becomes known. It takes a cascade to put all these parts together, but we do it unconsciously and effortlessly.

"Figure" and "out" are separated in the sentence. "It" is a pronoun, functioning here referring forward, via a binding circuit, to the clause "that Sally had left." "It" and the clause are separated by the word "out."

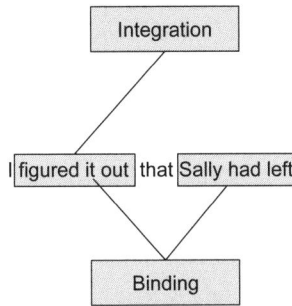

The integration and binding circuits overlap in the form of the sentence, but they function independently of each other. The meaning circuits are functioning on the basis of conceptual structure, integrating the meanings of "figure" and "out" while providing the meaning of "it" as something unknown to begin with, along with required reasoning. In the grammatical form of the sentence, "it" refers via binding to the meaning of the clause "that Sally had left." The circuitry combining "figure" and "out" is independent of the circuitry linking "it" and the following clause. However, there are sentence-level constraints on when an "it" + clause construction is needed versus when the construction can occur alone. You can say *I figured out that Sally had left* but not **I figured that Sally had left out.* A clause or other heavy noun phrase cannot come between a verb and a particle such as "out." But a short noun phrase can, as in *I can't figure Sally out.* These constraints apply to a grammatical frame characterizing the possible order of elements following a verb.[7]

INDEPENDENT CIRCUITS WITH A SHARED NODE

Consider the following examples:[8]

- *I drank the teapot dry.*
- *I washed the plate clean.*

In both cases, there is a result of an action: the teapot is dry as a result of drinking, and the plate is clean as a result of washing. The result shows a common grammar and semantics that apply in the construction in general. Here "the teapot" is the grammatical subject and the semantic theme of "dry," and "the plate" is the grammatical subject and the semantic theme of "clean."

But the semantics of the sentence grammar as a whole is different in these two cases. "The plate" is the direct object and patient of "wash": *I washed the plate*. But "the teapot" is not the direct object and patient of "drink": *I didn't drink the teapot*. What is a neural grammarian to make of this contrast?

First, we look at the ideas in both sentences. The concept of a teapot plays a role in a tea-drinking frame. The roles in the frame include the tea (a beverage); a teapot, in which the tea is brewed; the act of pouring tea from the teapot to the teacup; and a drinker (or more than one drinker), who drinks the tea from the teacup. This is characterized by an X-Net for performing, or simulating in imagination, the sequential actions of brewing, pouring, and drinking. This action sequence is commonly iterated, say, when the drinker wants to drink more than one cup. After a number of iterations, all the tea in the teapot will be gone, and the teapot will be dry.

This frame as well as the simulation of iterated drinking it defines is used to understand the sentence. In the frame, the tea is the patient of the drinking action; just as in the dishwashing frame, the plate is the patient of washing action. In the frame-based simulations of dishwashing and tea-drinking, the action-patient relation is the same.

Now consider the washing sentence. The plate is both (1) the direct object and patient of "wash" and (2) the subject and theme of "clean." This is because there are two circuits present that share a node (the plate) but are nonetheless independent outside of the construction. In addition, there is a cause-result X-Net circuit, with the washing of the plate as the cause and the plate being clean as a result.

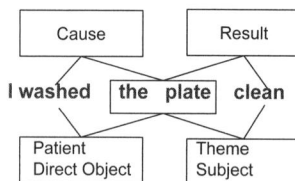

Finally, we need to bring up the issue of **null instantiation**, in which a role in a frame is left out of the sentence. Null instantiation was first studied in depth by Charles Fillmore (1976, 1982, 1985), the founder of the field of frame semantics and the discoverer of how framing works. It is regularly pointed out that "drink" and "eat" are classical null instantiation verbs that omit direct objects, as in the sentence *I drank after I ate*. From this perspective, the understood but unmentioned tea is not just a frame-based patient but also a null-instantiated direct object!

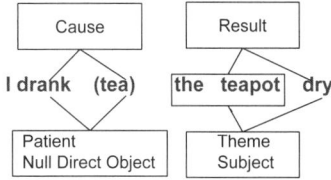

THE MORAL

Complex ideas develop prior to language. Language is meaningful; it links to ideas and uses the same neural circuitry as ideas. Grammar is not just about the elements of form in sentences. Grammar begins with meaning, with ideas, and inherently involves the pairing of meaning and form and *the linking of form to conceptual simulations*. Neural optimization often leads to a mismatch between *complex* neurally learned ideas and *simple* conventionalized linguistic form. One way that language deals with such a mismatch is via grammatical constructions. But grammatical constructions, as Charles Fillmore has taught us, include not only overt elements of form but also omitted (null-instantiated) elements of form, in this case the conceptually present but linguistically omitted tea.

Lexicon and Grammar

Every theory of language posits a mental lexicon, which contains what are called **lexical items**. They include words, morphemes, compounds, and idioms, each of which has a phonological structure (a mental characterization of sounds), a grammatical phrase structure, and a semantic meaning that is not predictable from the sound structure and the phrase structure alone.

Lexical items are neurally bound to grammatical roles in phrase structures. For example, "cup" is a noun. In using language, we make use of such neural circuitry as we unconsciously, automatically, and effortlessly fit lexical items (i.e., lexical constructions) to grammatical constructions.

The question we are considering here is how lexical and grammatical constructions fit together. The neural mechanisms for fitting ideas together include bindings, integrations, and cascades. How do those neural mechanisms work to combine lexical and grammatical constructions?

The example we will consider is the phrase *reusable wine cork*. "Reusable" is an adjective; "wine cork" is a noun, a compound noun. These terms fit into an adjective + noun phrase structure, where the adjective "reusable" is the modifier of the compound noun "wine cork," which is the head of the noun phrase. That is, a *reusable wine cork* is a special case of a wine cork.

"Reusable" is a whole word made up of these parts: prefix (*re-*), verb root (*use*), and suffix(*-able*). This type of word is constituted by a gestalt word circuit with the roles prefix, root, and suffix together with an X-Net, putting them in this linear order: prefix + root + suffix via neural bindings of the prefix to the start of the X-Net, the root to the center of the X-Net, and the suffix to the end of the X-Net.

A slightly different example would be the compound noun *wine cork*. It is constituted by a gestalt circuit with a gestalt node noun ("wine cork") and the two roles noun ("wine") and noun ("cork"). Here the noun "cork" is the head, since a wine cork is a special case of a cork. The noun "wine" is functioning as a modifier. But how?

The answer is metonymy, a mapping from one element of a frame to another element of the same frame.

The frame is the wine bottle frame, whose roles are (1) a tall glass *bottle* with a narrow opening on top; (2) *wine*, contained in the *bottle*; (3) the *cork*; and (4) the *use of the cork* to keep the wine in the bottle and to seal off the wine from the outside air to prevent it from turning into vinegar.

The metonymy is a mapping from the wine (2) to the use of the cork (4). In the compound *wine cork*, the wine stands via metonymy for the use, which modifies the cork by specifying the function of the cork in preserving the wine. That accounts for the role of "wine" as a modifier of the head "cork" in the compound noun *wine cork*.

WHAT ABOUT REUSABLE?

The phrase *reusable wine cork* is a noun phrase with the roles adjective ("reusable") and noun compound ("wine cork").

The two hierarchical structures, of course, can occur independently, as in the sentence *This wine cork is reusable.* "Wine cork" and "reusable" are two distinct items in our neural lexicon—a noun (with the structure of a compound) and an adjective—that are constituted by separate circuits. In addition, the noun phrase of the form adjective + noun is a distinct circuit in our neural grammar. To form the hierarchy, there must be two neural bindings, a binding of the lexical adjective "reusable" to the grammatical role adjective and a binding of the lexical compound noun "wine cork" to the grammatical role noun.

The situation is actually more neurally complex, since *re-, use, able, wine,* and *cork* are all independently existing lexical elements with meanings that have to be neurally bound to the listed lexical items "reusable" and "wine cork," with the meanings matching up properly. This is trickier than it sounds. For example, the morpheme *re-* has two related meanings: again (as in reuse and "redo") and back (as in revert, rebut, recoil, recrimination, and so on). Again and back are distinct meanings, although of course they are related by the fact that to move over a path from beginning to end again, you have to go back to the beginning. It is the *re-* with the meaning of "again" that occurs as the *re-* in "reusable."

Another such issue occurs with the "wine" in "wine cork." Here "wine" stands via metonymy for "wine bottle," since the cork goes in the bottle, not in the wine, and the cork functions with the bottle to contain the wine. Neurally such a metonymy is accomplished via a mapping circuit that links the contents of a bottle to the bottle within a conceptual frame that specifies *why a bottle containing wine needs a cork.* The neural circuitry for "wine cork" must make use of the neural circuitry for the wine to wine bottle metonymy, including all the above knowledge.

THE COMPOSITIONAL SEMANTICS

There is more. The meaning of the word "reusable" is more than just the meanings of its parts: *re-, use,* and *-able.* Its meaning includes the details of how those parts fit together. The term for this is **compositional semantics**.

Here's what is simple and fairly obvious:

- *Re-* means "again."
- In the again frame, "Again" modifies an event, indicating that an event with the same frame and fillers has occurred earlier.
- "Use" here is a verb root, with the meaning of an action. The concept of

use is characterized by a frame, the achieving a purpose frame, where "use" specifies the means of achieving the purpose.

THE PURPOSE X-NET

The purpose X-Net has the following aspects:

- Preconditions:
  - A desire by a person and a desired state for some entity,
  - A possible reason for desiring to reach the desired state,
  - The resources required for reaching the desired state, and
  - A course of action by the person desiring to reach the desired state.
- Center: The person desiring uses the resources in the course of action
- Problems: Possible difficulties in reaching the desired state
- Fulfillment: Reaching the desired state
- Afterward: Consequences of reaching the desired state

The verb root "use" is the form that is paired with "use" in this frame. The prefix *re-* has the meaning of the concept of again.

THE SEMANTIC COMPOSITION OF *RE-* AND *USE*

The lexical construction "reuse" is the integration of all of the following:

- Meaning:
  - The purpose X-Net
  - The again frame
  - A binding of use in purpose X-Net to the current event in the again frame
- Form:
  - The prefix + verb root word structure
  - *Re-* is bound to prefix
  - *Use* is bound to verb root
- Form to meaning linkages:
  - The neural doublet linking *re-* to the concept of again in the again frame
  - The neural doublet linking the word "use" to the concept of use in the purpose X-Net

When -*able* is suffixed to a verb, the result is usually given as passive in meaning. "Reusable" would be paraphrased as "able to be reused," where the subject is the patient of "reuse." Accordingly, the verbable construction has the constraints that

- the subject of the verbable is the patient of the verb, and
- the patient has the properties that allow the patient to undergo the action of the verb by the agent of the verb.

Thus, a *reusable wine cork* would have to have the properties that would allow it to be reused.

THE POINT OF THE EXAMPLE

Cognitive linguists for more than four decades have been pointing out the remarkable, highly structured complexities of grammatical and lexical constructions, many of which require conceptual frames, conceptual binding, and conceptual integration of the sort we have just given for *reusable wine cork*. These complexities are part of the cognitive unconscious, the 98 percent of thought and language that we have no conscious access to, since they are carried out via neural circuitry. Tens of thousands of such examples work pretty much the same way: unconsciously carried out by neural circuitry.

The point of the example is that we can understand the phrase *reusable wine cork* instantly, automatically, with no noticeable effort despite all of the complex circuitry for the frames, bindings, integrations, and doublets involved. This is possible because (1) we already have fixed circuitry for the meanings of *re-*, *use*, *-able*, *wine*, and *cork*; (2) we already have the circuitry for combining them, although we have not discussed how this circuitry comes about during learning; (3) once learned, the circuit functions very quickly, on the millisecond scale; and (4) the circuit functions unconsciously, since we have no conscious access to our neural systems.

Understanding language is incredibly complex yet incredibly fast and easy. By understanding how the neural system works, we can understand why incredibly complex language is incredibly fast to use and easy to understand.

There Is No Natural Language without Meaning

Natural language is meaningful. It allows us, at the very least, to name people, objects, actions, and places and mostly to communicate, express our thoughts and understand the thoughts of others, and even, in talking to ourselves, organize and clarify our own thoughts.

Meaning is expressed and communicated via form. "Form" is a technical term in linguistics that includes speech, gestures and signs (in signed languages), and writing, whatever can be communicated overtly, that is, not hidden from perception. Form is necessary for communication. Essential to form are the ways linguistic elements are put together: the structures of words, phrases, sentences, and discourses. But the form of language, devoid of meaning, is not natural language. Meaning is an essential part of natural language. This should be obvious.[9] People do not go around speaking word sequences that are not and are not supposed to be meaningful. Nor are books, newspapers, and newsletters published consisting of word sequences that make no sense and are not supposed to be meaningful.

CONSTRUCTIONS ORGANIZE THOUGHT

We pair meanings, with the forms that express them, in "constructions," formal structures that characterize how thought is expressed in natural language. Constructions thus allow thoughts to be communicated to others via linguistic form. But in doing so, constructions automatically play a remarkable role: they organize thought. Here's how.

In the neural theory we are proposing, each grammatical construction is a special kind of neural circuit in the brain of a language user. The more any neural circuit is used, the stronger its synapses become. Each construction links two parts: a form and a meaning. When neural construction circuits are used, the synapses in both their form circuits and meaning circuits become strengthened. This effect of the use of constructions on their meaning circuits is important.

Meanings tend to be complex, and conceptual complexity is made possible by combinatorial circuits, in particular,

- gestalts bringing together semantic roles,
- bindings creating identities,
- coordinations creating relations between circuits,
- integrations creating new circuits from old,

- conceptual metaphors applying the reasoning from one brain region to structure reasoning in another brain region, and
- cascades creating an overall unity out of disparate circuits across the brain.

Each complex meaning is a circuit with a control node that activates or inhibits that circuit depending on its input. Constructions are circuits that link linguistic form circuits to complex meaning circuits. When a linguistic form is used in communication—either said, heard, signed, or written—its complex meaning structure is activated and thus strengthened as a whole.

In short, the complex meaning structure in a construction forms a "packaged" meaning determined by how the construction's form activates its meaning. When constructions vary, sometimes considerably, from language to language and over time, meanings will be organized (packaged) differently in different languages and at different time periods in the same language. That is how, by their function in expressing meaning via form, different constructions may organize meanings differently in different languages, at different times, or over different dialects in the same language.

As we observed in chapter 2, embodied cognition experiments have shown that conceptual structure, including conceptual metaphor, leads to behavior that fits the concepts. This is expected, given our neural account of conceptual thought: neural circuitry for understanding language is also used to govern behavior. Given the link between language and thought, it is no surprise that previous theorists have attributed behavior differences to language differences.

The classic example of attributing differences in understanding and behavior to differences in linguistic form comes from Edward Sapir and Benjamin Lee Whorf. At the time they wrote, detailed studies of conceptual and neural structure were not available, so they resorted to the available explanatory mechanisms of the time: language "habits of the community" and social "agreements" according to social contract theory. Here are two typical examples of their ideas:[10]

> The fact of the matter is that the "real world" is to a large extent unconsciously built upon the language habits of the group. No two different natural languages are ever sufficiently similar to be considered as representing the same social reality. The worlds in which different societies live are distinct worlds, not merely the same world with different labels attached. We see, hear, and otherwise experience very largely as we do because the

language habits of our community predispose certain choices of interpretation. (Sapir 1958, p. 69)

We cut nature up, organize it into concepts, and ascribe significances as we do, largely because we are parties to an agreement to organize it in this way—an agreement that holds throughout our speech community and is codified in the patterns of our language. The agreement is, of course, an implicit and unstated one, but its terms are absolutely obligatory; we cannot talk at all except by subscribing to the organization and classification of data which the agreement decrees. (Whorf 1940, pp. 213–14)

There is a reason why Whorf speaks of the "organization" of concepts. Whorf, in teaching field methods courses in linguistics summer schools, gave out a list of more that 100 basic concepts that he took as universal. Students, in their fieldwork on Native American languages, were required to find out how those Whorfian "universal" concepts were organized and expressed in the languages they studied. This fact about Whorf's thought and practice is often ignored in inaccurate accounts of the Whorfian hypothesis, which is taken as claiming that there is no system of basic universals of conceptual systems.[11]

As we discussed in chapters 1 and 2, there is an extensive system of hundreds if not thousands of conceptual primitives. These are organized by grammatical constructions in each language. This can explain why translation is possible, even if inexact and clunky, in many useful cases, such as scientific papers, descriptions of products online, airline schedules, plots of movies, restaurant recommendations, driving routes, and Facebook entries.

The answer to the possibilities for translation rests partly with conceptual primitives—the building blocks of meaning—and with the commonalities of experience that make conceptual primitives possible (Lakoff 1987b, chap. 18). Moreover, as trade, travel, and communication become more global, experiences across cultures and the conceptual framings of them become more commonplace. Since words are defined relative to conceptual frames, increasingly common frames permit greater possibilities for translation. Examples abound: baseball (from the United States) in Japan and Latin America, yoga (from India) in the United States, French wine culture in China, lattes and bagels, and pizza and Big Macs everywhere in the United States and many other countries. The use of cellphones and computers spread the same images everywhere. Conceptual frames characterize social and cultural institutions and govern lived behavior. Conceptual frames make use of the universal conceptual primitives, with different frames organizing universal primitives in different and complex ways. The complex structures organizing those primitives in specific languages and cultures are different even though those primitives may be the same across languages and cultures.

It has been found empirically that some parts of the grammars of languages use more basic ideas than others, where "basic" refers to the use of conceptual primitives in the most common constructions in a language. We have seen that it is commonplace for constructions to build on more basic constructions. Part of what makes a construction basic is that its meaning uses conceptual primitives. Another part is that its form uses forms that are more "central" or "core" to the language. Although conceptual primitiveness and formal centrality are real and important in the constructional structure of a language, those constructions constitute only a relatively small part of the full constructional richness of natural languages. But basic cases are where we have to start in an overall discussion of constructions.

BASIC SEMANTICS AND SIMPLE CLAUSES

In 1965 two extraordinary young linguists, working independently, came up with essentially the same insight:

- The grammar of simple clauses reflects the semantic content of basic experiences and actions.

The linguists were Charles Fillmore (1968), who called the semantic roles "case roles," and Jeffrey Gruber (1965), who called essentially the same semantic roles "thematic roles." Their idea begins with everyday embodied experience as understood in terms of certain basic semantic roles. The semantic roles included agent, patient, experiencer, theme, recipient, benefactor, instrument, location, direction, source, goal, purpose, impediment, time, duration, causation, change, states, properties, aspect, and so on. The corresponding grammatical relations are subject, predicate, verb, adjective, direct object, indirect object, instrument, and so on. Their basic claim was that

- *each grammatical relation* (subject, predicate, direct object, indirect object, instrument, time, and so on) in the grammatical form of a sentence *has the meaning of a semantic role structuring a basic experience.*

With hindsight, one can see this as the beginning of a systematic embodied approach to language: looking at basic embodied human experiences, looking at how they are conceptualized, and taking that embodied semantic structure as basic to simple clause structure. Fillmore, after 1975, reframed the semantics of simple clauses to fit his theory of frame semantics, using the term "semantic role" in place of "case role."

With further hindsight, we can see the Fillmore-Gruber approach as bringing the concept of the frame into research on linguistic form, with subject, direct object, indirect object, and so on as grammatical roles in a frame for a clause. From a neural perspective, such hierarchical structures are characterized by gestalt circuits.

With still further hindsight, we can see this idea as explaining why there are more and less basic constructions and, correspondingly, (1) why less basic constructions are put together using the more basic constructions, as can be seen in the case of the unpassive adjective construction, the why not construction, and the amalgam constructions; and (2) why meaning—mainly pragmatic meaning—enters into these constructions.

THE UNPASSIVE ADJECTIVE CONSTRUCTION

Consider a situation in which there is a transitive event with a desired, expected, or normal outcome describable by a passive verb. Here are some examples: *The pregnancy was planned, The facts were accounted for, The bill was paid, The crime was solved, The coffee was poured, The book was read*, and so on.

Now consider the opposite case, where the desired, expected, or normal outcome does not occur. It can be expressed by the *un*+passive construction: *the unplanned pregnancy, the unaccounted for facts, the unpaid bill, the unsolved crime, the unfinished breakfast, the unread book*, and so on.

THE WHY (NOT) CONSTRUCTION

The form is why (not) verb phrase, with no subject or auxiliary. The understood subject is the addressee. The context is: the addressee either is (in the case of *why*), or is not (in the case of *why not*) carrying out or intending to carry out the content of the verb phrase:

- The *why* questions the intention for an action, suggesting that if the addressee does not have a good reason, the intended action should not be done. *Why paint your house purple?* suggests that you shouldn't do it unless you have a sufficient reason.
- *Why not* questions the lack of intention for an action, suggesting that

the action should be done unless you have a good reason for not doing it. *Why not paint your house purple?* suggests that you have no good reason for not doing it.

Note that *the positive **why** conveys a negative suggestion*, while *the negative **why not** conveys a positive suggestion*. The conveyed positive or negative suggestion has a grammatical effect, which can be seen with negative polarity items (such as "ever" and "a red cent"), which normally go with negatives but not positives. Normally the positive form (without a "not") conveys a negative suggestion, while the normally negative form (with a "not") conveys a positive suggestion.

Now an extremely important question arises. Which has a grammatical effect, grammatically present negatives (with "not" in the sentence) or indirectly conveyed negatives (without a "not" in the sentence)?

This question can be answered by a simple test. There are what are called **negative polarity items** that occur with simple negative sentences but not with the corresponding positive sentences. Common examples of negative polarity items are "ever" and "a red cent," as in

- *I did **not ever** see him there* but not **I **ever** saw him there.*
- *Harry did **not** give **a red cent** to charity* but not **Harry gave **a red cent** to charity.*

What governs the occurrence of negative polarity items is not the negative *grammatical form* of the sentence (with an overt *not*) but rather the *contextually conveyed negative* that functions grammatically (without an overt *not*). Note that a positive *why* question conveys a negative, while a negative *why* question conveys a positive:

- *Why **ever** go to an opera?* but not **Why not **ever** go to an opera?*
- *Why give him **a red cent**?* but not **Why not give him **a red cent**?*

Why do there exist such negative and positive polarity items?

One answer is that it is polarity items that constitute rhetorical devices for emphasis, which may be a way for a speaker to communicate emotions and degree of certainty. While this is true, it is not a sufficient explanation given the variety of such polarity items. More likely, the use of negative and positive polarity items is a speaker's signal to influence a decision by the hearer. "Why give John a red cent?" indicates that given the shared knowledge between the speaker and the hearer, the speaker does not want the hearer to give John any money at all. If the hearer wants to give John money, there better be a very good justification, which the hearer should then justify.[12]

In this construction, the main clause is a simple event with unfilled modifier roles. What fills the modifier roles are *exclamations about pragmatically conveyed extreme phrases*. To understand these cases better, let us start with the main clause schema, with blanks (__) where the modifiers go:

- *John invited __ (number of) people to a __ (kind of) party for __ (some) reason.*

Here is a classic amalgam sentence with the blanks filled in:

- *John invited **you'll never guess how many** people to **you can imagine** what kind of a party, for **God knows what** reason!*

Note (1) that the boldface subordinate clause fragments are normally fragments of main clauses and (2) that the main clause is understood as the content of the absent subordinate clause. Here are the corresponding sentences with the boldface fragments as main clauses:

- *You'll never guess how many people John invited to his party.*
- *You can imagine what kind of a party John invited a lot of people to.*
- *God knows why John invited a lot of people to a wild party.*

In our theory, this amalgam sentence could be formed by a cascade circuit consisting of three integration circuits. Each integration would consist of two meanings, two forms, together with two doublets, each linking a single meaning to a single form:

- The meanings:
  - Meaning 1 is an exclamatory speech act about an extreme value for a semantic role in some embedded content.
  - Meaning 2 is the content of a single simple statement, with the value of a role left unfilled.

- The forms:
  - Form 1 is the form of a single simple sentence for expressing meaning 2.
  - Form 2 is the form expressing meaning 1.
- The doublets:
  - Doublet 1 links meaning 1 to form 2.
  - Doublet 2 links meaning 1 to form 2.

These integration networks fit both speech production and speech understanding. However, production and understanding require different cascades. Each cascade is an X-Net with the structure start–center (a loop)–finish. In these cases, the loop operates three times: (1) one for each meaning to be produced in speech and one for each form to be understood in sentence processing of (2) inputs and (3) outputs.

In the general amalgam construction, the number of amalgamated meaning-to-form pairings is one or more.

The point of this example is to show the elements of meaning, form, and linking that we have discussed so far in this book come together naturally to form a construction that has vexed many traditional grammatical theorists.

THE BASEBALL EXAMPLE

Of course, one does not have to get so exotic to see the role of meaning in grammar. Consider a common sentence type in the language of baseball:

- *Lowrie doubled to left.*

Here "to left" is short for "to left field." Such a *to* phrase is a directional adverb. Directional adverbs go with someone or something moving in that direction, as in *Lowrie ran to left field*, where Lowrie is the mover, run is the motion, "left field" is the destination, and "to left field" is an adverb expressing direction.

But in *Lowrie doubled to left*, "double" is not a verb of motion, and Lowrie is not going to left field. The mover and the motion are part of the *meaning* of "double": it is the ball that is moving, but the ball is not mentioned overtly in the sentence and is understood via the meaning of "double." *The grammar of the directional adverb is determined not by the grammar of the sentence but instead by the meaning of the verb!* In addition, the shortening of "to left field" to "to left" only occurs within the baseball game frame. For example, you can say *The groundskeeper mowed left field* but not **The groundskeeper mowed left*.

Even such a simple sentence as *Lowrie doubled to left* shows the dependence of grammar on contextual meaning.

CONSTRUCTION GRAMMAR: MEANING INHERENT IN LANGUAGE

Following the initial evidence for embodied cognition in the mid-1970s, linguistics researchers who had been studying meaning in grammar since the 1960s joined the then-nascent cognitive science movement and began developing cognitively based approaches to meaning in grammar known as **construction grammar.**

The basic idea behind construction grammar is that we have conceptual systems characterizing conceptual thought, and language is a matter of pairing linguistic form with meanings from conceptual systems. In speaking, we express conceptual thought in linguistic form. In understanding, it is the reverse: we interpret linguistic form as communicating conceptual thought. The relation must go both ways: thought to form and form to thought. Hence, the idea of language as "pairing" conceptual thought and linguistic form seems natural. In addition, constructions themselves can form composites in new ways, allowing us to create new linguistic expressions and new ideas corresponding to them.

The first major papers on this were Lakoff (1984; 1987b, Case Study 3) and by Fillmore, Kay, and O'Connor (1988). To get a feel for why meaning is part of grammar, consider the following example:

- *Here comes the bus.*

The meaning includes a speech act frame with the following context:

- The speech act is *pointing out.*

In this context, there is a speaker, an addressee (or more than one addressee), a location, an entity in that location, and a motion. The location is the perceptual field of both speaker and addressee. The speaker is directing the attention of the addressee to the location. The entity (the bus) is moving toward the speaker and the addressee.

The meaning of the sentence determines constraints on the grammar of the sentence:

1. The sentence cannot be negated: *Here doesn't come the bus.
2. There is no question: *Does here come the bus? *Comes here the bus?
3. There is no past or future tense: *Here came the bus. *Here will come the bus.
4. There is no tag question: *Here comes the bus, doesn't here. *Here comes the bus, doesn't it.

Since (1) the motion of the bus is present in the context, a negation would contradict the context. Since (2) the speaker recognizes that it *is* happening, a question as to whether it is happening would not fit the context. Since (3) it is happening now, a past or future would be ruled out by the context. And since (4) it is certain in context, a question tag indicating uncertainty would not fit.

THE GOLDBERG CASES

The fullest account of constructions by a linguist came in Adele Goldberg's (1995) book *Constructions*. In her classic study, Goldberg points out that the pairing of meaning with form in constructions makes a subtle but very real distinction:

- The caused motion construction has the meaning: an exertion of force on an entity causes the entity to move to (or from) a location.

The caused motion construction contains the form:

- verb–noun phrase–prepositional phrase

In the construction, the verb expresses the exerted force. The noun phrase expresses the entity that the force is exerted on. And the preposition phrase expresses the path of motion. An example is *Sally pushed Harry into the kitchen*.

Goldberg points out that this construction, which includes its meaning, is not determined just by its form. The same form can occur with a different construction, the direct object construction, which has a verb expressing the exertion of force on an entity but with a following prepositional phrase that does not indicate motion, as in *Sally kicked Harry in the kitchen*.

It follows that the caused motion construction and the direct object construction are different constructions and that the caused motion construction is not just the direct object construction with some added preposition phrase that does not characterize caused motion.

Purely syntactic theories of grammar, based on form alone and without meaning-laden constructions, would claim that it is syntactic transitivity that allows a verb to occur in positions before a noun phrase regardless of whether caused motion is involved.

A theory of grammar incorporating meaning via constructions claims the opposite, that *the caused motion in the meaning of the construction* would distinguish the caused-motion case of the form verb–noun phrase–prepositional phrase from a non–caused-motion case.

The matter, Goldberg observed, can be decided by finding a verb that is intransitive, that is, a verb that normally occurs without a direct object and normally cannot occur with a direct object but is understood as describing an exertion of force.

"Sneeze" is such a verb. In *She sneezed*, "sneeze" can occur as an intransitive verb without a direct object. In normal contexts and simple sentences, "sneeze" does not take a direct object. When you sneeze, you emit a nasal fluid. But the name of that fluid cannot occur as a direct object of "sneeze":

- *He sneezed snot.*

Even though a sneeze actually does expel air containing nasal fluid, that fact is not expressed overtly in linguistic form:

- *He sneezed air with nasal fluid in it.*

But when you sneeze, you forcefully release air from your nose. So, with a very light object, the force of sneezing can move that light object, as in

- *She sneezed the tissue off the table.*

It is the semantic property—the meaning—of "sneeze" (forcefully releasing air from the nose), not its syntactic intransitivity (failure to take a direct object such as "snot"), that determines its grammatical function in the sentence! Note the ungrammaticality of the simple direct object construction:

- *She sneezed the tissue.*

The intransitivity of "sneeze," together with the lack of an expression of caused motion, rules this case out.

Why is "sneeze" intransitive in form? Because normally objects are not moved around by sneezing! In the overwhelming range of experienced cases, there can be no direct object that would make sense. In most cases—virtually all—"sneeze" would be used intransitively, and via neural optimization (overall least energy use), "sneeze" would be learned as *intransitive* in form.

Neural optimization also applies to caused motion construction cases. Caused motion is so very common that the pairing of the idea of caused motion with its formal expression is learned separately.

Goldberg's tissue example does not just work for tissues. Her general principle is productive, as shown by new cases that would work by her principle, made-up cases where sneezing causes motion:

- *Paul Bunyan was so big that he once sneezed a forest into the neighboring river.*
- *Superman accidentally sneezed the Golden Gate Bridge into San Francisco Bay.*

But this does not work for simple intransitivity with a simple direct object, where form constraints, not semantic constraints, matter:

- **Paul Bunyan sneezed a forest.*
- **Superman sneezed the Golden Gate Bridge.*

Investigators using the techniques of psychological experimentation and statistics have no direct way to study neural optimization. In such experiments, what we see as neural optimization effects are known as cases of **statistical preemption**. That is, *considerably greater normal use in one structure overrides uses in a minimally competing structure*. Thus, the overwhelming use of "sneeze" as intransitive in form preempts the possibility of its being used as a transitive in form (followed by just a direct object, as would occur in the ill-formed **She sneezed the tissue*). But this does not preempt the possibility of the meaning of "sneeze" as a producer of force; instead, a sneeze is insufficient to effect a change of form or location in almost all everyday situations. Goldberg's claim is that statistical preemption rules out **She sneezed the tissue*.

A CRUCIAL DISTINCTION

Meaningful thought and linguistic form are grounded in different ways, as we have seen. Each is characterized via neural circuitry and is subject to neural optimization and generalizations. But linguistic form alone is not language. The distinction is crucial.

Neural optimization applies to meaning alone, but meaning alone is not language. Neural optimization applies to form alone. But form alone is not language.

In language, form is paired with meaning: form expresses meaning, and meaning is expressed via form. This two-way relation is carried out by constructions, which pair meaning and form. *Neural optimization applies separately to constructions, since they constitute a separate class of neural circuits.* In short, constructions have a neural life of their own.

A GENERAL POINT ABOUT NEURAL OPTIMIZATION

Neural optimization applies in the learning of circuits, that is, in the process of neural recruitment. Once circuits are recruited and made permanent, they can be activated with minimal neural resources. And the more they are activated, the stronger their synapses become, and more likely they are to become permanent and to continue to be activated with minimal neural resources.

This has a very important consequence for natural language. Languages are passed on to children generation after generation. Children have a great deal of conceptual fixed knowledge (e.g., primitive concepts, basic frames, primary conceptual metaphors) *before* they learn language. They also learn language piecemeal, as Michael Tomasello (2003) observed in his classic verb island research: the grammar of verbs is learned at first one verb's grammar at a time and then generalized. As Dan Slobin showed in even earlier child language research, bilingual children (in Serbo-Croatian and English) learn grammatical morphemes earlier than grammatical words with the corresponding meaning. Thus, there is a stage at which such a child will know the meaning of the Serbo-Croatian morpheme before learning the corresponding English word with the same meaning. For such English words, meaning is learned prior to form and prior to form-meaning constructions. Research by Jean Mandler (1992) also indicates that children learn image schemas well before they learn language for them.

This means that in the learning of constructions, when the form is paired with an earlier-learned meaning, the form can be neurally optimized, that is, omitted or shortened on the basis of the already optimized meaning. This leads to cases that linguists have called "unmarked," "uninflected," "deleted," "zeroed," or "null." We will discuss such cases as we go along.

As we discussed in chapter 3 in the section "Lateral Inhibition and Winner-Takes-All Circuits," a comparison circuit, or winner-takes-all circuit, is one where the nodes mutually inhibit one another. The largest input inhibits all smaller competing inputs and is the only node remaining active.

Similarly, there can be convergence zones with competing, mutually inhibitory inputs. Imagine such a convergence zone, where each input has a different output. If the inputs occur at the same time, the strongest wins, and its output is chosen. The strongest could have become strongest via neural optimization.

Grammar involves competitiveness. For example, adjectives can occur in the predicative construction (*the boy is sleepy*) or in the modifier construction (*the sleepy boy*). These are minimally different options for adjectives. But some adjectives can only occur in the predicative construction: *the boy is asleep* but not **the asleep boy*. Other adjectives have the opposite distribution: *the mere child* but not **the child is mere*. Either these two constructions have subtle semantic differences that appear to determine where the adjectives can occur, or the adjectives act as exceptions to everyday constructions, with "asleep" and "mere" being exceptions to different constructions.

NEURAL MODELING ENTERS LINGUISTICS

The idea that a grammar using semantic roles is a "processing grammar" operating over time is an old idea first introduced in a computational form by computational linguist William Woods (1970) in his "Transition Network Grammars for Natural Language Analysis" and adapted by Lakoff and Thompson (1975) in what they called "cognitive grammars." The neural version of embodied cognition entered the picture in 1992 with Narayanan's research on coordinated control, X-Nets, and linguistic aspect.[13] It was not until the late 1990s with Narayanan's dissertation (1997a) that neural computation and simulation semantics entered in a serious way. These ideas have been integrated together in recent years in a computationally sophisticated way by Feldman et al. (2009), Bergen and Chang (2005), and Bryant (2008), what they have called "embodied construction grammars." This research has become even more sophisticated with incorporation of Sullivan's (2013) insights on where in grammatical constructions the sources and targets of conceptual metaphors can occur.

NEURAL INTEGRATION IN CONSTRUCTION GRAMMAR

Conceptual integration was brought to the fore in cognitive linguistics by Gilles Fauconnier and Mark Turner (2002) in their classic book *The Way We Think*. In each of the cases we will be discussing, *neural* integration is necessary in order to state certain generalizations about *conceptual* integration.

George Lakoff (1963) in his first paper, "Toward Generative Semantics," considered the following cases of integration:

- *The liquid is cool.* Coolness is a state. "Cool" is an adjective that expresses a state.
- *The liquid cooled.* Cooling is a change of state, ending in coolness. "Cool" is an intransitive, inchoative verb, expressing a change to a state.
- *The scientist cooled the liquid.* Here we have causation. The scientist is causing the cooling change in the liquid, which results in the liquid being in a cool state. It is an example of a caused change to a state.

There are two cases of semantic integration. In *The liquid cooled*, the cooling—the change to a state—is understood as an integrated whole. In *The scientist cooled the liquid*, the act of cooling by the scientist—the cause of a change to a state—is also seen as an integrated whole.

Lakoff analyses cause, change, and state as conceptual primitives that semantically combine in two separate processes: a change and a cause of a change. To get the integrated understandings of the sentences, he posits two "transformations" that perform integrations that are both conceptual and linguistic at the same time. The first creates an integrated change to a state, with an integrated intransitive verb "cool" formed from the adjective "cool." The second creates an integrated cause of a change to a state expressed by the transitive verb "cool" formed from the intransitive verb "cool." The "transformations" operate generally on the conceptual level: state to change to a state and change to a state to caused change to a state, independently of any particular state. The choice of the lexical item, for example *cool* or *warm*, can occur at the state level, with the same words with augmented meanings carried along by the integrations state with change and state with change with cause.

The idea was that that the integration transformations were fully general and occurred at the conceptual level, creating integrated concepts from separate conceptual primitives. "Kill" could then be seen as an integrated form of cause to change to be dead, made up of the three primitives cause, change, and state, with dead being the state in question. Jerry Morgan (1969), considerably later, argued in favor of this analysis by showing that the adverb "almost" can pick out the individual primitives cause, change, and dead. His examples were like the following:

> **Almost cause**: *I almost killed him, but my shot just missed him.*
> **Almost change**: *I almost killed him by poisoning him, but he happened to have an antidote to the poison he drank, and the poison had no effect.*
> **Almost dead**: *I almost killed him with a shot to the heart, but the doctors operated on him and saved him.*

In the theory of neural integration that we are proposing, the integrated idea and the unintegrated ideas that comprise it are all present: the integrated idea is formed by integrating neural bindings with an integration node. A state is a single occurrence, as in *The tea is cool*. In *The tea cooled*, the same word ("cool") is used. The same single word is used for a single event: the cooling. Conceptualizing (via integration) the X-Net as a single overall integrated event makes "cool" the verb expressing the event.

When the change is caused via an X-Net, the progressive cooling is in the process of being caused at each stage of the X-Net. At each stage there is a causal result in progress: the progressive cooling. The cause of change is conceptualized via integration as a single event, just as the change is. "Cool" is the best-fitting verb (taking the least energy, since the same form is used for change of state as for state) to express the cause of the change of state.

When a neural integration occurs, the separate unintegrated ideas correspond to the circuits being integrated. "Kill" names the integrated node, which activates all the unintegrated nodes of causation, change, and the final state: dead. "Almost kill" is a gestalt structure in form with two roles, the adverb "almost" and the verb "kill," which is the head of the construction. As Morgan (1969) observed, "almost" can modify either the integrated causation, the change, or the state, all of which are present in the integration.

INTEGRATION AND INCORPORATION

Another early example of integration occurred in Jeffrey Gruber's (1965) master's thesis "Studies in Lexical Relations." Gruber's form of integration came in lexical rules, in which two or more primitive concepts could be "incorporated" into a single word. For example, he considers the word "pierce" as in *The knife pierced the cardboard.* He argued that the verb "pierce" incorporates both the concept "pierce" and the concept "through." He also argues that the verb "cross" as in "He crossed the street," incorporates the concepts "across" and "motion." Gruber's theoretical idea of incorporation is *a mapping to a word from a prelexical sequence of concepts.* For Gruber, the verb "ascend" incorporates motion plus up and on. If you ascend a staircase or a mountain, you are not just going up but are also being supported by the staircase or mountain and are therefore on it.

NEURAL INTEGRATION IN GOLDBERG'S *CONSTRUCTIONS* BOOK

Adele Goldberg (1995) in her *Constructions* book assigns to the word "give" the "semantically fused" integrated meaning "cause-receive." In chapter 2, we discussed this basic construction as an example of language-mediated integration. Here we will provide a more detailed analysis of the construction. For *She gave Harry the book*, Goldberg provides the following pairing of semantic roles and grammatical relations as the cause-receive construction:

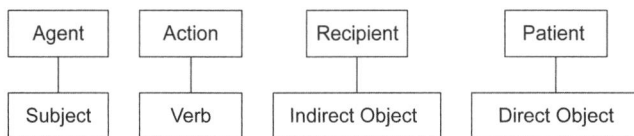

Agent	Action	Recipient	Patient
Subject	Verb	Indirect Object	Direct Object

The pairings of meanings and form are indicated by vertical lines in the cause-receive construction. Not shown is the integration node I that integrates all of the form-meaning pairs to create the construction as a whole.

Goldberg proposes that the prototypical meaning of "give" is X causes Y to receive Z. She then seeks to explain why many apparently disparate verbs occur with the same verb–direct object syntax as the verb "give," verbs such as diverse as "promise," "permit," "refuse," "bake," and "bequeath."

THE CENTER IS GENERAL, AND THE RADIAL IS SPECIFIC

Goldberg goes on to use the principle that specific cases of a general case activate the general case. But she adds a new idea: We know that radial categories are made up of variations on a central prototypical case. She proposes that *in a radial category, the central prototype is the general case and the radial varia-*

tions are specific cases. Hence the radial variations will activate the prototypical general case and the predicates that occur in the prototype.

Goldberg then observes that the predicates that occur in the prototypical case (cause and receive) also occur with meanings that are radial variations on the central prototypical case, namely "give" as it occurs in example 1 below:

1. X causes Y to receive Z (central sense).
   Example: *Joe gave Sally the bracelet.*

2. Conditions of speech act satisfaction with the speech act of promising, namely that the promise is fulfilled, imply X causes Y to receive Z.
   Example: *Joe promised Bill a car.* The promise would be fulfilled if Joe gave Bill a car.

3. X enables Y to receive Z (enabling makes future causation possible).
   Example: *Joe permitted Chris an apple.* Joe's permission enabled Chris to get an apple.

4. X causes Y not to receive Z.
   Example: *Sam refused Bob a cookie.* Bob's receipt of the cookie depends on Sam's permission, which is denied.

5. X intends to cause Y to receive Z.
   Example: *Mary baked Bob a cake.* Mary intended the cake for Bob.

6. X acts to cause Y to receive Z at some future point in time.
   Example: *Joe bequeathed Bob a fortune.* Joe specified in his will that after his death Bob would receive a large amount of Joe's money.

Goldberg then gives a polysemy explanation implicitly invoking the radial category analysis of polysemy from Lakoff (1987b), Talmy's (1988) "causes are forces," metaphor and Fillmore's (1982) frame semantics.

- In example 2 above, there is a commitment, and in the framing of commitments, the social norms exert a metaphorical moral "force" to satisfy the commitment.
- Example 3 exemplifies Talmy's force-dynamic account of enablement as removing a blocking force, adding a propulsive force, or both.

- Example 4 exemplifies the well-known fact that negating a frame activates the frame, as in the title of Lakoff's (2004) book *Don't Think of an Elephant!*, which makes you think of an elephant.
- In regard to example 5, Talmy observed that an *intention* is conceptualized metaphorically as *an internal force acting to move a potential actor toward carrying out an action*.
- Regarding example 6, in Gilles Fauconnier's (1985) theory of mental spaces, a future event is a real event in a future mental space, which, in our approach, is a partial simulation of a future situation in which the event occurs. A full theory of mental spaces in terms of simulation semantics has not yet been developed at the time of this writing.

In each of these cases Talmy's force-dynamic account of causation occurs, with a special frame using force to expand upon what "force" means in each situation. The radial category approach to such variations appears in Lakoff (1987a).

Goldberg then included conceptual metaphors, applying them to semantic roles from frame semantics. For example, the conceptual metaphor "applying force to an entity is the transfer of a force to that entity," when applied to the constructional semantics of the first construction above, yields expressions such as *give someone a punch, a kick, a pinch, a push*, and so on.

In the metaphor "causing someone to be in a state is transferring an object to that person," *the state caused is the object as transferred*. When applied to the same constructional semantics, this metaphor yields examples such as *The Affordable Care Act **gave** poor patients a hope of recovery* and *His magic tricks **gave** the children pleasure*.

These analyses are in the service of maintaining certain general claims:

1. "X causes Y to receive Z" is the meaning that is paired with grammatical relations (subject, indirect object, and direct object) in the cause-receive construction.
2. The grammatical relations in a construction can be preserved even when the semantics of the construction undergoes variations.
3. The central case in a radial category is a general case with the radial variations as specific cases.
4. The general case is activated by each specific case and incorporated as part of the meaning of each specific case.
5. The verb "give" is an integration of cause and receive, and its grammar follows from that integration.

Each of the basic neural mind ideas that we have hypothesized has three aspects:

- Concepts can display hierarchical organization, sequencing, general-ization, mapping, integration, forming simulations, and so on.
- Concepts work via neural mechanisms to carry out those functions, namely proposed circuitry types in the computational model that can carry out a given function.
- Kinds of concepts are types of ideas.

Table 7 is a list of how these three aspects of neural mind theory come together in each of many cases. Examples of each will be given below as we discuss examples presented above from Adele Goldberg's (1995) book *Constructions*.

TABLE 7.   The Goldberg analysis

CONCEPTUAL FUNCTION	NEURAL REALIZATION	CONCEPTUAL CONTENT
perceptual-motor grounding	embodied schemas	primary concepts
conceptual organization	gestalt circuits	conceptual frames
conceptual coordination of processes	coordination circuits (X-Nets) to form cascades	coordinated activation of cascades and complex concepts
conceptual identity	neural binding circuits	composite concepts
conceptual mapping	mapping circuits	conceptual metonymy and metaphor
conceptual generalization	neural optimization: best fit to existing circuitry	higher-level concepts
conceptual variation	structured minimal additions and changes to existing circuitry	radial categories and polysemy
conceptual specialization	activation of general circuitry with specialized changes and additions	A more specific concept
conceptual integration	integration circuits	holistic concepts and imaginative ideas
conceptual simulation	the filling in of semantic roles in gestalt circuits and the coordinated activation of cascades	conceptualized situations and mental spaces

- Perceptual-motor grounding is used in the embodied understanding of **cause** (via our X-Net analysis in chapter 1) and **receive** (taking a transferred object into one's possession).
- This analysis inherently uses frame semantics to characterize semantic roles. Frames are **conceptual organizations**, formed *via* **gestalt circuits**.
- Giving, the action of transferring an object, is **sequential**, starting with the giver's possession, then the transfer, and finally the reception (taking into one's possession). This is carried out via a coordination circuit, which is an X-Net.[14]
- **Neural binding** occurs in the giving frame: Binding makes the object given the same entity as the object received. Neural binding also occurs in the integration in that frame, making the causing with an effect and the receiving into one single, integrated event via binding of the effect and the receiving.
- **Conceptual metaphors** neurally map from the central sense to metaphorical senses.
- The central sense is a **generalization**.
- The radial senses are **conceptual variations**.
- The variations are seen as **special cases** that incorporate the general sense and elaborate it with specific cases.
- Cause-give is a **conceptual integration**.
- The filling of the semantic roles with individuals (e.g., Joe, the ball, Sally) is a use of **simulation**. The conceptual understanding of the example sentences indicate the use of **simulation semantics**.

One of the most important topics that Goldberg takes on is the existence of intermediate generalizations: constructions that are special cases of more general constructions and, at the same time, generalize over other more specialized constructions. This provides a hierarchy of generalizations.

The generalization hierarchy also fits the remarkable finding we mentioned in chapter 1 on generalization. The findings by Badre (2008) and Nee and D'Esposito (2016) show that general rules activate regions in the prefrontal cortex, and when there is a hierarchy of generalization, the more general the rules are, the closer to the front of the brain they are. Under our interpretation of the Nee and D'Esposito results, basic-level concepts (e.g., the basic-level verb "give" as opposed to the superordinate "cause to receive") are controlled from the midlateral prefrontal cortex and in the middle of the generalization hierarchy, which Rosch (1973, 1977) describes as the "basic level." The actual prefrontal circuitry layout is that most general is most frontal, intermediate generalization is middle frontal, and least general (most specific) is least frontal.

This general layout in the brain corresponds to other findings:

- The middle position in the generalization hierarchy fits Rosch's basic-level in the category hierarchy.
- The intermediate position in the generalization hierarchy corresponds to Goldberg's intermediate generalizations.

There is a moral here. We have been isolating separate conceptual functions and separate neural mechanisms. But in real brains, they do not apply in isolation; they apply together, since the neural mechanisms are constantly engaged and carry out those conceptual functions.

We have only discussed one of the examples from Goldberg's book *Constructions*. Her book is filled with many such examples with deep and informative discussions. *Constructions* was not written from a neural perspective. But from our perspective, the neural mind issues we have just discussed seem to apply throughout Goldberg's prescient book.

At this point, we turn to three case studies of difficult problems in the theory of language and how a neural theory can provide new insight into such problems.

Case Study 1: One-Anaphora

Anaphora is the use of a pronoun (e.g., "she," "it," "one") to refer to a prior use of a noun or noun phrase, as in

(a) Sam bought *a portrait of the queen*, and Sara borrowed *it* (same portrait).
(b) Sam bought *a portrait of the queen*, and Sara bought *one* too (different portrait).

In the 1960s and into the 1970s, it was common for grammarians to use phrase structure trees, which had just hierarchical and sequential structure without semantic roles. The exceptions were Charles Fillmore (1976) and Jeffrey Gruber (1965). The use of phrase structure trees changed with the advent of construction grammar. The difference between sentences as in the examples above was assumed to lie in the difference between (a) *identity of reference* (same entity) and (b) *identity of sense* (same type of entity).[15] This division was complicated by sentences such as

(c) I bought *a portrait* of the queen, and Sara bought *one* of Princess Di.

In (c), "one" refers only to a portrait. The natural proposal, given phrase structure grammar, was that "a portrait of the queen" had a phrase structure such as [a [[portrait] of the queen]] and that *one* can refer to either [[portrait] of the queen] or just [portrait]. This would also work for (d):

(d) I bought an expensive *portrait*, and Sara bought a cheap *one*.

This assumes the phrase structure [an [expensive [portrait]]]. Presumably, *one* could refer to either the phrase [expensive portrait] or just [portrait].

But this solution breaks down in the following examples:

(e) I bought an expensive portrait of the queen, and Sara bought one too.
(f) I bought an expensive portrait of the queen, and Sara bought a cheap one.
(g) I bought an expensive portrait of the queen, and Sara bought one of Princess Di.
(h) I bought an expensive portrait of the queen. and Sara bought a cheap one of Princess Di.

Examples (f) and (g) would require phrase structure analyses such as

(f) [expensive [[portrait] of the queen]] (*one* = portrait of the queen)
     and
(g) [[expensive portrait] of the queen] (*one* = expensive portrait).

Examples (f) and (g) are contradictory phrase structures; you can't have both at once and maintain the theory that "one" refers to a single constituent in a phrase structure tree.

These cases become more complicated with more information.

1. Sam bought a large expensive portrait of the queen by Jones, and Sara bought *one* of Princess Di.
2. Sam bought a large expensive portrait of the queen by Jones, and Sara bought a small *one*.

3. Sam bought a large expensive portrait of the queen by Jones, and Sara bought a small *one* of Princess Di.

4. Sam bought a large expensive portrait of the queen by Jones, and Sara bought a small cheap *one*.

5. Sam bought a large expensive portrait of the queen by Jones, and Sara bought a cheap *one* of Princess Di.

6. Sam bought a large expensive portrait of the queen by Jones, and Sara bought a cheap *one* by Wainwright.

7. Sam bought a large expensive portrait of the queen by Jones, and Sara bought a cheap *one* of Princess Di by Wainwright.

8. Sam bought *a large expensive portrait of the queen by Jones*, and Sara bought *one* too.

9. Sam bought *a large expensive portrait of the queen by Jones*, and Sara bought a small cheap *one*.

10. Sam bought *a large expensive portrait of the queen* by Jones, and Sara bought *one* by Wainwright.

11. Sam bought *a large expensive portrait* of the queen by Jones, and Sara bought a small cheap *one* of Princess Di by Wainwright.

The antecedents of "one" in these sentences include

1. a large expensive portrait by Jones,
2. an expensive portrait of the queen by Jones,
3. an expensive portrait by Jones,
4. a portrait of the queen by Jones,
5. a large portrait by Jones,
6. a large portrait of the queen,
7. a large portrait,
8. a large expensive portrait of the queen by Jones,
9. a portrait of the queen by Jones,
10. a large expensive portrait of the queen, and
11. a portrait.

How can we characterize a general rule that accounts for exactly what the antecedents of "one" are in these sentences?

To approach the problem, think not of a grammatical tree for the first noun phrase but instead of a semantic frame for portraits that includes the following semantic roles: size, cost, the person painted, and the painter. Size and cost are type roles, filled by type values: large and small for size, expensive and cheap for cost. The portrait frame's semantic roles include painter and the person painted.

In sentence 11 above, there is maximal difference in the fillers of the frame roles:

1. Sam bought a large expensive *portrait* of the queen by Jones, and Sara bought a small cheap *one* of Princess Di by Wainwright.

antecedent—**size**: large; **cost**: expensive; **subject**: the queen; **painter**: Jones

anaphor—**size**: small; **cost**: cheap; **subject**: Princess Di; **painter**: Wainwright

In the other examples, there is only a partial difference or none at all, as in sentence 8.

Here is what is true of all examples:

1. The frame of the "one" phrase has the same semantic roles as the frame of the antecedent. That is, it is the same frame, not counting the fillers of the roles.
2. Where the fillers of the "one" phrase frame are different from the antecedent frame, the fillers of the "one" phrase take priority and stay.
3. The conceptual fillers of the "one" phrase frame are copies of the conceptual fillers of the antecedent frame. Those conceptual copies, being redundant, do not appear in linguistic form; that is, they are linguistically null.

This statement covers all the cases, but *why* is it true? How would we characterize this truth neurally, and would a neural account explain why this occurs?

Anaphora is an energy-saving phenomenon that uses a short linguistic form (e.g., "it," "one") to express some sameness or similarity of thought between a prior thought and one being processed right now. Processing at the moment requires more attentional resources (hence, more neural firing) than what is past or anticipated. Speakers and writers signal reuse of the previous frame bindings *as much as possible*. To express the same referent (i.e., the *same individual*), identity-of-reference pronouns such as "he," "she," "it," and "they" are used. To express identity of sense (i.e., the *same type* as the antecedent), either one-anaphora or an omission is used. A typical omission occurs in sentences such as, *Bill bought a cheap portrait of the queen, and so did I / just as I had / and I did so too.*

The neural version of one-anaphora is as follows:

1. Previously activated frames remain somewhat active (with a decay over time) and are primed (i.e., partly active) for a while, although they are less active than those that one is currently paying attention to. This is typical of short-term memory and is useful for many recency-based phenomena in language and thought.

2. In the default case, "one" evokes *the same frame as the antecedent*, and the values for the antecedent frame are still active but a bit decayed.

3. Mutually exclusive values for a frame role participate in a winner-takes-all network, which is defined by mutual inhibition circuitry and has the effect of prohibiting two different fillers from occurring simultaneously for the same role.

4. The *most recent fillers* for semantic roles are the most active and *will win*, while the competing values among the antecedent values will lose.

5. For example, the explicit value provided in the "one" phrase will be at a higher activation level for the particular slot (such as the "cheap" in sentence 4 in the previous section) than the primed (decayed) value ("expensive" in sentence 4). The anaphoric value will naturally inhibit the earlier value (now having decayed and therefore less activation) for the cost role in the frame.

6. The rest of the frame fillers (the part not inhibited by the new role values) will now be conceptually activated with the "one" anaphor but will not appear in linguistic form, since they are redundant and the expenditure of energy on them is not needed.

The point is that the neural theory we are proposing explains all of this.

From the perspective of our theoretical approach, one-anaphora looks like the polysemy of "give" that we have just described in the Goldberg case. In each, there is a default case:

- For Goldberg, the default is the central sense of "give" as X causes Y to receive Z.
- For this one-anaphora case, the default is the antecedent frame with its fillers:

antecedent—**size**: large; **cost**: expensive; **painted**: the Queen; **painter**: Jones

As Goldberg observes, what we call the "default" acts like a generalization in an inheritance hierarchy, and the variations on the default act like special cases, that is, using as much of the generalization as possible in its meaning and substituting part for part only when the special cases are incompatible with the general case.

In the Goldberg case, *the substitutions are variations on the central, general case* use of force dynamics for cause to receive: for example, "bequeath" as a future use of cause to receive, and "promise" as a commitment to cause to receive. With one-anaphora, *the antecedent frame with its fillers is the default and functions like a general case, and differences with anaphoric **one** are variations on the default case.* When special case anaphoric fillers differ from those of their antecedents, they substitute for the corresponding general case fillers: for example, "cheap" for "expensive" and "Princess Di" for "the queen."

The neural account is the same for both and explains both cases. *Defaults and variants function just as do general cases and their special case variations.*

Why do the filler values for anaphoric "one" replace the corresponding antecedent filler values? Because different fillers for the same role in the same frame are mutually inconsistent. Neurally, they are mutually inhibitory. Anaphoric "one" neurally activates its antecedent's frame. The antecedent frame then acts as "one's" frame except for the mutually inhibitory cases!

Why do "one's" fillers win and override the antecedents conflicting fillers?

1. They have present attention and hence somewhat greater activation than the prior frame, which has past attention and hence less activation due to activation decay.
2. Conflicting fillers for the same role are conceptually mutually exclusive and neurally mutually inhibitory.
3. In a circuit defined by mutual inhibition, the winner is the node with the greatest activation. The result is the antecedent's frame and frame roles with the "one" frame's fillers substituted in place of the inhibited frames' fillers that have decayed from lack of present attention.

The explanation and comparison is somewhat long and complicated. But from a neural perspective, this is simple and happens automatically within milliseconds.

Very complex linguistic behavior such as one-anaphora can arise from very basic low-level neural phenomena such as activation decay and mutual inhibition.

Case Study 2: Cascades of Metaphors

IDEAS ARE STRUCTURED

One of the themes of this book is that thought is structured, and that the neural circuitry constituting ideas and thinking is correspondingly structured. Such structure comes from elementary forms of embodiment, that is, from primary schemas, X-Nets and coordination circuits, the physiology of emotions, and so on. The structure of thought also arises from the structured neural compositional mechanisms for forming complex neural circuitry: neural binding, neural integration, and cascades.

A corresponding theme is that neural integration, which is needed for consciousness, makes much of the neural structure of ideas inaccessible to consciousness once they are integrated into wholes with inaccessible parts. Moreover, the speed and ubiquity of neural integration means that *most conscious intentional behavior is the result of the integration of complex circuits* and that *it is relatively rare for structural neural **elements** to occur isolated—by themselves—in normal, everyday consciously intended and controlled behavior.*

However, language is a major form of behavior in which elementary conceptual structure does make an appearance. This is not always an obvious or straightforward appearance but enough of one so that cognitive linguists have been able to disentangle them sufficiently that the conceptual structures in language become accessible.

The task of the following section is to provide some examples of metaphor composition and disentanglement and to show how the combinatorial complexity of ideas that is unconscious and usually goes unnoticed can appear in language.

In 2012, the 111th US Congress failed to pass a federal budget. Given existing laws, unless something was done there would have been an automatic broad across-the-board cut in federal spending and a tax increase. Ben Bernanke, chair of the US Federal Reserve, predicted a severe economic decline, which he pictured as going over a "fiscal cliff."

The American people became focused on this metaphor, understanding it as Bernanke intended. Although it sounds simple enough, the "fiscal cliff" metaphor is actually complex, since it makes use of other metaphoric concepts neurally bound together. This is a simple case of the kind of phenomenon that a theory of neural cascades has to deal with.

Let's take a look at the metaphorical complexity of "fiscal cliff" and how the metaphors that comprise it fit together. The simplest is the metaphor named "more is up," which we hypothesize is a neural circuit linking two distinct brain regions, one for verticality and one for quantity. It is a primary metaphor widespread throughout the world and occurs in a vast number of sentences such as *Turn the radio up*, *The temperature fell*, and so on.

The economy is metaphorically seen as moving forward, from the past toward the future, and either moving up, moving down, or staying level. Here verticality metaphorically indicates the value of economic indicators such as the gross domestic product (GDP) and a stock market average. These are indicators of economic activity such as overall spending on goods and services and the sale of stocks. Why is economic activity conceptualized as motion? Because a common conceptual metaphor is being used: "activity is motion," as in sentences such as *The project is moving along smoothly*, *The remodeling is getting bogged down*, and so on. The common metaphor "the future is ahead" accounts for why the motion is "forward."

In a diagram of changes over time in a stock market or the GDP, the metaphor used is "the past is left and the future is right," which is why the diagram goes from left to right when the economy is conceptualized as moving "forward."

When Federal Reserve chief Bernanke spoke of the "fiscal cliff," he undoubtedly had an mind a graph of the economy moving along, left to right, on a slight incline and then suddenly dropping way down, which looks like a line drawing of a cliff from the side view. Such a graph has values built in via the metaphor "good is up." Literally going over a cliff usually results in great harm. Via the metaphor that economic harm is physical harm (as in *The stock market took a hit*), going over the fiscal cliff would be very harmful.

The administration has the goal of increasing GDP. Here common metaphors apply: "success is up" and "failing is falling." Hence, going over the fiscal cliff would be a serious failure for the administration and harm for the populace.

These metaphors are part of a cascade circuit in which the metaphors fit together tightly with the graph of changes in a linear scale over time. This cascade circuit allows us to understand a wide range of graphs. In this case, the linear scale characterizes the GDP, the metric of national economic activity. From the neural perspective, these metaphors form a tightly integrated neural cascade, so tightly integrated and so natural and normal that we barely notice them if we notice them at all.

There is, of course, more content to the "fiscal cliff." Imagine driving toward a cliff with the possibility of going over. The car you are in is out of control. The cliff is a feature of the natural environment. If the car goes over, everyone in it would be harmed or killed. Thus, if the economy is a vehicle moving forward without control toward the cliff, there is great and immediate danger, and so the "fiscal cliff" metaphor engenders fear. Thus, knowledge about driving out of control toward a cliff, together with the metaphors cited above, characterizes the implications of the "fiscal cliff" metaphor. We get these implications through the normal process of doing a mental (imaginative) simulation of going over the cliff.

Yet, for all this complexity, the metaphor was understood by hundreds of millions of Americans. How? That's how cascades of metaphors work; *they are learned as a single cascade of metaphors.*

And that complex understanding can be triggered by only two words: *fiscal cliff.*

SOME METAPHOR SYSTEMS AND METAPHORS

We will begin with four conceptual metaphor examples and how they combine to form complexes. We will the use plus sign (+) to indicate composition. The four conceptual metaphor examples are

1. Linear scales as paths of motion
2. The spatial event structure system
3. The object event structure system
4. More is up, less is down[16]

The first and fourth examples are the most basic:

1: Linear scales are paths of motion; more is ahead, and less is behind.

Examples: *Jon's intelligence goes way beyond Bill's. Jon is ahead of Bill in intelligence. Bill is lagging behind Jon in achievements. Bill's grades are catching up to Jon's. In the past year, Bill's grades have zoomed ahead of Jon's. Bill improved for a while, but now he's going backward. Bill has a long way to go to catch up with Jon.*

4: "More is up, less is down" is a specific metaphor using verticality.

Examples: *Jon is at the top of the class; Bill is at the bottom. Bill's grades are rising; Jon's are falling.*

1+4:

Examples: *Jon's grades have risen above Bill's. Bill has fallen behind Jon. Bill's grades have been dropping. Jon's grade point average this year started at 2.6 and has gone up 6 points to 3.2 but is unlikely to reach 4.*

The examples of 1 alone (without 4) indicate motion but not verticality. Example 1 has *ahead* and *behind* but not *above* and *below*. Conceptual metaphors 1 and 4 both have quantity (with more and less) in their targets. They differ in their conceptual domains (motion and verticality). But their conceptual domains are compatible, and their targets are identical. Therefore, both metaphor mapping circuits can be active at once. Since each circuit is gated, both gates can be open simultaneously.

We can see this in words. In "go up," "go" indicates motion, and "up" indicates verticality. The verbs "fall" and "drop" are both lexical integrations of motion and downward. The combination "fall behind" works because both "fall" and "behind" mean "less" but for different metaphorical reasons.

COMPLEXITIES

The second and third examples add complexity:

2: The spatial event structure system, in which states are locations, changes are movements (to or from locations), causation is forced movement, purposes are desired locations (destinations), action is motion, progress is distance increase from source to destination, difficulties are impediments to motion, lack of progress is lack of motion forward, and regress is motion backward.

Examples: *How far along are you on your thesis? Is the going slow? Are you stuck? Are you close to finishing? Is there pressure on you to make progress? Is there anything standing in your way? The finish line isn't that far away. Push yourself hard!*

In this metaphor system, the subject matter—the target—is purposeful action. The conceptual purposeful action frame that defines the target includes a desire, a purpose, actions (defining a course of action), progress, and possible difficulties. There are separate individual metaphors that make up this metaphor system. Each is a primary metaphor. Each comes into being around the world, person-by-person, by virtue of repeated correspondences in experience, given the nature of human beings and the nature of the world. In each case, we hypothesize circuits in two distinct brain regions that are activated by the distinct experiences. The activation spreads, forming a circuit. The repetition leads to synaptic strengthening by Hebbian learning. Spike-timing-dependent plasticity determines the direction of the strengthening and hence which is source and which is target in primary metaphors. The source-to-target orientation of primary metaphors is preserved in composites of those metaphors.

In daily life, satisfying purposes that arise from desires requires literally going to some destination. Whether it is an infant having to crawl to where her favorite toy is, a guy having to go to the fridge for a cold beer, a family having to drive to the shore for a day at the beach, or someone having to go to the bathroom to, we regularly have to go to some destination to achieve a purpose. Actions require motion: moving the body in place or from place to place. A difficulty is something that gets in the way of achieving a purpose. In getting to destinations, difficulties are anything that impedes motion and/or limits progress. Each of the separate primary metaphors making up the system arises by the general neural mechanism that gives rise to primary metaphors. The result is called the **spatial event structure system**. Again, all of the targets fit together in a consistent frame. And all of the sources fit together into a consistent frame. This means that any of the metaphors in the system can apply together with no problem of interference. When two or more apply together, we get a complex metaphor.

1+2: Linear scales are paths of motion and event structure.

>   **Examples:** *He's made very little progress on his thesis. He's ahead of schedule. He's behind schedule.*

There is a target frame that includes both the amount of progress required and the amounts of time and energy needed to achieve the purpose. That target frame—the amount required frame—is a binding of two frames: the purposeful action frame (the target for metaphor 4) and a frame for relative amounts (a linear scale that is the target for metaphor 1). The mapping circuit for metaphor 1 has connections to the amount part of the amount required frame, and the mapping circuit for metaphor 2 has connections to the purpose part of the amount required frame.

The context is defined by which frames are evoked and used. The context could include only the purpose frame or the amount required frame, with its neural binding active. Both are possibilities that could occur, depending on whether the gate for the binding that links the two frames is open and activated or is inhibited. In language understanding, the understood context is shown in the language used. "Schedule" in the examples above *indicates where on the path you are supposed to be at a given time.* "Ahead of schedule" and "behind schedule" indicate that you have either made more progress than required by the schedule and are "ahead" of schedule or have made less progress and are "behind" schedule. "A lot of effort" suggests more effort than is required to be on schedule and thus that one is "ahead" of schedule. "Little progress" suggests not enough to be ahead or on schedule, and thus one is behind schedule. The word "schedule" is typically used with the amount required frame applying to effort for progress, in which case the amount of progress depends on whether the amount of effort required is more or less than what is needed.

In a competition there are two or more competitors, but only one can achieve the given purpose of winning. In the race metaphor for competition, the competitors are in a race. Only one can win, the one who is ahead at the finish line. The race metaphor is usually used for a long competition in which there is a regular concern with who is ahead and by how much and who has what amount of resources left to use in the competition.

1+2+4:

>   **Examples:** *Bill's income has fallen far behind Jon's. The Dow climbed above its previous high. Stephen Vogt's batting average slipped so much that he was traded.*

More is up (example) adds a vertical dimension to 1+2 that comes with words such as "fallen," "slipped" and therefore "fell behind," and "climbed," which are all lexical integrations of 2+4. The use of "income," "stock prices," and "batting average" adds a linear scale (metaphor 1) of amounts and comparisons.

3: The object event structure system, in which attributes are possessions, changes are movements of possessions (acquisitions or losses), causation is transfer of possessions (giving or taking), and purposes are desired objects.

Examples: *He **has** a headache. He **has** a lot of intelligence. That noise **gives** people headaches. He **got** a headache. The aspirin **took away** his headache. His headache **went away**. Dementia **robbed him** of his intellect. He slowly **lost** his intellectual abilities.*

2+3:

Examples: *The table **has** a book **on** it. He **has** a lot of strength **in** his legs. He **hasn't** an idea **in** his head. He **got to** Chicago. She **got into** Harvard. John **got** a crazy idea **into** his head. **Get** me **to** the church on time.*

These are all integrations, commonplace integrations. In the two event structure systems, the targets are the same. Changes, causes, and purposes are obviously the same. A state and an attribute are both something about a person: their happiness, their confidence, their energy level, and so on. The words "state" and "attribute" are etymologically derived from their corresponding conceptual metaphors: a bounded region of space (for "state") and an object given ("tribute" plus "at," from "ad-," meaning "to" in Latin). "Properties" are metaphorically understood as possessions that one can *have, gain,* or *lose.*

Examples 2 and 3 have the same targets and different sources (location in space or object possession). There is similarity in some of the metaphorical sources. In both cases, causation is forced motion (bodily movement or object transfer). Also in both cases, a purpose is something desired, and purpose achieved for a person is being in a resulting location (via movement) or acquiring a possession (via object transfer). The phrase "get to" is *an integration of two different metaphors* for achieving a purpose: (1) a change of possession ("get") and (2) a change of location ("to"). Both are used to indicate a goal, namely achieving a purpose by either (1) metaphorically acquiring a desired object ("get") or (2) metaphorically reaching a destination ("to").

This identity of metaphorical targets for two different metaphors allows for integrations, both conceptual and lexical. English has a construction of the form "X has Y located at X," a conceptual integration of "X has Y" and "Y is located at X." The grammatical form contains both "have" and a preposition indicating a location associated with X, hence *The table **has** a book **on** it* and *He **has** a lot of strength **in** his legs*. "On" is the literal location between the book and the table, while "in" is the (metaphorical) location between his strength and his legs.

The construction that combines "get" with "to" or "into" is a particularly interesting form of integration. The metaphor used with "get" is "achieving a purpose is getting a desired object." In *He got to Chicago*, there is a subject matter (target) meaning that he was going to Chicago, which was a purpose that he achieved. The "to Chicago" is literally about motion to Chicago. The "get" indicates the achievement of a purpose, which was going to Chicago. "Get to" integrates these two meanings linguistically to form a single linguistic expression that conceptually integrates the meanings of the two linguistic expressions.

This integration works metaphorically as well. Consider *He got to the end of the book*, in which the "process is motion" metaphor applies to the process of reading. The "to" is used metaphorically, not about literal motion but instead about the metaphorical "end point" of the process (conceptualized metaphorically as motion) of reading the book (where finishing reading is *reaching the end* of the *process*).

The metaphor yielding the achievement sense of "get" has a metaphorical source domain of active acquisition. But there is also a source sense of "get" as being a passive recipient of an acquisition, not an active acquirer, as in *I got your letter*. The metaphorical version of this, via the conduit metaphor, is *Jon happened to get a crazy idea into his head* and *I got the joke*. In other words, a crazy idea metaphorically "entered" his head, with Jon as the passive metaphorical recipient. The same happens in the phrase "received ideas." Here the recipient is not doing anything active to create the ideas but instead is only receiving ideas created by others.

The cause-change integration circuit discussed above has, as a special case, the causal use of this "get." We can see it in sentences such as *Get me to the church on time*. It is an instruction to bring about the achievement of my purpose of being at the church on time.

Any single cascade circuit for both of these uses at once would use integrative circuitry, integrating two or more metaphors.

The expression "the glass ceiling" was coined by feminists to describe the widely known limits placed on women in professions that keep them from reaching full professional success. When the expression was coined, it was understood instantly by millions of Americans. Why does it have the meaning that it has? And how could it have been instantly understood by so many millions of people?

As we have seen, the integration of metaphors 1+2+4 places achieving a purpose in a competitive situation (winning) at the top of a vertical linear scale, as in *The American women's soccer team came out **on top** in the World Cup. Who is the **top** home run hitter of all time? What will be the **top** movie of the year at the Oscars?*

In any professionally run large organization, there is a hierarchy of authority. The primary metaphor it fits is the metaphor "control is up," as in *She's on top of the situation, How did Putin rise to power?*, and *The government was toppled in Egypt.* In such a professional organization, the degree of success is measured by how high one has gotten in the hierarchy. The hierarchy is commonly formalized by having "steps," which is expressed in the metaphor *climbing the ladder of success.* In a university, faculty who have access to such "steps" are called "ladder faculty."

"The glass ceiling" assumes the entire metaphorical structure of the "ladder of success." The idea is that while men can climb all the way to the top of the organization, there is a "ceiling" for women, a blockage they cannot get past, as if they hit their heads on a metaphorical ceiling. Why "glass"? Because you can see through glass. The metaphor used here is knowing is seeing. Women climbing the ladder and the public in general can see the ladder to the top that women are blocked from reaching. And glass has another attribute: it is brittle. There can be cracks in the glass ceiling, and a major goal of fairness in our society is to break a hole in the glass ceiling. Each woman who creates a crack in the glass ceiling comes closer to creating such a metaphorical "hole" in the "ceiling."

"The glass ceiling" is understood via primary metaphor mapping circuits that are already there in our brains, with existing cascade circuits forming metaphorical complexes such as 1+2+4+**knowing is seeing**. Each of those metaphors is gated and can be activated in milliseconds. The issue of limitation on the rise of American women in the professions has been discussed all over America for decades. By the time "the glass ceiling" was coined, it could be understood instantly by millions of people.

How is it possible for human conceptual thought, especially abstract thought, to expand and be understood without bound and for that unbounded range of complex thought to be expressed in language? Here is what this section contributes to a possible answer.

Structured embodied meanings arise via many connections between the brain and the body: primary schemas, X-Nets and coordination circuits, the physiological connections that form emotions, etc. Embodied idea circuits in the brain are formed via gestalt circuits, mapping circuits, binding circuits, coordinating circuits, integration circuits, and physically realized cascades of such neural circuits. Binding circuits, integration circuits, and cascades create neural control structures throughout the brain, which can be extended over time. The ability to form composite neural circuitry applies to ideas, constructions, lexical items, and linguistic forms.

This section has shown how the ability to form composites of ideas, especially metaphorical ideas, can extend indefinitely to create an unlimited range of new meanings, ultimately grounded in connections to the body.

A moral: The metaphors we see in figurative language—linguistic metaphorical expressions such as "the fiscal cliff" and "the glass ceiling"—are anything but simple; they are conceptually complex. The same is true of expressions that on the surface don't seem metaphorical, such as *She got into Harvard* and *The table has a book on it*. A thorough analysis of primary metaphor systems is required in order to understand not just language that is obviously metaphorical but also much of everyday language that looks literal yet is conceptually metaphorical, with the metaphors functioning conceptually and unconsciously.

Case Study 3: Looking Over without Overlooking

"OVER'S" RADIAL SEMANTICS

There are over 100 meanings of the term "over." Brugman (1981, 1988) first argued that "over" has a radial semantics: a central sense linked by general processes to a radial network of systematically related senses branching out from the central sense. An extended theoretical discussion of Brugman's research, in the context of prototype theory, was provided by Lakoff (1987b, Case Study 2) and focused on the details of the general processes forming the radial category.

What the Brugman and Lakoff analyses missed was the fact that the radial senses of "over" are distributed over several grammatical categories. To get a sense of the problem, consider the following examples:

- *Look **over** the manuscript, but don't **over**look anything.*

Here the "over" in "look over" is a verb particle, while the "over" in "overlook" is a prefix. To see that there are many senses of "over" for a given grammatical usage, consider these sentences. "Over" is a verb particle in all of them:

- *Turn the pancake **over**.*
- *Turn the page **over**.*
- *Turn the evidence **over** in your mind.*
- *Turn the evidence **over** to the FBI.*

To get an idea of how senses of "over" are grammatically distributed, take a look at this list:

Preposition: *over the fence, over her divorce, over lunch, over the hill, over the hole.*

Prefix: *overshoot, overthrow, overflow, overindulge, overwork, overcook, overlook, overestimate*

Verb particle: *send it over, think it over, turn it over, look it over, take it over, do it over*

Final state: *The play is over, He's over!* (after he has walked over the Golden Gate Bridge), *Over and out!* (as said by walkie-talkie or ham radio operators)

Noun: *an oversight, an overlook, the overflow, the leftovers, apple turnovers*

Adjective: *overworked, overcommitted, overweight, overcooked, overdetermined, leftover* (as in leftover vegetables)

From the point of view of their phonology—their sound structure—and their status as words, verb particles are instances of prepositions: *over, in, out, up, down,* and so on. Semantically, verb particles are integrated with the meanings of verbs to create new meanings. Grammatically, they either immediately follow the verb or occur right after a direct object (or the first of a double-object pair). Their ordering depends on the size of the direct object. If the direct object is short, the particle follows (as in *He looked it over*). If the direct object is long, the particle precedes it (as in *He looked over the documents found by the FBI*). And if the direct object is medium-size (say, two words long), it can occur either before or after (as in *He turned over the papers to the FBI, He turned the papers over to the FBI*.)

What kind of cases are there where a particle follows the first of a double-object pair, where the first object has the meaning of an indirect object? There are plenty of common cases. For example, *I'll **cook you up** a delicious meal for your birthday. Sonny Gray **served Dozier up** a fastball, and Dozier hit it out of the park. I **handed the students out** their graded homework papers. Rockefeller **threw the workers down** buckets full of dimes, and my father told me how he picked up a few of those dimes. The nurse **brought him in** a cake baked by his wife. **Send me over** some students who need a summer job.*

The technical question raised is how our approach to the neural mind can account for the grammar of these cases while still showing how the radial senses are linked to central sense. This, of course, is a problem for any theory, neural or not.

CENTRAL "OVER"

The central sense of "over" is "above" and "across." Recall the meaning of "across" from chapter 2. "Across" is a composite of (1) the opposite schema, in which opposite entities are understood as if they were people facing each other; (2) the opposite side schema, with a landmark adjacent to two nonoverlapping opposite side regions; and (3) the source-path-goal schema, with the source on one side and the goal on the opposite side.

The path is *across* the landmark. In *Jack swam across the river*, Jack is the trajector, the swimming occurs on the path, the landmark is the river, and the path extends from one side (a region) of the landmark river to the opposite side.

The central sense of "over" assumes the meaning of "across," with the path above the landmark as in *He jumped over the fence.*

LIKE THE GOLDBERG CASE

Adele Goldberg (1995) hypothesized, as noted in the sections above, that the central case of a radial category acts like a general case and that the radial variations based on the central case act like specific variations based on a general case. This works for the case of "over."

To approach the problem, we need a clear idea of the complex conceptual metaphors that are used. A good place to start is the collection of metaphors for seeing and looking, which have been well studied.

- Seeing is touching/eyes are limbs. The extended metaphorical limbs are called a "gaze."
  Examples: *Don't make eye* **contact**. *Their eyes* **met**. *Her eyes* **picked out** *every detail of the pattern. She* **ran** *her* **eyes over** *my body. She* **undressed** *him with her eyes. My* **gaze** *is* **out over** *the bay. I can* **see** *San Francisco Bay* **from** *my window.* **From** *my window, I can* **see all the way to** *Mt. Tamalpais.*

The sentence *She can see over the fence* uses the seeing is touching/eyes are limbs metaphor. The metaphorical "limbs" are a visual "gaze" that extends above the fence from one side to the other.

- Knowing is seeing.
  Examples: *I* **see** *what you mean. He* **pulled the wool over** *my eyes. She has* **blinders** *on. Keep him* **in the dark** *about the new project.*

- Making certain is seeing (as an action resulting in perceptual seeing).
  Examples: *See that he does his homework. See that the doors are locked. See to it that the garbage is taken out.*

Len Talmy's (1996) fictive motion metaphor, as in *The path runs through the woods* and *The road meanders through the valley*, reifies a path of motion, making it a thing (such as a person) that can run or meander. Applying that metaphor to the central sense of "over," the path of motion becomes a long thin object extending above and across a landmark, as in *He looked over the fence at the house next door* and *I gazed over the bay at Mount Tamalpais from my home in the Berkeley hills.*

There is a "covering" sense of "over," as in *The tapestry hung over the wall, He sprayed paint over the wall*, and a variant with the word "all" in which a path can go through points or regions that "cover" a surface, as in *Guards were posted all over the hill* and *Ants crawled all over the wall.*

In a sentence such as *Look over the manuscript*, the seeing is touching metaphor imposes a gaze that touches the manuscript and moves above and across the writing, metaphorically touching (i.e., is looking at) each word as the gaze moves all across each line. The knowing is seeing metaphor combines with the seeing is touching metaphor, indicating that each part of the manuscript should be understood by the reader if the reader looks it over.

Now, consider *Don't overlook any typos*. Words such as "overlook," "overcook," "overshoot," "overstay," "overflow," "overestimate," "overpopulate," and so on use the "above" and "across" meaning but add a context:

- There is motion toward a destination and a purpose to be served by just reaching, or almost reaching, that destination.[17]
- But on the other side of that destination point, there is a region to be avoided, and moving into that region undercuts the purpose.[18]

For example, if you are filling your coffee cup, it is okay to fill it near the lip of the cup, but if you *overfill* it, the coffee will *overflow*, creating a mess. Suppose a river runs through the middle of a town, and the townspeople have built a wall to protect the town from the river during a flood season. If the river rises less than the height of the wall, the town is protected, but if the river *overflows* the wall, the town gets flooded, which negates the purpose of the wall.

That is true of real motion, but it works for metaphorical motion as well. Time is metaphorically understood in terms of motion. When you cook string beans, there is a length of cooking time that will leave the string beans cooked through but juicy and with a slight crunch. If you cook them too long, they will go limp and become tasteless. That is called "overcooking," that is, beyond the optimum cooking time. The same will happen if you *overestimate* how long to cook the string beans. On your birthday, you may want to go to an excellent restaurant to celebrate by indulging yourself, eating rich food that is generally not good for you. There may be a certain amount for food that will make you feel indulged but unharmed. But if you go beyond that, you may *overindulge* and feel sick afterward.

The point is that this *excess* meaning of the prefix *over-* activates the central above and across sense but adds to it the extra context cited above. And with metaphors applying to that situation, you get the excess sense of the prefix *over-*.

However, there is another meaning of the prefix *over-* that can be seen in sentences such as *The chef de cuisine **oversees** the cooking staff*, *He **overpowered** the intruder*, and *The citizens **overthrew** the dictator*.

The metaphor "making certain is seeing" assumes the "knowing is seeing" metaphor. Suppose your job is to make sure that certain norms are met: that the doors are locked at night, that others do their assigned tasks, that dangers are averted, and so on. If you see that some norm is or might be violated, then by knowing is seeing, you will know it and your job will be to "see to it" that that norm is met.

The sense of "over" in "look over" (with seeing is touching and fictive motion) provides the sense of "covering" everything looked at. The *over-* adds the sense of being "above" in the social situation where the norm is to be maintained. That is why "oversee" means "supervise" from a superior social position (metaphorically above others), while the metaphorical gaze metaphorically touches those supervised and (via knowing is seeing) allows whoever is overseeing to know what those who are supervised (being overseen) are doing.

"Overlook" has the gaze going above and across and not touching and therefore means to miss seeing, that is, to miss metaphorically knowing.

"Look over," "overlook," and "oversee" each use the seeing is touching metaphor, which creates a limb-like gaze that reaches out and sees and therefore knows (via knowing is seeing) whatever the gaze "touches." In "look over" and "oversee," the gaze "touches," sees, and therefore knows. In *look over the manuscript*, all lines are completely touched and therefore seen. In "oversee," there is a supervisor monitoring whatever is being supervised: people, processes, or things. In "overlook," the gaze goes above and across, not touching, and thereby metaphorically not seeing and not knowing what the gaze fails to touch.

WHAT IS NEEDED NEURALLY?

We hypothesize that each of these verbs, with prefix or particle, works neurally via cascades that are partially different and partially identical. In each case, the cascade activates the seeing is touching and knowing is seeing metaphors but with differently framed contexts and details, as we have just discussed.

In addition, the prefix and particle meanings of "over" require *grammatical* integrations but of different kinds. The prefix *over-* creates a new verb when attached before a verb. The particle "over" follows its verb and remains a separate word, attached semantically and grammatically to create a two-word verb, with the grammatical constraints as described above.

In each of the prefix and particle cases, frames and metaphors are applied to make use of both the above and across senses of the central case of "over" but in different ways.

Language Types and Embodiment

We have pointed to two major ways in which concepts arise from embodiment, that is, from the ways our bodies work and the ways we use our bodies.

1. Our brains have X-Nets (neural networks that "execute," that is, perform actions) for coordination and control, which characterize events and actions and the manner in which they occur and unfold.
2. Embodied image schemas characterize our structuring of vision and motion, have topological and force-dynamic properties, and also allow for the shifting of viewpoint.

The dynamics of X-Nets capture the manner, aspectual, and dynamic features of the action, while the dynamics of image schemas draw attention to spatial and structural state changes (e.g., exiting or entering a state).

The major forms of conceptual embodiment have an effect in shaping the possible structures of grammar in the world's languages.

The Two-Type Language Contrast

Research by Leonard Talmy (2000b) and Dan Slobin (2006) has distinguished two types of languages, what might best be called **manner-verb languages** versus change-verb languages. English is a manner-verb language and expresses spatial change mostly via what Talmy calls "satellites." We just discussed two examples of "satellites" in English: the prefix *over-* and the verb-particle "over," which share a linguistic form that is differently integrated grammatically and semantically with the verb.

Another example is the sentence *Harry ran in.* "Run," the main verb, expresses a manner of motion, and "in" expresses the spatial change from an exterior to an interior, that is, across a boundary. In a language such as Spanish, the spatial change would be in the main verb and the manner of motion in a modifier. The Spanish equivalent would be translated literally as *Harry entered running*. The same thought elements are present in *Harry ran in* and *Harry entered running*, but they are distributed grammatically in different ways. The language difference corresponds to a difference between how neural doublets link meaning and form, in other words, to a difference in constructions.

To further illustrate the difference, here are examples taken from a classic paper by Dan Slobin (2004), "The Many Ways to Search for a Frog: Linguistic Typology and the Expression of Motion Events." Slobin discusses how the emergence of an owl in a story is expressed grammatically in five languages categorized by Talmy as *change-verb languages* (Spanish, French, Italian, Turkish, and Hebrew). In these languages, narrators of all ages almost always describe the appearance of the owl with a single path verb meaning "exit":

- **Spanish**: *Sale un buho* (Exits an owl)
- **French**: *D'un trou de l'arbre sort un hibou* (From a hole of the tree exits an owl)
- **Italian**: *Da quest' albero esce un gufo* (From that tree exits an owl)
- **Turkish**: *Oradan bir baykus çıkıyor* (From there an owl exits)
- **Hebrew**: *Yaca mitox haxor yanšuf* (Exits from inside the hole owl)

By contrast, many manner-verb language narrators, at all ages, use some kind of manner verb together with a path satellite to add some sort of dynamic information about the owl's emergence. For example, consider English, German, Dutch, Russian, and Mandarin:

- **English**: *An owl popped out.*
- **German**: . . . *weil da eine Eule plötzlich raus-flattert* (. . . because there an owl suddenly out-flaps)
- **Dutch**: . . . *omdat er een uil uit-vliegt* (. . . because there an owl out-flies)
- **Russian**: *Tam vy-skočila sova* (There out-jumped owl)
- **Mandarin**: *Fei1-chu1 yi1 zhi1 mao1 tou2 ying1* (Fly out one owl)

THE THIRD TYPE

Since the early Talmy-Slobin research, a third type of language has been added to their typology. These are languages that make the two-type distinction but where neither manner nor change takes precedence in grammar. Rather, both manner and change are expressed with equal grammatical prominence. Slobin (2006) calls these "'equipollently-framed' languages."

Slobin points to three subtypes of equipollently framed languages, based on morphological criteria:

1. Serial-verb languages in which it is not always evident which verb in a series, if any, is the "main" verb: Niger-Congo, Hmong-Mien, Sino-Tibetan, Tai-Kadai, Mon-Khmer, (some) Austronesian.

2. Bipartite verb languages, such as the Hokan and Penutian languages described by DeLancey (1996), in which the verb consists of two morphemes of equal status, one expressing manner and the other path. Talmy (2000a, 2000b) provides a similar description of Nez Perce manner prefixes, such as *ququ'-láhsa*, "gallop-ascend" (Aoki 1970). Richard Rhodes (personal communication, 2003) reports that such constructions are typical of Algonquian, Athabaskan, Hokan, and Klamath-Takelman. Huang and Tanangkingsing (2004) report that at least one Austronesian language, Tsou, has apparently developed bipartite manner-path verbs from serial-verb constructions.

3. Generic verb languages, such as the Australian language Jaminjung (Schultze-Berndt 2000), with a very small verb lexicon of about twenty-four "function verbs." For encoding motion events, one of five verbs is used, expressing a deictic or aspectual function: "go," "come," "fall," "hit," "do." These verbs are combined with satellite-like elements, "coverbs," that encode both path and manner in the same fashion. In such a language, neither path nor manner is unequivocally the "main" element in a clause (Slobin 2006, p. 64).

What is extraordinary here is that the major bifurcation in embodiment circuitry types (X-Nets versus image schemas) is reflected in the grammars of the world's languages.

How Ideas Spread

MEMES ARE NEURAL CIRCUITS

Viruses spread through the air. Ideas don't.

How, then, do ideas spread? They spread via neural recruitment and strengthening through use. That's the short answer. But there's more to it.

Richard Dawkins (1976) introduced the idea of the meme in his book *The Selfish Gene*, seeing the meme as the conceptual equivalent of the gene. The meme concept has since spread enough to make it into Wikipedia, which cites the Merriam-Webster dictionary definition as "an idea, behavior, or style that spreads from person to person within a culture." Dawkins saw memes not as metaphorical but instead as physical, living structures in the brain spreading by natural selection: by competition, inheritance, and mutation.

There seems to be something important and right about the meme idea. What Dawkins calls a "meme" is basically anything that can be understood and reproduced, what we would call a neural circuit for ideas or behaviors. And like many neural circuits for ideas, the neural circuit for an idea that Dawkins called a "meme" can be realized in behavior of all kinds.

From our perspective, ideas—neural circuits for thoughts—can spread and either become popular, die out, or just hang around. Are they physical? Yes, as are all neural circuits for ideas. Do they "live" in the brain? Well, all neural circuits that constitute ideas are made up of living cells. Does natural selection operate? That is the metaphor applied by neuroscientists such as Gerald Edelman (1992). Neural recruitment and strengthening apply. If there are neurons in the right configuration, they can be recruited through use and "strengthened" by repeated use. If the synapses are not strengthened, the idea will not come into existence. If an idea is not used, it can die off. "Natural selection" is a reasonable metaphor for that process.

REFLEXIVITY: HOW IDEAS SPREAD

An idea is constituted by a neural circuit in the brain. If it happens to be activated and "crosses your mind," say unconsciously, once with no significant inferences or other effects, it may "disappear"; that is, the synapses along the circuit will not be strengthened, and the connections within the circuit will cease to exist. There is no synaptic strengthening, no lasting idea. If an idea matters to you, if it has important consequences, and if the idea and its consequences become conscious, you may well think of it again, and the synapses along the circuit may be strengthened. Once strengthened, the idea may be expressed in behavior or in language, which is a particular form of behavior.

Suppose you communicate the idea in language or through a recognizable behavior. Since language and recognizable behavior activate frames and metaphors in the brains of others and since those frames and metaphors are constituted by neural circuitry, the neural circuitry will be activated and thus strengthened through understanding the language or behavior. With enough repetition of the language or behavior in social situations, the idea will gradually spread. If the language and/or recognizable behavior keeps being repeated, the idea's neural circuitry will become permanent in the brains of many people, and the more the idea is expressed, the more the idea will keep spreading and become stronger in the brains of others. As the idea becomes stronger in the brains of others, it becomes still more likely to spread: a *self-reinforcing positive feedback loop*. This works especially well when the ideas are unconscious. Since they are not consciously expressed, they are less likely to be challenged.

Reflexivity in communication is the self-reinforcing communication of an idea through positive feedback in either language or other behavior. Can the reflexive spread of an idea be halted? Yes, when there are sufficiently strong "competing" ideas that spread with greater effect.[19]

GEORGE SOROS ON REFLEXIVITY

The concept of reflexivity has been popularized by George Soros (2008), who learned it from his university tutor in England, the great philosopher Karl Popper. Soros has applied the concept of reflexivity to financial markets. Financial behavior is noticed and understood. Serious investors keep track of their profits. Market fundamentalism, the religion of Wall Street, seeks to maximize profits in as short a time as possible. This is justified as increasing efficiency, where "efficiency" is seen as *optimally using capital to increase investors' profits*. Investors look for maximal efficiency, that is, getting rich as much and as fast as possible.

Market fundamentalism makes a claim: markets move toward an equilibrium of supply and demand. Soros points out that the claim is false in many cases. He cites the housing bubble. Rising prices attract buyers, whose buying keeps driving prices higher, which attracts more buyers and so on in a **feedback loop**.

What is the reflexive idea for buyers? Housing prices are rising so high and so quickly that people can always profit greatly by buying a house and selling it soon. The assumption is that no matter what the cost of the house when you buy, you will make a lot of money soon when you sell. Therefore, you should borrow as much as you can, because you can always afford to pay it back and still make money.

This is unsustainable. There is no equilibrium. Buyers soon owe more on the mortgages for their houses than the houses are worth. The banks bundle mortgages and set values on the bundle that are unrealistically high. At some point the feedback loop reverses, and a new reflexive idea comes into being: the faster you sell, the less you will lose. Selling breeds more selling, which breeds more selling. There is a crash.

The system needs to be regulated to prevent the feedback loops.

Politics is about morality. Politicians put forth policies on the assumption that the policies are morally right, not wrong or morally irrelevant. Conservatives and progressives have very different, often opposite, views of morality, of what is right.

Most people want to think of themselves as doing what is right. Their ideas of what is right, like all ideas, are physically constituted by circuits in their brains. Competing ideas have different—and competing—brain circuits. How are brain circuits strengthened? By being activated repeatedly. Conservative language activates conservative ideas; progressive language activates progressive ideas. Whoever gets their language to dominate public discourse will have their ideas of what is right dominate in society.

This is reflexivity at work in political communication. There is a feedback loop: the more the public hears political language from one side, the more the political ideas from that side will be activated in the brains of those who hear the language. And the more those ideas are activated, the stronger their synapses will become and the more likely they will be to think in terms of those ideas. And if they are significantly more likely to think using the circuitry for the ideas of one side, the more they are likely to speak and behave using that side's ideas. This is a feedback loop, which results in the ongoing strengthening and spreading of particular ideas.

Reflexivity in the use of political language in public discourse if unchecked can lead to political domination. Simply negating or arguing against political ideas may well just make them stronger. Recall that the book title *Don't Think of an Elephant!* makes you think of an elephant. If an idea is discussed regularly and loudly and repeated over and over, whether discussed positively or negatively, that idea will become physically represented in the brains of speakers.

Neural Phonology

Phonology is the study of sound—how words and morphemes sound—and applies to both speaking and hearing. As a science, phonology seeks the generalizations governing pronunciation and understanding. In our neural theory, those generalizations are stated as phonological constructions, linking the *mental phonemic level in the mind with the pronounced and heard phonetic level*. We use the abbreviations "P" for phonemic and "F" for phonetic.

The elements at the phonemic level are called "phonemes," and those at the phonetic level are called "phones."

In grammatical constructions, the **meaning and form levels** are linked via gestalt doublets: circuits going in both directions, from meaning to form and from form to meaning. The circuits are neurally fixed in the brain and in use can operate in both directions.

Phonological constructions work in a similar way. *Gestalt doublet circuits link the corresponding elements at the phonemic P level and the phonetic F level.* In the notation we will be using, each doublet circuit will be represented by a line between the P and F levels.

Neurally, the linear ordering of phonemes and phones will be characterized by X-Nets, neural networks that control speaking and the understanding of speech. In our very simplified notations, we will just use left-to-right ordering.

In modern phonology, as initiated by Roman Jakobson (1968), each P-level phoneme and F-level phone is to be understood as a bundle of their phonological properties, called "distinctive features," properties that distinguish one phoneme or phone from another. They are often called just "features" for short. *The distinctive features are embodied and multimodal, with each feature linking the mouth in pronunciation with the acoustic cortex in hearing.* The plus sign (+) indicates the presence of a feature; the minus sign (−) indicates its absence. For example, +voicing indicates the presence of vocal cord vibration, and −voicing indicates the absence of vocal cord vibration. Jakobson intended such embodied distinctive features to be universally valid for all natural languages.

In our approach to neural phonology, we will consider each phoneme or phone to be characterized by an integration circuit capable of activating all of that phoneme's or phone's distinctive features at once, allowing it to function as a single entity, an integrated phoneme or phone. In our notation, the integration circuit will be notated as a bracketing of the features, just a list putting the features together, or a single symbol (e.g., *t*) for the phone or phoneme that has the properties of being a voiceless dental stop consonant.

WHAT COGNITIVE PHONOLOGY REPLACES

The old theory of generative phonology (Chomsky and Halle 1968) assumed that there were generative phonological rules that "applied" in typically long sequences, not in real time but abstractly. Although the rules, when formulated insightfully as they often were, seemed to say something real about phonology, the generative theory in which they were formulated took them out of the realm of neural possibility.

One of our authors, George Lakoff, was a student of both Chomsky and Halle in the 1960s and had worked through their theory, appreciating the insights that seemed right but frustrated by the neural impossibility of the theory. In 1993 while working with both Jerome Feldman and Srini Narayanan on structured neural computation, Lakoff published a paper titled "Cognitive Phonology" that anticipated the neural mind theory in many ways (Lakoff 1993). He used ideas from the theory of grammatical constructions, which he had worked out with Charles Fillmore and Paul Kay in the 1980s (see Lakoff 1987b, Case Study 3).

What Lakoff took from the Chomsky-Halle theory was the insight of a phonological element to be changed in a given context. The context, in the input to the rule, is a sequence of phonological elements to the left or right or both of the element to be changed, or features within the element to be changed, together with the features to be introduced by the rule. The output of the rule consists of the features resulting from the change. Rules can also delete or insert whole phonemes, that is, bundles of features. The context can also contain higher-level morphological or word-level information.

The basic insight of cognitive phonology is that phonological constructions replace the old Chomsky-Halle phonological rules. In the Chomsky-Halle generative theory, the context always applies to the input to the rule. In Lakoff's "Cognitive Phonology" and in our present neural phonology, the context can apply at either the phonemic or phonetic level. In place of a change to a different phonological element from the input to the output of the rule, cognitive phonology has the equivalent of the Chomsky-Halle "input" at the phonemic P level paired bidirectionally with the Chomsky-Halle "output" at the phonetic F level.

In the Chomsky-Halle theory, rules are ordered top to bottom (cyclically) with the context in the input of each rule. However, Chomsky-Halle rules cannot apply left to right, that is, earlier to later in the utterance. The Chomsky-Halle approach thus misses how real utterances unfold over time, which is possible in cognitive phonology notation, that is, how they apply left to right (earlier to later) in cognitive phonology.

As it turns out, cognitive phonology can handle the same range of cases as the Chomsky-Halle theory without generative phonological rules, rule ordering, or cycles.

The "Cognitive Phonology" paper, written with the help of the noted phonologists Larry Hyman and John Goldsmith, went through the basic classical cases of phonological problems that appeared to require generative phonology and then showed that cognitive phonology could handle those cases without sequences of generative rules (Lakoff 1993).[20]

FROM COGNITIVE TO NEURAL PHONOLOGY

In general, cognitive phonology analyses can now be reinterpreted as neural phonology analyses in the same way that cognitive grammar constructions can be reinterpreted as neural grammar constructions.

In each neural grammar construction, as we have seen, there are two levels: meaning and form. Elements of meaning are bidirectionally linked to corresponding elements of form via gestalt doublets. Thus, a meaning element, when activated, will activate the corresponding form element during language production because they are linked via a gestalt doublet. During language understanding, the activation of a form element will activate the corresponding meaning element because they are linked via the same gestalt doublet. All of the individual correspondences are organized by a gestalt circuit, with the gestalt nodes of each of the gestalt doublets being roles in the gestalt circuit.

As in cognitive phonology, lexical items in neural phonology are constructions. In phonological representations of lexical items, the form level is characterized by a P-level phonological representation. Neural phonology links the lexical P-level phonemic representations to F-level phonetic representations. The cognitive distinctive features of cognitive phonology are replaced by neural features. Bidirectional pairing of P-level neural feature bundles with F-level feature bundles are carried out via gestalt doublets, as in neural grammar.

The left-to-right, start-to-finish processing in both language production and language understanding is carried out by X-Nets (neural networks that perform operations), as in neural grammar.

The examples accounted for in the "Cognitive Phonology" paper would thus be accounted for as well by a neural phonology, given neural characterizations of phonological features (Lakoff 1993).

Of course, the huge gap here is in neural phonetics, a field seriously under study but by no means completed as of this writing.

The Takeaways

THE MAIN POINTS

Once you take a neural perspective, language looks different, and linguistics changes utterly. Disembodied generative linguistics and formal semantics are gone. Meaningless formal syntax is gone. All language is meaningful. All meaning is embodied, so meaningful language must be embodied. It is embodied neurally. Language learning becomes neural learning and neural recruitment, which depends on prior neural structure in both body and brain. *The neural basics of meaningful thought are learned before language is learned. This includes circuitry for X-Nets, embodied primitive schemas, logic schemas, basic frames, primary conceptual metaphors, bindings, integrations, and cascades.*

Meaning does not simply *fit* reality, which is often claimed by philosophers. In crucial ways *the mechanisms of meaning **create** reality as consciously experienced*, as with the *experienced reality* of colored objects when we know *there is no color in the external world* and that in the brain color and shape are computed in separate brain regions and are brought together via neural integration. *Our real experience of colored objects is created via our neural systems.*

Simulation—of both immediate reality and imaginative simulation—plays a major role in thought. So does viewpoint—called "deixis" by Charles Fillmore (1977)—and viewpoint shift. What has been called the "protagonist role" picks out a viewpoint from which one is viewing an event. The neural mechanism for viewpoint shift is substitution of one viewpoint for another, that is, a viewpoint substitution of one person, time, or situation for another. (Consider how a Democrat might try to understand a political situation from the viewpoint of a Republican or vice versa.)

Although agent subjects of active sentences and patient subjects of passive sentences have different semantic roles (agent and patient), as subjects both can bear the protagonist role. For example, the difference between seller and buyer in a commercial exchange is defined by the viewpoint taken: who is the protagonist; that is, who is thought of as doing the buying or selling.

Viewpoint is neurally complex, using both the mechanisms of place (presumably via place and grid cells) and empathy (circuitry linking motor control, vision, mental imagery, and the embodiment of emotion, that is, the somatosensory system, the amygdala and insula, and the corticostriatal reward system).

The aspect of language that is conscious is small relative to the cognitive and linguistic unconscious. Their timescales are utterly different. *The neural unconscious functions at the millisecond timescale, while the conscious use of language functions at the hundred millisecond timescale* and requires successful *neural integration of disparate unconscious circuitry, often from different brain regions*, and requires a longer timescale.

The idea of a "language module" or "language organ," as proposed by

Noam Chomsky, is dead, killed off by our understanding of the neural mind. So is the "warehouse theory" that memory and knowledge are "stored" in a single local region of the brain. Embodied conceptual structure may be *triggered* locally, perhaps prefrontally, but embodied meaning is *carried out* globally by neural cascades going through multiple convergence zones across the brain and connecting multiple brain-to-body brain regions.

Different, often incompatible, cultural worldviews develop across cultures and subcultures. Communication is worldview-dependent because we can only understand what our brains allow us to understand, namely what fits our worldviews. The reason is that worldviews are carried out by neural circuitry, and we can only understand what our neural circuitry allows us to understand. In short, *our neural circuitry can be seen as a filter on our understanding of reality.*

Finally, language that is repeated can change worldviews over time. The reason is that morphemes, words, and grammatical constructions have embodied meanings carried out by neural circuits. The more those circuits are activated, the stronger they become and the more they activate the worldviews being used. The more a given worldview is activated, the stronger it becomes, and if a minor worldview is activated enough, it can gradually become major. A change in political worldviews, for example, requires repetition, either external repetition or repetition in one's own mind, that is, repeated activations of the neural structures characterizing our worldviews.

THE POINT-BY-POINT TAKEAWAY

1. Since language expresses thought, it uses all the neural mechanisms of thought. Those neural mechanisms appear to be sufficient for language; for example, gestalt circuits can characterize hierarchical structure, X-Nets can characterize linear structure, binding and integration circuits can provide simple constructions, and cascades can characterize complex constructions.
2. The general concept of event structure, causation, and purpose, as they are used in constructions, arises from the structure of action coordination and control and is expressible through X-Nets.
3. Simple clauses are embodied via the framing of everyday actions and events. Each framing has conceptual "roles." These have been characterized in linguistics both by Charles Fillmore's (1968) case roles and semantic roles and by Jeffrey Gruber's (1965) thematic roles. We

owe a great intellectual debt to Fillmore and Gruber. Such integrated ideas are the locus of intellectual complexity.

4. Grammatical relations (subject, direct object, etc.) get their meanings from semantic roles via *gestalt nodes linking meaning and form*, that is, gestalt doublets.

5. The mechanism of integration is used everywhere in grammar. Examples include change-state constructions (*The milk cooled*) and cause-change constructions (*He cooled the milk*) as well as cause-receive constructions (e.g., *I gave Harry a book* and *I gave a book to Harry*).

6. Minor rules—rules limited to specific lexical items and constructions—can be modeled by having a control node for a construction inhibited by a gating node, which can be disinhibited in context by the activation of specific lexical items or specific constructions.

7. Words can have very complex meanings, which can be characterized via neural integration over neural cascades. Overall and largely unconsciously lexical cascades integrate a great many kinds of embodied circuitry.

8. "Over" and other prefixes and verb particles have simple central meanings that are extended via frames and/or mappings, metaphoric and metonymic. Prefixes and verb particles are semantically and grammatically integrated with verbs and verb meanings via cascade circuitry.

9. One-anaphora works via general mutual inhibition among role fillers, neural decay, and least energy use (also called "neural optimization" and "best fit").

10. The main mechanisms of embodiment are image schemas and X-Nets. Languages differ according to which of these, if either, is given priority in the choice of verbal structures.

11. Languages are enormously complex, and there are thousands around the world, many undescribed. This section can only begin to outline an approach to how language is able to be constituted by the relatively small range of neural circuitry discussed in this book.

12. Human conceptual systems and language depend in many respects on genetically inherited aspects of brain anatomy and neural circuitry, which shapes a great deal of our conceptual systems. The rest is "emergent" and is learned via neural recruitment under condi-

tions of best fit, that is, neural optimization. *Language and thought cannot be totally emergent ("usage-based"), because neural recruitment depends on prior structure, not just language use at particular times.*

We began by asking whether the circuitry types and neural mechanisms for ideas that we have been discussing are sufficient for grammar and morphology as characterized by constructions. We have, of necessity, left much undiscussed. But for the wide range of examples considered in this chapter, the answer seems to be *yes!* Idea circuitry types seem to be sufficient for grammar.

If so, a "universal grammar" need not be hypothesized and claimed to be genetically encoded, as Noam Chomsky has proposed, *without any idea of what the gene-to-language mechanism would be and how the huge gap between genes and grammars could be bridged.*

Instead, grammars can be neurally recruited. This is a major advantage of the neural theory of thought and language. No magical gene-to-grammar mechanism is required by the neural theory presented here.

BEYOND ENGLISH

There are several thousand languages around the world, most barely described in any detail. The questions to ask for each language are these:

1. Are the neural mechanisms for ideas in all languages the same as for English (even if the specific ideas are different in many details)?
2. Are the proposed neural mechanisms for thought also sufficient for the characterization of linguistic form in each language (grammar, morphology, phonology)?
3. Are those neural mechanisms sufficient for the constructions linking form and meaning (including pragmatics) in those languages?

Such questions, as is common in science, will not be answered soon or without debate. One of our goals in this book has been to be detailed enough so that such questions can at least be asked, solutions can be offered, and, quite likely, our theory can be extended or changed.

A full understanding of how our neural systems can constitute thought and language is the object of a great scientific quest. We hope that this book has been a significant step on that quest, and we welcome others to continue the quest wherever it may lead.

George Lakoff's interest in phonology and its biological basis came from working closely with the great linguist Roman Jakobson in 1961 while George was an undergraduate at MIT and Jacobson was visiting there from Harvard. It was Jakobson who had come up with the idea of distinctive features in phonology based on how the basic phonological structures came from the physical structure of the mouth. George worked closely with Jakobson on the phonological analysis of Yeats's poetry and in the process learned an enormous amount from Jakobson about how to do phonological analysis.

In 1989 while working with Jerome Feldman and Srini Narayanan on structured neural computation, Lakoff he wrote a paper titled "Cognitive Phonology" that looked forward to ideas in this book. This paper became a chapter in *The Last Phonological Rule* (Goldsmith 1993). With the help of phonologists Larry Hyman and Goldsmith, George showed how long sequences of phonological rules in generative phonology could be replaced by simple phonological constructions, akin to grammatical constructions as devised by Charles Fillmore.

Grammatical constructions link meaning and syntactic form, while phonological constructions link the phonemic level (called "P") to the phonetic level (called "F"). The neural linking mechanism is the same: gestalt doublets. A line linking a phonemic to a phonetic property in the "Cognitive Phonology" paper would be interpreted as a gestalt doublet doing the neural linking in neural phonology.

Each neural phonological construction is a fixed circuit capable of linking the P and F levels, indicating the differences and correspondences. Whereas generative phonological rules have contextual conditions in their inputs alone, neural phonology would have contextual conditions at either the P level or the F level. This eliminates the need for the rule orderings required in generative phonology but accomplishes the same effects in a simpler, straightforward way.

FEEDING VERSUS BLEEDING

In generative phonology there is a difference between, for example, "feeding order" and "bleeding order" for two rules; call them A and B. In feeding order, rule A creates the contextual condition for the application of rule B. In bleeding order, rule A removes the contextual condition for the application of rule B, which then cannot apply if rule A has applied.

In neural phonology, feeding and bleeding effects occur for constructions A with a contextual condition at level P (feeding) and for construction B with its contextual condition at level F (bleeding). The feeding effect occurs when contextual conditions at level P in construction A and at level F for construction B are compatible so that both fit. The bleeding effect occurs when there is an incompatibility between the contextual conditions in constructions A and B. This means that the phonetic condition of B does not fit with the phonemic condition of A. Hence, where A occurs at level P, there is no corresponding fit for B at the phonetic level.

Afterword

The Neural Mind versus Deep Learning AI

We are concerned with the neural mind, with how real human brains think.

Ours is a scientific enterprise focused on embodied cognition where the goal is to study how cognitive phenomena arise from the brain and body interacting in a social and physical environment. Neural computation and the models we introduce constitute a technique using computers for modeling how the neural mind works. Our present research therefore is part of the human sciences rather than a form of artificial intelligence (AI), which is concerned with building intelligent artifacts.

AI is concerned with building systems that learn patterns from human-generated data and cultural artifacts (e.g., text, audio, photos, videos) and use these learned patterns to perform a variety of human tasks (e.g., conversation, answering questions, playing games, generating images). Research in AI is in no way limited to how real human brains think.

Indeed, there has been a revolution in AI spearheaded by the amazing success of a decades-old technique called deep learning. Deep learning uses a network of artificial neural network nodes arranged in multiple layers (deep) between the input and output and a stochastic version of gradient descent (e.g., backpropagation) for updating the weights between nodes connecting all the intermediate layers. At every input (or batch of inputs), the gradient descent algorithm calculates the sensitivity of the different weights across all the deep layers to the prediction error at the output layer. The gradient descent algorithm then adjusts the weights of the different cross-layer connections in a direction, in addition to the amount, to minimize the error.

The huge success of deep learning has come from recent advancements in both the architecture of the network and the ability to efficiently process and learn from massive amounts of sequential data (e.g., text). The architectural advancement comes in the form of decoder networks that predict and produce the next token (words or parts of words) trained on vast amounts of web data. Some AI systems both encode the input (encoder) and then produce the output (decoder) as deep network towers, each with multiple layers or stages. The learning effectiveness comes from mechanisms of cross-input attention in a technique called "transformer networks." Transformers are able to learn and attend to the relevant part of the input by processing all of the input (which consists of a sequence of tokens such as words in a text) and calculating the relevant parts of the context by learning attention weights between parts of the input in successive layers. This allows much larger networks to be trained effectively in parallel and is particularly useful for processing sequential data such as text, video, and speech signals.

Transformer-based deep generative networks such as GPT-X from Open AI and the Google Gemini network are trained on large amounts of web text to predict and generate (hence "generative") the next word as the output when presented with a sequence of input text. Such networks learn from web-scale data to both generalize and memorize patterns and sequences of words, phrases, and constructions. Their success, given their goals, has been remarkable. For example, a novel written by AI was shortlisted for a literary prize. Self-driving cars recognize images and then make decisions in real time. A new method combining deep networks with a reward-based learning mechanism known as deep learning has made huge advances in computer game playing (e.g., the Google AlphaGo) and more recently in protein folding (AlphaFold), one of the major open problems in biochemistry. Governments and large corporations are rapidly picking up on the potential of deep learning technology, and AI has become a strategic priority at multiple levels of society.

There is an obvious question we can ask about how picking up statistical correlations and generalized input patterns in massive amounts of data including text, images, speech, and video and sets of these sources simultaneously (in multimodal networks) is related to human understanding. The two processes appear completely different in multiple ways. As we've shown in this book, human understanding is fundamentally embodied and cannot be divorced from emotions, social cognition, and action. As we explored in chapter 1 and chapter 2, the brain is not a general-purpose computer but instead is a specific device that has evolved to monitor and control the body. Language and the brain have coevolved so that brain designs and constraints through evolution are matched by linguistic structures and devices. Thus, as we discussed, perception, motor action, decision-making, and language all rely crucially on the architecture, specific brain circuits including timing details, and the functional interaction of these circuits to monitor and control the body. AI is simply a different enterprise than the neural mind, defining a different field of research that is important for other reasons, practical reasons outside of the human sciences.

Some proponents of deep learning AI may want to claim a connection to real human thought. But the better way to serve the technology and enterprise would be to remove the implicit or explicit goal of mimicking human cognition and instead focus on the systems engineering and the enormous practical benefit of the models, which have superhuman performance in data-rich statistically based applications in the business, medical, scientific, and commercial sectors.

In short, it is important to separate our study of the neural mind in the human sciences from the practical need for massive data processing in vital data-rich applications. They are very different enterprises, and both are vital. In this book, we have been concerned only with the neural mind, and computers are of interest only to the extent that they can accurately model aspects of the neural mind.

Acknowledgments

Writing a book requires remarkable cooperation from family members on a daily basis. We owe enormous debts to George's wife, artist and writer Kathleen Frumkin, and Srini's partner, law professor Andrea Büchler, and their daughter, Tara.

Jerome Feldman is the major intellectual influence behind this book. He has contributed fundamental ideas and support throughout. This work would not have been possible without his contributions.

The Neural Theory of Language Project at the International Computer Science Institute (ICSI) was the crucible for the ideas in this book over many years. We would like to thank Lisa Aziz-Zadeh, David Bailey, John Bryant, Nancy Chang, Matt Gedigian, Joseph Makin, Eva Mok, Shweta Narayan, Terry Regier, Steve Sinha, Nathaniel Smith, and Andreas Stolcke.

The members of the ICSI MetaNet research team did crucial work in the development of cascade theory, embodied construction grammar, and empirical research, especially Lisa Aziz-Zadeh, Ben Bergen, Lera Boroditsky, Oana David, Ellen Dodge, Jisup Hong, Teenie Matlock, Karie Moorman, Elise Stickles, and Eve Sweetser.

Vittorio Gallese spent several months visiting ICSI and was crucially important in integrating our work with neuroscience. Thanks also to Tim Verstynen and Rich Ivry.

Several of these ideas were crystalized while Srini was a fellow of the Institute for Advanced Study in Berlin (Wiko Berlin). Thanks go to Lisa Aziz-

Zadeh, Holk Cruse, Tom Metzinger, Rafael Núñez, Luc Steels, and Robert Trivers.

Discussions of neuroscience with Robert Knight and Antonio and Hannah Damasio have been invaluable. David Presti contributed important details of neurochemistry.

Shin Shimojo helped enormously with the results on neural integration prior to consciousness.

Lara Krisst has helped greatly by rooting out relevant neuroscience papers.

George's research assistants, Eleanor Garret and Kayla Briones, helped with discussions of early versions and with editing. Members of George's Linguistics 205 class at the University of California, Berkeley, read and critiqued an earlier draft of the manuscript.

George's seminar students helped with regular insightful comments and questions, especially Helen de Andrade Abreu, Elizabeth Amazing, Jongmin Baek, Travis Bartley, Brittany Blankenship, Michael Greenberg, Laura Henderson, Levon Korganyan, Joy MacKay, Sergio Nieves, John Sokolov, and Ozge Ugurlu.

Roy Deckard helped with the computer version of this manuscript. His help has been invaluable. Roy has been a great resource, a great teacher, and a delight to work with.

Notes

Chapter One

1   For an excellent review of these and other complexities, see Conway et al. (2010).
2   For interesting examples of the sensory worlds created by different organisms, see Yong (2022).
3   See Gilbert et al. (2005).
4   For facts and figures about neuron size, we recommend "Neurons" at the website "Brain Facts and Figures," Washington University, http://faculty.washington.edu /chudler/facts.html#neuron. For some speculations on the profound implications of this potential gradient, see Lane (2022).
5   Note that in traditional cognitive science or logic, the part-whole relation is treated as a semantic relation to be interpreted by a reasoning algorithm. While this view is consistent, it is not sufficient for research on the neural mind. The idea of a circuit requires us to be precise about the way the relation operates as a dynamic system and to the extent possible in a neurally plausible manner, primarily through activation or inhibition within a frame or between frames. In this way, reasoning and representation become tied together in circuits that, when activated, do such things as actions and reasoning.
6   See Raichle (2015).
7   See Badre et al. (2008), Bhandari and Badre (2020) and Koechlin et al. (2003). It has since been discovered that there is a second "hierarchical" prefrontal neural organization of connections but with the *top* of the connection hierarchy in the *midprefrontal* region because it is in the middle of the generalization hierarchy. We will discuss this phenomenon below in the section on basic-level concepts.
8   See Koechlin et al. (2003).
9   See Dehaene (2014) and Koch (2012).
10 See also Hubel (1995, p. 43).

11 See Man, Kaplan, Damasio, and Meyer (2012); Man, Damasio, Meyer, and Kaplan, (2015); Damasio (1989a, 1989b); Damasio and Damasio (1992); Damasio, Damasio, et al. (1990); and Damasio, Tranel, and Damasio (1990).

12 See Damasio, Tranel, and Damasio (1990).

13 See Averbeck et al. (2014); Jarbo and Verstynen (2015); Meyer and Damasio (2009); Man, Kaplan, Damasio, and Damasio (2013); Meyer, Kaplan, Essex, Webber, Damasio, and Damasio (2010); Man, Kaplan, Damasio, and Meyer (2012); and Man, Damasio, Meyer, and Kaplan, 2015.

14 See Rosch et al. (1976); Rosch and Mervis (1981); and Lakoff (1987a, chap. 2).

15 See Gallese and Lakoff (2005).

16 See Dentico et al. (2014).

17 See Gallese et al. (1996).

18 Personal conversation with George Lakoff.

19 Incidentally, C. C. Pratt in 1930 suggested that "high" and "low" coded height in the cochlea, from the apex to the base. The mouth is a better bet. We all, even infants, have regular conscious interactions with our mouths. Few people even know they have a cochlea with an apex and a base.

20 David Huron (n.d.), music theorist at Ohio State University, notes that there are other metaphors for pitch around the world. Huron cites Larry Zbikowski, the noted music theorist at the University of Chicago, as pointing out that what we call "high" and "low" pitch is metaphorized as "sharp" and "heavy" by the ancient Greeks, as "small" and "large" in Bali and Java, and as "young" and "old" by the Suyá of the Amazon. Dolscheid et al. (2014) observe that in Persian and Turkish, the metaphors are "thick" and "thin." None of this is surprising, given what we know of the role of culture in conceptual metaphor as studied by Kövecses (2005).

21 See Krisst et al. (2015).

Chapter Two

1 Although as we will see in chapter 3, recent findings suggest that some of the synergies include premotor cortex and may be stereotypical but complicated movements, which may result from combining multiple basic synergies.

2 See Chang et al. (1998).

3 See Ramachandran and Blakeslee (1999).

4 There are two important places to start. The first is Regier (1996, pp. 93–107), which suggests that spatial image schemas could be computationally modeled using comparisons between models of topographic maps. Neurally, this would amount to cascades within and across maps. The other is Silver and Kastner (2009), which reports on the existence of maps with internal radial and rotational structures that can do some of the things that can be done by image schemas (although that was not the point of their research).

5 Consistent with our hypothesis, Badre (2008) and Badre et al. (2009) have shown experimentally that in the case of general versus specific rules, generalization circuits for the general rules are located in the forward (rostral) part of the prefrontal cortex.

6   For a magnificent book on embodied simulation, read Benjamin Bergen's (2012) delightful and very thorough book *Louder Than Words*.

7   The great sociologist Erving Goffman, one of the major contributors to the theory of frames with his classic book *Frame Analysis* (1974), first brought up the importance of asking what happens when a frame "breaks." Goffman used questions as a means of studying the limits of a given frame.

8   For the FrameNet research, see FrameNet, https://framenet.icsi.berkeley.edu/.

9   See Lakoff and Johnson (1999, pp. 380–82).

10 See Lakoff and Johnson (1999).

11 See Lakoff and Johnson (1999, chap. 12).

12 One example of a new idiom is "hookup."

13 See Plaster and Polinsky (2007).

Chapter Three

1   See Narayanan (1997a, 2003a, 2003b), Feldman (2006), and Barrett (2010).

2   In general, the model is called a leaky integrate and fire model and is modeled by a linear differential equation based on an electrical circuit model called the RC circuit. The interested reader can find more information in this online textbook *Neuronal Dynamics* by Gerstner et al. (2014). The more realistic Hodgkin-Huxley model is a second-order differential equation with the resistive and inductive cable properties of the axon taken into account.

3   See Markram et al. (2011).

4   The term "backpropagation" also has a very different neuron-internal use. Recent research has shown that, when firing occurs in certain neurons, a weak secondary signal, sometimes called an "echo," propagates from the end of the axon to the cell body and through the cell body to and through the dendrites. Precise details are presently unclear. But there is no evidence at all that the process works as it does in artificial neural networks as an error-correction mechanism propagating back and forth across neurons.

5   An account of learning of these circuits is outside the scope of this book. For some initial steps, see Hoerzer et al. (2014).

6   See Rizzolatti et al. (1996, 2004).

7   As an example of how quickly research develops in neuroscience, a world of new research has been developing as we write. The research concerns two different kinds of inhibition beyond the simple case of inhibition by chloride ions entering the dendrites of a cell and subtracting from the positive charge inside the cell.

Optogenetic techniques have led to the discovery of two special types of inhibition via (1) somatostatin and (2) parvalbumin inhibitory synapses in certain brain regions. The neuroscience research is going on as we write, and a full understanding of the results of this discovery does not exist. These discoveries do not appear to alter the basic theory presented in this book, since they concern the neural mechanisms of memory and homeostasis but not the neural mechanisms for characterizing ideas and language.

An excellent example of this research is reported in Horn and Nicoll (2018). Computational modeling research has not caught up to the neuroscience here.

8   See Gisiger and Boukadoum (2011).

9   For a wonderful very readable account of the neuroscience behind these discoveries, see Graziano (2018).

10   See Kahn et al., 2013.

11   For an interesting and provocative perspective on the current state of our knowledge on consciousness and the mind-body problem, see Feldman (2006). The central idea is that the neural system is part of the mind, and its functioning is what we experience of the workings of the mind.

12   Feldman (2006) contains a good description of the different approaches that have been tried to date.

13   For the complex structure and relationships among mental spaces, see Fauconnier (1985) and Fauconnier and Turner (2002).

14   See Constantinescu et al. (2016).

15   Buzsáki (2019, p. 108).

Chapter Four

1   Human Connectome Project, USC Mark and Mary Stevens Neuroimaging and Informatics Institute, https://www.humanconnectomeproject.org.

2   Older linguists will recall theories by Zellig Harris and his student, Noam Chomsky, in which there are "kernel" sentences (in the form of active statements) and "transformations" linking them to passives, questions, and so on. Those were purely "syntactic" theories based only on form, consisting only of abstract symbols rather than on speech acts and other aspects of pragmatics—presuppositions, viewpoint shifts, emotional effects, and so on—that would have to be taken into account in a neural theory. These needs take us beyond transformational grammar, which is purely syntactic and does not involve meaning at all.

3   The linguist Zellig Harris used the term "kernel sentence." Harris took the term "kernel" from abstract algebra, which he was trying to apply to syntax.

4   The central role of pragmatics in grammar is anything but new. A vast range of examples was uncovered in the generative semantics period from 1963 to 1975 by George Lakoff, Haj Ross, James D. McCawley, Robin Lakoff, and many others.

5   In what follows, "1" stands for first person (i.e., the speaker), "2" stands for second person (i.e., the hearer), and "3" stands for third person (i.e., what is spoken about).

6   In the field of linguistics, the asterisk indicates that something is ill-formed.

7   This grammatical frame, or "template," was discovered by John Robert Ross (1967) in his PhD dissertation.

8   With thanks to Professor Beth Levin of Stanford University.

9   There are, however, linguists, computer scientists, and even psychologists who see natural language as purely formal, just a matter of meaningless symbol sequences. The view that natural language is not fundamentally meaningful does not seem to us to make sense, and we will not discuss that view further.

10   We are grateful for these citations to the Ask a Linguist FAQ service at the Linguist List website, https://linguistlist.org. For a detailed discussion of the complexities and empirical issues raised by the Whorfian hypothesis," see Lakoff (1987b, chap. 18).

11   Incidentally, there is currently a responsible neo-Whorfian movement as represented in the work of great linguistic scholars such as Dan Slobin and Leonard Talmy. Their work is discussed in the section titled "The Two-Type Language Contrast."

12   For a superb book-length discussion of the details of negative and positive polarity items, see Israel (2011).

13   See Narayanan (1997b).

14   See chapter 2.

15   See Lakoff (1968).

16   These metaphors are abbreviated here but are elaborated in detail in Lakoff and Johnson (1999).

17   The metaphor cascade used in these examples includes *Purposes are destinations*, *Linear scales are paths*, and *More is up*, which makes linear scales vertical.

18   Compare the meaning of *over-* with the meaning of *under-* in *undercook, underperform, underestimate, undervalue, underexpose, underinvest*, and so on. In these cases, there is a metaphorical region before reaching the destination, and if you stop short of the destination (*below* that region on the linear scale), the purpose is unrealized.

19   See Lakoff (2014a).

20   Elaborate examples can be found in Goldsmith (1993).

References

Aoki, H. (1970). *Nez Perce Grammar*. University of California Press.

Arbib, M. A. (1998). Schema theory. In M. A. Arbib (ed.), *The Handbook of Brain Theory and Neural Networks* (pp. 830–34). MIT Press.

Austin, J. L. (1962). *How to Do Things with Words*. J. O. Urmson (ed.). Clarendon Press.

Averbeck, B. B., Lehman, J., Jacobson, J. M., and Haber, S. N. (2014). Estimates of projection overlap and zones of convergence within frontal-striatal circuits. *Journal of Neuroscience 34*(29), 9497–9505.

Badre, D. (2008). Cognitive control, hierarchy and the rostro-caudal organization of the frontal lobes. *Trends in Cognitive Science 12*(5), 193–200.

Badre, D., Hoffman, J., Cooney, J. W., and D'Esposito, M. (2009). Hierarchical cognitive control deficits following damage to the human frontal lobe. *Nature Neuroscience 12*(4), 515–22.

Bailey, D. (1997). *A Computational Model of Embodiment in the Acquisition of Action Verbs* (doctoral dissertation). University of California, Berkeley.

Bargh, J. A., Chen, M., and Burrows, L. (1996). Automaticity of social behavior: direct effects of trait construct and stereotype-activation on action. *Journal of Personality and Social Psychology 71*, 2, 230–44.

Bargmann, C. I. (2012). Beyond the connectome: How neuromodulators shape neural circuits. *Bioessays 34*, 458–65.

Barrett, L. (2010). *An Architecture for Structured, Concurrent, Real-Time Action* (doctoral dissertation). University of California, Berkeley.

Benoit, R. G., Szpunar, K. K., and Schacter, D. L. (2014). Ventromedial prefrontal cortex supports affective future simulation by integrating distributed knowledge. *Proceedings of the National Academy of Sciences of the USA 111*, 16550–55.

Bergen, B. (2001). *Of Sound, Mind, and Body: Neural Explanations for Noncategorical Phonology* (doctoral dissertation). University of California, Berkeley.

Bergen, B. (2004). The psychological reality of phonaesthemes. *Language 80*(2), 290–311.

Bergen, B. (2012). *Louder Than Words*. Basic Books.

Bergen, B., and Chang, N. (2005). Embodied construction grammar in simulation-based language understanding. In J.-O. Östman and M. Fried (eds.), *Construction Grammars: Cognitive Grounding and Theoretical Extensions* (pp. 147–90). Benjamins.

Bernshteïn, N. A. (1967). *The Co-ordination and Regulation of Movements*. Pergamon.

Bhandari, A., and Badre, D. (2020). Fronto-parietal, cingulo-opercular and striatal contributions to learning and implementing control policies. *bioRxiv*. https://doi.org/10.1101/2020.05.10.086587.

Bienenstock, E. L., Cooper, L. N., and Munro, P. W. (1982). Theory for the development of neuron selectivity: Orientation specificity and binocular interaction in visual cortex. *Journal of Neuroscience 2*(1): 32–48.

Bode, S., He, A. H., Soon, C. S., Trampel, R., Turner, R., and Haynes, J.-D. (2011). Tracking the unconscious generation of free decisions using ultra-high field fMRI. *PLoS ONE 6*(6), e21612.

Boroditsky, L. (2009). How does our language shape the way we think? In Brockman (ed.), *What's Next? Dispatches on the Future of Science*. Vintage Press.

Brugman, C. (1988). *The Story of Over: Polysemy, Semantics, and the Structure of the Lexicon*. Garland.

Buckner, R. L., Andrews-Hanna, J. R., and Schacter, D. L. (2008). The brain's default network: Anatomy, function, and relevance to disease. *The Year in Cognitive Neuroscience, Annals of the New York Academy of Sciences 1124*, 1–38.

Buzsáki, G. (2019). *The Brain from Inside Out*. Oxford University Press, 2019.

Carpenter, P. A., and Eisenberg, P. (1978). Mental rotation and the frame of reference in blind and sighted individuals. *Perception and Psychophysics 23*, 117–24.

Carpenter, G. A., and Grossberg, S. (1988). A massively parallel architecture for a self-organizing neural pattern recognition machine. *Computer Vision Graphics and Image Processing 37*, 54–115.

Cassanto, D., and Boroditsky, B. (2005). Writing direction influences spatial cognition. *Proceedings of the Twenty-Seventh Annual Conference of the Cognitive Science Society 27*, 412–17.

Chang, N., Gildea, D., and Narayanan, S. (1998). A dynamic model of aspectual composition. In *Proceedings of the Twentieth Annual Conference of the Cognitive Science Society*. Routledge.

Chang, N., Narayanan, S., and Petruck, M. (2002). Putting frames in perspective. In *Proceedings of the Nineteenth International Conference on Computational Linguistics 1*, 1–7.

Chang, E. F., Raygor, K. P., and Berger, M. S. (2015). Contemporary model of language organization: An overview for neurosurgeons. *Journal of Neurosurgery 122*(2), 250–61.

Chomsky, N., and Halle, M. (1968). *The Sound Pattern of English*. Harper and Row.

Churchland, P. S. (1986). *Neurophilosophy: Toward a Unified Science of the Mind-Brain*. MIT.

Cienki, A., and Müller, C. (2008). *Metaphor and Gesture*. John Benjamins.

Citron, F. M., and Goldberg, A. E. (2014). Metaphorical sentences are more emotionally engaging than their literal counterparts. *Journal of Cognitive Neuroscience 26*, 2585–95.

Constantinescu, A. O., O'Reilly, J. X., and Behrens, T. E. J. (2016). Organizing conceptual knowledge in humans with a gridlike code. *Science 352*, 1464–68.

Conway, B. R., Chatterjee, S., Field, G. D., Horwitz, G. D., Johnson, E. N., Koida, K., and Mancuso, K. (2010). Advances in color science: From retina to behavior. *Journal of Neuroscience 30*(45), 14955–63.

Coulson, S. (2001). *Semantic Leaps: Frame-Shifting and Conceptual Blending in Meaning Construction*. Cambridge University Press.

Dahl, C. D., and Adachi, I. (2013). Conceptual metaphorical mapping in chimpanzees (*Pan troglodytes*). *eLife 2*, e00932.

Damasio, A. (1989a). Time-locked multiregional retroactivation: A systems-level proposal for the neural substrates of recall and recognition. *Cognition 33*, 25–62.

Damasio, A. (1989b). The brain binds entities and events by multiregional activation from convergence zones. *Neural Computation 1*, 123–32.

Damasio, A. (1994). *Descartes' Error*. Penguin.

Damasio, A. (1996). The somatic marker hypothesis and the possible functions of the prefrontal cortex. *Philosophical Transactions of the Royal Society 351*, 1413–20.

Damasio, A., and Damasio, H. (1992). Brain and language. *Scientific American 267*, 88–95.

Damasio, A., Damasio, H., Tranel, D., and Brandt, J. P. (1990). Neural regionalization of knowledge access: preliminary evidence. In *Cold Spring Harbor Symposia on Quantitative Biology* (vol. 55, pp. 1039–47). Cold Spring Harbor Laboratory Press.

Damasio, A., and Tranel, D. (1993). Nouns and verbs are retrieved with differently distributed neural systems. *Proceedings of the National Academy of Sciences of the USA 90*, 4957–60.

Damasio, A., Tranel, D., and Damasio, H. (1990). Face agnosia and the neural substrates of memory. *Annual Review of Neuroscience 13*, 89–109.

Dancygier, B., and Sweetser, E. (2014). *Figurative Language*. Cambridge University Press.

David-Barrett, T., and Dunbar, R. (2016). Language as a coordination tool evolves slowly. *Royalty Society Open Science*, 3(12). https://doi.org/10.1098/rsos.160259.

Dawkins, R. (1976). *The Selfish Gene*. Oxford University Press.

De Brigard, F., Spreng, R. N., Mitchell, J. P., and Schacter, D. L. (2015). Neural activity associated with self, other, and object-based counterfactual thinking. *NeuroImage, 109*, 12–26.

Dehaene, S. (1997). *The Number Sense: How the Mind Creates Mathematics*. Oxford University Press.

Dehaene, S. (2014). *Consciousness and the Brain: Deciphering How the Brain Codes Our Thoughts*. Penguin Books.

DeLancey, S. (1996). The bipartite stem belt: Disentangling areal and genetic correspondences. In *Proceedings of the Twenty-Second Annual Meeting of the Berkeley Linguistics Society: Special Session on Historical Topics in Native American languages*, 37–54.

Dentico, D., Cheung, B. L., Chang, J. Y., Guokas, J., Boly, M., Tononi, G., and Van Veen, B. (2014). Reversal of cortical information flow during visual imagery as compared to visual perception. *NeuroImage 100*, 237–43. https://doi.org/10.1016/j.neuroimage.2014.05.081.

Dixon, R. M. W. (1972). *The Dyirbal Language of North Queensland*. Cambridge University Press.

Dodge, E., and Wright, A. (2002, August). Herds of wildebeest, flasks of vodka, heaps of trouble: An embodied construction grammar approach to English measure phrases. In *Proceedings of the Annual Meeting of the Berkeley Linguistics Society* (pp. 75–86). Berkeley Linguistics Society.

Dolscheid, S., Hunnius, S., Casasanto, D., and Majid, A. (2014). Prelinguistic infants are sensitive to space-pitch associations found across cultures. *Psychological Science, 25*(6), 1256–1261.

Dowty, D. R. (1979). *Word Meaning and Montague Grammar: The Semantics of Verbs and Times in Generative Semantics and in Montague's PTQ*. Springer Netherlands.

Edelman, G. (1992). *Bright Air, Brilliant Fire*. Basic Books.

Ekman, P., Levenson, R. W., and Friesen, W. V. (1983). Autonomic nervous system activity distinguishes among emotions. *Science 221*, 1208–10.

Enfield, N. (2022). *Language vs. Reality: Why Language Is Good for Lawyers and Bad for Scientists*. MIT Press.

Farah, M. J. (1989). The neural basis of mental imagery. *Trends in Neuroscience 12*, 395–99.

Fauconnier, G. (1985). *Mental Spaces: Aspects of Meaning Construction in Natural Language*. MIT Press.

Fauconnier, G., and Turner, M. (2002). *The Way We Think*. Basic Books.

Fedorenko, E., Terri, L. S., Brunnerd, P., Coond, W. G., Pritchett, B., Schalkd, G., and Kanwisher, N. (2016). Neural correlate of the construction of sentence meaning. *Proceedings of the National Academy of Sciences of the USA 113*(41), E6256–62.

Feldman, J. (1982). Dynamic connections in neural networks. *Biological Cybernetics 46*(1), 27–39.

Feldman, J. (2006). *From Molecules to Metaphors: A Neural Theory of Language*. MIT Press.

Feldman J. (2013). The neural binding problem(s). *Cognitive Neurodynamics 7*(1), 1–11. https://doi.org/10.1007/s11571-012-9219-8.

Feldman, J. A., and Ballard, D. H. (1982). Connectionist models and their properties. *Cognitive Science 6*(3), 205–54. https://doi.org/10.1207/s15516709cog0603_1.

Feldman, J., Dodge, E., and Bryant, J. (2009). Embodied construction grammar. In *Oxford Handbook of Linguistic Analysis* (pp. 111–38). Oxford University Press.

Fillmore, C. (1968). The case for case. In Bach and Harms (eds.), *Universals in Linguistic Theory* (pp. 1–88). Holt, Rinehart, and Winston.

Fillmore, C. (1976). Frame semantics and the nature of language. *Annals of the New York Academy of Sciences: Conference on the Origin and Development of Language and Speech 280*, 20–32.

Fillmore, C. (1977). *Lectures on Deixis*. CSLI Publications. (Originally distributed as *Santa Cruz Lectures on Deixis, 1971*, by the Indiana University Linguistics Club, 1975.)

Fillmore, C. (1982). Frame semantics. In Linguistic Society of Korea (ed.), *Linguistics in the Morning Calm* (pp. 111–37). Hanshin.

Fillmore, C. (1985). Frames and the semantics of understanding. *Quaderni di Semantica 6*, 222–53.

Fillmore, C., Kay, P., and O'Connor, M. C. (1988). Regularity and idiomaticity in grammatical constructions: The case of let alone. *Language 64*(3), 501–38.

Gallese, V. (2001). The "shared manifold" hypothesis: From mirror neurons to empathy. *Journal of Consciousness Studies 8*(5–7), 33–50.

Gallese, V., Fadiga, L., Fogassi, L., and Rizzolatti, G. (1996). Action recognition in the premotor cortex. *Brain 119*, 593–609.

Gallese, V., Fogassi, L., Fadiga, L., and Rizzolatti, G. (2002). Action representation and the inferior parietal lobule. In W. Printz and B. Hommel (eds.), *Attention and Performance XIX* (pp. 247–66). Oxford University Press.

Gallese, V., and Lakoff, G. (2005). The brain's concepts: The role of the sensory-motor system in conceptual knowledge. *Cognitive Neuropsychology 22*(3), 455–79.

Gilbert, A. L., Regier, T., Kay, P., and Ivry, R. B. (2005). Whorf hypothesis is supported in the right visual field but not the left. In *Proceedings of the National Academy of Sciences of the USA 103*(2), 489–94. https://doi.org/10.1073/pnas .0509868103.

Gilbert, S. J., Frith, C. D., and Burgess, P. W. (2005). Involvement of rostral prefrontal cortex in selection between stimulus-oriented and stimulus-independent thought. *European Journal of Neuroscience 21*, 1423–31.

Gisiger, T., and Boukadoum, M. (2011). Mechanisms gating the flow of information in the cortex: What they might look like and what their uses may be. *Frontiers in Computational Neuroscience 5*, 1. https://doi.org/10.3389/fncom .2011.00001.

Goffman, E. (1974). *Frame Analysis: An Essay on the Organization of Experience*. Harper and Row.

Goldberg, A. E. (1995). *Constructions: A Construction Grammar Approach to Argument Structure*. University of Chicago Press.

Goldsmith, John. (1993). *The Last Phonological Rule: Reflections on Constraints and Derivations*. University of Chicago Press.

Grady, J. (1997). *Foundations of Meaning: Primary Metaphors and Primary Scenes* (doctoral dissertation). University of California, Berkeley.

Grafton, S. T., Aziz-Zadeh, L., and Ivry, R. B. (2009). Relative hierarchies and the representation of action. In M. Gazzaniga (ed.), *The Cognitive Neurosciences IV* (pp. 641–55). MIT Press.

Graybiel, A. M., and Smith, K. S. (2014). Good habits, bad habits. *Scientific American 310*, 38–43.

Graziano, M. S. A. (2018). *The Spaces between Us: A Story of Neuroscience, Evolution, and Human Nature*. Oxford University Press.

Graziano, M. S. A., and Gross, C. G. (1995). The representation of extrapersonal space: A possible role for bimodal visual-tactile neurons. In M. S. Gazzinga (ed.), *The Cognitive Neurosciences* (pp. 1021–34). MIT Press.

Graziano, M. S. A., Hu, X., and Gross, C. G. (1997). Visuo-spatial properties of ventral premotor cortex. *Journal of Neurophysiology 77*, 2268–92.

Gruber, J. S. (1965). *Studies in Lexical Relations* (master's thesis). MIT. http://www.ai.mit.edu/projects/dm/theses/gruber65.pdf.

Hebb, D. O. (1972). *The Textbook of Psychology*. W. B. Saunders.

Hoerzer, G. M., Legenstein, R., and Maass, W. (2014). Emergence of complex computational structures from chaotic neural networks through reward modulated Hebbian learning. *Cerebral Cortex 24*(3), 677–90.

Horn, M. E., and Nicoll, R. A. (2018). Somatostatin and parvalbumin inhibitory synapses onto hippocampal pyramidal neurons are regulated by distinct mechanisms. *Proceedings of the National Academy of Sciences of the USA 115*(3), 589–94. https://doi.org/10.1073/pnas.1719523115.

Huang, S., and Tanangkingsing, M. (2004). Reference to motion events in six Western Austronesian languages: Towards a semantic typology [Unpublished paper]. National Taiwan University.

Hubel, D. H., and Wiesel, T. N. (1962). Receptive fields, binocular interaction and functional architecture in the cat's visual cortex. *Journal of Physiology 160*, 106–54.

Huron, D. (n.d.). The metaphor of "high" and "low" in pitch. Archive.org. https://web.archive.org/web/20100702204154/http://www.musiccog.ohio-state.edu/Music829C/Notes/high_low.html.

Israel, M. (2011). *The Grammar of Polarity: Pragmatics, Sensitivity, and the Logic of Scales*. Cambridge University Press.

Iriki, A., and Taoka, M. (2012). Triadic (ecological, neural, cognitive) niche construction: a scenario of human brain evolution extrapolating tool use and language from the control of reaching actions. *Philosophical Transactions of the Royal Society of London. Series B, Biological Sciences 367*(1585), 10–23. https://doi.org/10.1098/rstb.2011.0190.

Itti, L., and Arbib, M. A. (2006). Attention and the minimal subscene. In M. A. Arbib (ed.), *Action to Language via the Mirror Neuron System* (pp. 289–346). Cambridge University Press.

Itti, L., Koch, C., and Niebur, E. (1998). A model of saliency-based visual attention for rapid scene analysis. In *IEEE Transactions on Pattern Analysis and Machine Intelligence 20*(11), 1254–59. doi:10.1109/34.730558.

Jakobson, R. (1968). *Child Language, Aphasia and Phonological Universals*. De Gruyter Mouton. https://doi.org/10.1515/9783111353562.

Jarbo, K., and Verstynen, T. D. (2015). Converging structural and functional connectivity of orbitofrontal, dorsolateral prefrontal, and posterior parietal cortex in the human striatum. *Journal of Neuroscience 35*(9), 3865–78.

Johnson, M. (1987). *The Body in the Mind: The Bodily Basis of Meaning, Imagination, and Reason*. University of Chicago Press.

Johnson, M. (2007). *The Meaning of the Body: Aesthetics of Human Understanding*. University of Chicago Press.

Kahneman, D. (2013). *Thinking, Fast and Slow*. Farrar, Straus and Giroux.

Kant, I. (1965). *Critique of Pure Reason*. Norman Kemp Smith (trans.). St. Martin's.

Kay, P., and Kempton, W. (1984). What is the Sapir-Whorf hypothesis? *American Anthropologist 86*, 65–79.

Koch, C. (2012). *Consciousness: Confessions of a Romantic Reductionist*. MIT Press.

Koechlin, E., Ody, C., and Kouneiher, F. (2003). The architecture of cognitive control in the human prefrontal cortex. *Science 302*, 1181–85.

Kövecses, Z. (2000). *Metaphor and Emotion: Language, Culture, and Body in Human Feeling*. Cambridge University Press.

Kövecses, Z. (2005). *Metaphor in Culture: Universality and Variation*. Cambridge University Press.

Krisst, L., Montemayor, C., and Morsella, E. (2015). Deconstructing voluntary action: Unconscious and conscious component processes. In B. Eitam and P. Haggard (eds.), *The Sense of Agency*. Oxford University Press.

Lakoff, G. (1963). Toward generative semantics. MIT Mechanical Translation Project Report. Reprinted in J. D. McCawley, *Notes from the Linguistic Underground*, Academic Press, 1977.

Lakoff, G. (1965). *On the Nature of Syntactic Irregularity*. Harvard University, Computation Laboratory. Reprinted as *Irregularity in Syntax*, Holt, Rinehart, and Winston, 1970.

Lakoff, G. (1968). Pronouns and reference. A Harvard Mimeo, distributed by the

Indiana Linguistics Club. Reprinted in J. D. McCawley (ed.), *Notes from the Linguistic Underground*, Academic Press, 1977.

Lakoff, G. (1970). *Irregularity in Syntax*. Holt, Rinehart, and Winston, 1970.

Lakoff, G. (1984). *There-Constructions: A Study in Prototype Theory and Grammatical Construction Theory*. Cognitive Science Report, No. 18, University of California, Berkeley.

Lakoff, G. (1987a). Prototype theory and cognitive models. In U. Neisser (ed.), *Concepts and Conceptual Development: Ecological and Intellectual Factors in Categorization*. Cambridge University Press.

Lakoff, G. (1987b). *Women, Fire, and Dangerous Things: What Categories Reveal about the Mind*. University of Chicago Press.

Lakoff, G. (1993). Cognitive phonology. In *The Last Phonological Rule*. Retrieved from https://escholarship.org/uc/item/45n1m8xk.

Lakoff, G. (1996). *Moral Politics: How Liberals and Conservatives Think*. University of Chicago Press.

Lakoff, G. (2003). Part II: The embodied mind, and how to live with one: In A. Sanford (ed.), *The Nature and Limits of Human Understanding: The 2001 Gifford Lectures at the University of Glasgow* (pp. 47–108). T and T Clark.

Lakoff, G. (2004). *Don't Think of an Elephant!* Chelsea Green.

Lakoff, G. (2006a). The neuroscience of form in art. In M. Turner (ed.), *The Artful Mind*. Oxford University Press.

Lakoff, G. (2006b). *Whose Freedom?* Farrar, Strauss and Giroux.

Lakoff, G. (2008). *The Political Mind*. Viking Penguin.

Lakoff, G. (2014a). *The ALL NEW Don't Think of an Elephant! Know Your Values and Frame the Debate*. Chelsea Green.

Lakoff, G. (2014b). Mapping the brain's metaphor circuitry: Metaphorical thought in everyday reasoning. *Frontiers in Human Neuroscience 8*. doi:10.3389/fnhum.2014.00958.

Lakoff, G., and Johnson, M. (1980). *Metaphors We Live By*. University of Chicago Press. [Updated version, 2002.]

Lakoff, G., and Johnson, M. (1999). *Philosophy in the Flesh*. Basic Books.

Lakoff, G., and Núñez, R. (2000). *Where Mathematics Comes From: How the Embodied Mind Brings Mathematics into Being*. Basic Books.

Lakoff, G., and Thompson, H. (1975). Introduction to cognitive grammar. *Proceedings of the Berkeley Linguistics Society 1*, 295.

Lakoff, G., and Turner, M. (1989). *More Than Cool Reason*. University of Chicago Press.

Lakoff, G., and Wehling, E. (2012). *The Little Blue Book*. Free Press.

Lane, N. (2022). Transformer: The deep chemistry of life and death. United Kingdom: Profile.

Lazzaro, J., Ryckebusch, S., Mahowald, M. A., and Mead, C. (1988). Winner-take-all networks of O(N) complexity. *Neural Information Processing Systems*.

Lindeman, L. M., and Abramson, L. Y. (2008). The mental simulation of motor incapacity in depression. *Journal of Cognitive Psychotherapy, 22*(3), 228–51.

Loftus, E. F. (2003). Our changeable memories: Legal and practical implications. *Nature Reviews Neuroscience 4*, 231–34.

Luntz, F. (2007). *Words That Work: It's Not What You Say, It's What People Hear.* Hyperion.

Maass, W. (2000). On the computational power of winner-take-all. *Neural Computation 12*, 2519–35.

Man, K., Damasio, A., Meyer, K., and Kaplan, J. T. (2015). Convergent and invariant object representations for sight, sound, and touch. *Human Brain Mapping 36*(9), 3629–40. https://doi.org/10.1002/hbm.22867.

Man, K., Kaplan, J. T., Damasio, A., and Meyer, K. (2012). Sight and sound converge to form modality-invariant representations in temporoparietal cortex. *Journal of Neuroscience, 32*(47), 16629–36. https://doi.org/10.1523/JNEUROSCI.2342-12.2012.

Man, K., Kaplan, J., Damasio, H, and Damasio, A. (2013). Neural convergence and divergence in the mammalian cerebral cortex: From experimental neuroanatomy to functional neuroimaging. *Journal of Comparative Neurology 521*(18). https://doi.org/10.1002/cne.23408.

Mandler, J. M. (1992). How to build a baby: II. Conceptual primitives. *Psychological Review 99*(4), 587–604. https://doi.org/10.1037/0033-295X.99.4.587.

Mandler, J. M., and Cánovas, P. C. (2014). On defining image schemas. *Language and Cognition 6*(4), 510–32.

Markram, H., Gerstner, W., and Sjöström, P. J. (2011). A history of spike-timing dependent plasticity. *Frontiers in Synaptic Neuroscience 3*(4), 1–24.

McGurk, H., and MacDonald, J. W. (1976). Hearing lips and seeing voices. *Nature 264*, 746–48.

Mervis, C. B., Catlin, J., and Rosch, E. (1976). Relationships among goodness-of-example, category norms, and word frequency. *Bulletin of the Psychonomic Society 7*(3), 283–84 https://doi.org/10.3758/BF03337190.

Mervis, C., and Rosch, E. (1981). Categorization of natural objects. *Annual Review of Psychology 32*, 89–115.

Meyer, K., and Damasio, A. (2009). Convergence and divergence in a neural architecture for recognition and memory. *Trends in Neurosciences 32*(7), 376–82. https://doi.org/10.1016/j.tins.2009.04.002.

Meyer, K., Kaplan, J. T., Essex, R., Webber, C., Damasio, H., and Damasio, A. R. (2010). Predicting visual stimuli on the basis of activity in auditory cortices. *Nature Neuroscience 13*, 667–68.

Morgan, J. L. (1969). On the treatment of presupposition in transformational grammar. *Proceedings from the Annual Meeting of the Chicago Linguistic Society 5*(1), 167–77.

Morsella, E., Godwin, C. A., Jantz, T. K., Krieger, S. C., and Gazaley, A. (2016). Homing in on consciousness in the nervous system: An action-based synthesis. *Behavioral and Brain Sciences, 39*, e168. https://doi.org/10.1017/S0140525X15000643.

Narayanan, S. (1997a). *KARMA: Knowledge-Based Action Representations for Metaphor and Aspect* (doctoral dissertation). University of California, Berkeley.

Narayanan, S. (1997b). *Talking the Talk Is Like Walking the Walk: A Computational model of Verbal Aspect* (paper presentation). Proceedings of the 19th Cognitive Science Society Conference.

Narayanan, S. (2003a). Cortico-subcortical loops and cognition: A computational model and preliminary results. *Neurocomputing 52–54*, 605–14.

Narayanan, S. (2003b). *Simulation Semantics: A Neurally Motivated Computational Framework for Metaphor* (paper presentation). International Cognitive Linguistics Conference, La Rioja, Spain.

Nee, D. E., and D'Esposito, M. (2016). The hierarchical organization of the lateral prefrontal cortex. *eLife 5*, e12112. https://doi.org/10.7554/eLife.12112.

Papadimitriou, C. H., Vempala, S. S., Mitropolsky, D., and Maass, W. (2020). Brain computation by assemblies of neurons. *Proceedings of the National Academy of Sciences of the USA 117*(25), 14464–72.

Paulus, W., and Kröger-Paulus, A. (1983). A new concept of retinal colour coding. *Vision Research 23*(5): 529–40. PMID:6880050.

Pfeiffer, B. E., and Foster, D. J. (2013). Hippocampal place-cell sequences depict future paths to remembered goals. *Nature 497*(7447), 74–79. doi:10.1038/nature12112.

Plaster, K., and Polinsky, M. (2007). Women are not dangerous things: Gender and categorization. Harvard Working Papers in Linguistics 12.

Pratt, C. C. (1930). The spatial character of high and low tones. *Journal of Experimental Psychology 13*, 278–85.

Raichle, M. E. (2015). The brain's default mode network. *Annual Review of Neuroscience 38*, 433–47.

Ramachandran, V. S., and Blakeslee, S. (1999). *Phantoms in the Brain: Probing the Mysteries of the Human Mind*. HarperCollins.

Ravignani, A., and Sonnweber, R. (2017). Chimpanzees process structural isomorphisms across sensory modalities. *Cognition 161*, 74–79. doi:10.10.16/j.cognition.2017.

Reddy, M. L. (1979). The conduit metaphor: A case of frame conflict in our language about language. *Metaphor and Thought 2*, 164–201.

Regier, T. (1996). *The Human Semantic Potential: Spatial Language and Constrained Connectionism*. MIT Press.

Rhodes, R., and Lawler, J. (1981). Athematic metaphor. In R. Hendrick, C. Masek, and M. F. Miller (eds.), *Papers from the Seventeenth Regional Meeting of the Chicago Linguistic Society* (pp. 318–42).

Riesenhuber, M., and Poggio, T. (1999). Hierarchical models of object recognition in cortex. *Nature Reviews Neuroscience 2*(11), 1019–25. https://doi.org/10.1038/14819.

Rizzolatti, G., Fadiga, L., Gallese, V., and Fogassi, L. (1996). Premotor cortex and the recognition of motor actions. *Cognitive Brain Research 3*, 131–41.

Rizzolatti, G., and Wolpert, D. M. (2005). Motor systems. *Current Opinion in Neurobiology 15*(6), 623–25.

Rocco, C., and Rich, A. (2012). Cross-modality correspondence between pitch and spatial location modulates attentional orienting. *Perception 41*, 339–53.

Rock, A. (2005). *The Mind at Night: The New Science of How and Why We Dream*. Basic Books.

Rosch, E. (1973). Natural categories. *Cognitive Psychology 4*(3): 328–50. doi:10.1016/0010-0285(73)90017-0.

Rosch, E. (1975). Cognitive reference points. *Cognitive Psychology 7*(4), 532–47.

Rosch, E. H. (1977). Human categorization. In N. Warren (ed.), *Advances in Cross-Cultural Psychology* (vol. 1, pp. 1–72). Academic Press.

Rosch, E. H., Mervis, C. B., Gray, W. D., Johnson, D. M., and Boyes-Braem, P. (1976). Basic objects in natural categories. *Cognitive Psychology 8*(3), 382–439.

Rosenblum, L. D. (2005). The primacy of multimodal speech perception. In D. Pisoni and R. Remez (eds.), *Handbook of Speech Perception* (pp. 51–78). Blackwell.

Ross, J. H. (1967). *Constraints on Variables in Syntax* (doctoral dissertation). Massachusetts Institution of Technology. https://dspace.mit.edu/handle/1721.1/15166.

Rumelhart, D. E. (1975). Notes on a schema for stories. In D. G. Bobrow and A. Collins (eds.), *Representation and Understanding: Studies in Cognitive Science*. Academic Press.

Sapir, E. (1958). *Culture, Language and Personality*. University of California Press.

Schacter, D. L., Addis, D. R., and Buckner, R. L. (2009). Constructive memory and the simulation of future events. In M. S. Gazzaniga (ed.), *The Cognitive Neurosciences IV* (pp. 751–62). MIT Press.

Schultze-Berndt, E. (2000). *Simple and Complex Verbs in Jaminjung: A Study of Event Categorization in an Australian Language*. Katholieke Universiteit Nijmegen.

Searle, J. (1979). *Expression and Meaning*. Cambridge University Press.

Shimojo, S. (2014). Postdiction: Its implications on visual awareness, hindsight, and sense of agency. *Frontiers in Psychology 5*, 196. https://doi.org/10.3389/fpsyg.2014.00196.

Silver, A. M., and Kastner, S. (2009). Topographic maps in human frontal and parietal cortex. *Trends in Cognitive Science 13*(11), 488–95.

Slobin, D. (2004). The many ways to search for a frog: Linguistic typology and the expression of motion events. In S. Strömqvist and L. Verhoeven (eds.), *Relating Events in Narrative*. Vol. 2, *Typological and Contextual Perspectives* (pp. 219–57). Erlbaum.

Slobin, D. (2006). What makes manner of motion salient? Explorations in linguistic typology, discourse, and cognition. In M. Hickmann and S. Robert (eds.), *Space in Languages: Linguistic Systems and Cognitive Categories* (pp. 59–81). John Benjamins.

Soros, G. (2008). *The New Paradigm for Financial Markets: The Credit Crisis of 2008 and What It Means*. PublicAffairs.

Spreng, R. N., Sepulcre, J., Turner, G. R., Stevens, D. W., and Schacter, D. L. (2013). Intrinsic architecture underlying the relations among the default, dorsal attention, and frontoparietal control networks of the human brain. *Journal of Cognitive Neuroscience 25*, 74–86.

Sullivan, K. A. (2007). *Grammar in Metaphor: A Construction Grammar Account of Metaphoric Language* (doctoral dissertation). University of California, Berkeley.

Sullivan, K. A. (2013). *Frames and Constructions in Metaphoric Language*. John Benjamins.

Talmy, L. (1988). Force dynamics in language and cognition. *Cognitive Science 12*(1), 49–100.

Talmy, L. (1996). Fictive motion in language and "ception." In P. Bloom, M. Peterson, L. Nadel, and M. Garrett (eds.), *Language and Space* (pp. 211–75). MIT Press.

Talmy, L. (2000a). *Toward a Cognitive Semantics*. Vol. 1, *Concept Structuring Systems*. MIT Press.

Talmy, L. (2000b). *Toward a Cognitive Semantics*. Vol. 2, *Typology and Process in Concept Structuring*. MIT Press.

Timble, O. C. (1934). Localization of sound in the anterior-posterior and vertical dimensions of "auditory" space. *British Journal of Psychology 24*, 320–34.

Tomasello, M. (2003). Constructing a language: A usage-based theory of language acquisition. Harvard University Press.

Tomasello, M. (2022). *The Evolution of Agency: Behavioral Organization from Lizards to Humans*. MIT Press.

Vendler, Z. (1957). Verbs and times. *Philosophical Review 66*(2), 143–60. https://doi.org/10.2307/2182371.

Walker, P., Bremner, J. G., Mason, U., Spring, J., Mattock, K., Slater, A., and Johnson, S. P. (2010). Preverbal infants' sensitivity to synaesthetic cross-modality correspondences. *Psychological Science 21*, 21–25.

Whorf, B. L. (1940). Science and linguistics. *Technology Review, 35*, 229–31, 247–48.

Woods, W. A. (1970). Transition network grammars for natural language analysis. *Communications of the ACM 13*(10), 591–606.

Yang, H. *A Priming Circuit Controls the Olfactory Response and Memory in Drosophila* (doctoral dissertation). Harvard University, Graduate School of Arts and Sciences.

Yong, E. (2022). *An Immense World: How Animal Senses Reveal the Hidden Realms around Us*. Random House.

Bibliography

Adolphs, R., Damasio, H., Tranel, D., Cooper, G., and Damasio, A. R. (2000). A role for somatosensory cortices in the visual recognition of emotion as revealed by three dimensional lesion mapping. *Journal of Neuroscience 20*(7), 2683–90.

Aziz-Zadeh, L., Cruse, H., Metzinger, T., Narayanan, S., Nunez, R., and Steels, L. (eds.). (2009). *Workshop Report: Toward the Neural Basis of High-Level Cognition: The Embodiment of Compositional Meaning.* Wissenschaftskolleg zu Berlin.

Aziz-Zadeh, L., and Damasio, A. (2008). Embodied semantics for actions: Findings from functional brain imaging. *Journal of Physiology 102*, 35–39.

Aziz-Zadeh, L., Fiebach, C., Narayanan, S., Feldman, J., Dodge, E., and Ivry, R. (2006). *Modulation of FFA and PPA by Language Related to Faces and Places* (paper presentation). Human Brain Mapping Conference, Italy.

Aziz-Zadeh, L., Fiebach, C. J., Naranayan, S., Feldman, J., Dodge, E., and Ivry, R. B. (2008). Modulation of the FFA and PPA by language related to faces and places. *Social Neuroscience 3*, 229–38.

Aziz-Zadeh, L., and Ivry, R. B. (2009). The human mirror neuron system and embodied representations. In D. Sternad (ed.), *Progress in Motor Control* (pp. 355–76). Springer.

Aziz-Zadeh, L., Wilson, S., Rizzolatti, G., and Iacoboni, M. (2006). Congruent embodied representations for visually presented actions and linguistic phrases describing actions. *Current Biology 16*, 1–6.

Barsalou, L. W. (2005). Continuity of the conceptual system across species. *Trends in Cognitive Sciences 9*, 309–11.

Bergen, B. K., Lindsay, S., Matlockc, T., and Narayanan, S. (2007). Spatial and linguistic aspects of visual imagery in sentence comprehension. *Cognitive Science 31*(5), 733–64.

Bergen, B., Medeiros-Ward, N., Wheeler, K., Drews, F., and Strayer, D. (2013). The crosstalk hypothesis: Why language interferes with driving. *Journal of Experimental Psychology: General 142*(1), 119–30.

Bergen, B., and Wheeler, K. (2005). Sentence understanding engages motor processes. In *Proceedings of the Twenty-Seventh Annual Conference of the Cognitive Science Society 27*, 238–43.

Bergen, B., and Wheeler, K. (2010). Grammatical aspect and mental simulation. *Brain and Language 112*, 150–58.

Berlin, B., and Kay, P. (1969). *Basic Color Terms: Their Universality and Evolution*. University of California Press.

Bolinger, D. (1977). *Meaning and Form*. Longman.

Boroditsky, L. (1997). Evidence for metaphoric representation: Perspective in space and time. In M. G. Shafto and P. Langley (eds.), *Proceedings of the Nineteenth Annual Conference of the Cognitive Science Society*.

Boroditsky, L. (2000). Metaphoric structuring: Understanding time through spatial metaphors. *Cognition 75*, 1–28.

Boulenger, V., Hauk, O., and Pulvermuller, F. (2009). Grasping ideas with the motor system: Semantic somatotopy in idiom comprehension. *Cerebral Cortex 19*, 190–91.

Boulenger, V., Shtyrov, Y., and Pulvermuller, F. (2012). When do you grasp the idea? MEG evidence for instantaneous idiom understanding. *NeuroImage 59*, 3502–13.

Brachman, R. J., and Schmolze, J. G. (1985). An overview of the KL-One knowledge representation system. *Cognitive Science 9*(2), 171–216.

Brockman, M. (2009). *What's Next? Dispatches on the Future of Science: Original Essays from a New Generation of Scientists*. Vintage Books.

Brown, T. L. (2003). *Making Truth: Metaphor in Science*. University of Illinois Press.

Brugman, C. (1981). *The Story of over* (master's thesis). University of California, Berkeley.

Brugman, C. (1983). The use of body-part terms as locatives in Chalcatongo mixtec. In *Report No. 4 in the Survey of California and Other Indian Languages* (pp. 235–90). University of California, Berkeley.

Buccino, G., Binkofski, F., Fink, G. R., Fadiga, L., Fogassi, L., Gallese, V., Seitz,

R. J., Zilles, K., Rizzolatti, G., and Freund, H. J. (2001). Action observation activates premotor and parietal areas in a somatotopic manner: An fMRI study. *European Journal of Neuroscience 13*, 400–404.

Bryant, J. (2008). *Best-Fit Constructional Analysis* (doctoral dissertation). University of California, Berkeley.

Casad, E., and Langacker, R. W. (1985). "Inside" and "outside" in Cora grammar. *International Journal of American Linguistics 51*, 247–81.

Casasanto, D. (2008). Similarity and proximity: When does close in space mean close in mind? *Memory and Cognition 36*, 1047–56.

Chang, N. (2009). *Constructing Grammar: A Computational Model of the Emergence of Early Constructions* (doctoral dissertation). University of California, Berkeley.

Damasio, H., Grabowski, T. J., Tranel, D., Hichwa, R. D., and Damasio, A. (1996). A neural basis for lexical retrieval. *Nature 380*(6574), 499–505.

Damasio, H., Grabowski, T. J., Tranel, D., Ponto, L. L. B., Hichwa, R. D., and Damasio, A. (2001). Neural correlates of naming actions and of naming spatial relations. *NeuroImage 13*, 1053–64.

Dayan, P. (2009). Goal-directed control and its antipodes. *Neural Networks 22*(3), 213–19.

Decety, J., Sjoholm, H., Ryding, E., Stenberg, G., and Ingvar, D. (1990). The cerebellum participates in cognitive activity: Tomographic measurements of regional cerebral blood flow. *Brain Research 535*, 313–17.

Dehaene, S. (2009). *Reading in the Brain*. Penguin Viking.

Desai, R. H., Binder, J. R., Conant, L. L., Mano, Q. R., and Seidenberg, M. S. (2012). The neural career of sensorimotor metaphors. *Journal of Cognitive Neuroscience 23*(9), 2376–86.

Desai, R. H., Conant, L. L., Binder, J. R., Park, H., and Seidenberg, M. S. (2013). A piece of the action: Modulation of sensory-motor regions by action idioms and metaphors. *NeuroImage 83*, 862–69.

D'Esposito, M., and Postle, B. R. (2015). The cognitive neuroscience of working memory. *Annual Review of Psychology 66*, 115–42.

DeValois, R. L., and DeValois, K. (1975). Neural coding of color. In E. C. Careterette and M. P. Friedman (eds.), *Handbook of Perception*. Vol. 5, *Seeing*. Academic Press.

de Vignemont, F., and Singer, T. (2006). The empathic brain: How, when and why? *Trends in Cognitive Science 10*(10), 435–41.

Dodge, E. K. (2010). *Constructional and Conceptual Composition* (doctoral dissertation). University of California, Berkeley.

Dodge, E., Hong, J., and Stickles, E. (2015). MetaNet: Deep semantic automatic metaphor analysis. In *Proceedings of the Third Workshop on Metaphor in NLP* (pp. 40–49). Association for Computational Linguistics.

Doya, K. (2002). Metalearning and neuromodulation. *Neural Networks 15*, 495–506.

Doya, K. (2008). Modulators of decision making. *Nature Neuroscience 11*, 410–16.

Duce, D. A., and Ringland, G. A. (1988). *Approaches to Knowledge Representation: An Introduction*. Research Studies Press.

Ekman, P. (1999). Basic emotions. In T. Dalgleish and M. Power (eds.), *Handbook of Cognition and Emotion*. Wiley.

Fadiga, L., and Rizzolatti, G. (2001). Cortical mechanism for the visual guidance of hand grasping movements in the monkey: A reversible inactivation study. *Brain 124*, 571–86.

Farah, M. J. (1985). Psychophysical evidence for a shared representational medium for visual images and percepts. *Journal of Experimental Psychology: General 114*, 93–104.

Farah, M. J. (1988). Is visual imagery really visual? Overlooked evidence from neuropsychology. *Psychological Review 95*, 307–17.

Farah, M. J. (1989). Semantic and perceptual priming: How similar are the underlying mechanisms? *Journal of Experimental Psychology: Human Perception and Performance 15*, 188–94.

Farah, M. J. (2000). The neural bases of mental imagery. In M. S. Gazzinga (ed.), *The Cognitive Neurosciences* (2nd ed.). MIT Press.

Feldman, J., and Narayanan, S. (2004). Embodied meaning in a neural theory of language. *Brain and Language 89*(2), 385–92.

Feldman, J., and Narayanan, S. (2008). *Simulation and Cognition in Language Understanding* (paper presentation). Proceedings of Exciting Biologies, Chantilly, France.

Feldman, J., and Narayanan, S. (2011). *Simulation Semantics, Embodied Construction Grammar, and the Language of Events* (paper presentation). Workshop proceedings of the conference of the American Association for Artificial Intelligence: Language Action Tools for Cognitive Artificial Agents, San Francisco.

Filimon, F., Nelson, J. D., Hagler, D. J., and Sereno, M. I. (2007). Human cortical representations for reaching: Mirror neurons for execution, observation, and imagery. *NeuroImage 37*(4), 1315–28.

Fillmore, C. (1977). Scenes and frames semantics. In A. Zampolli (ed.), *Linguistic Structures Processing, Fundamental Studies in Computer Science*. North Holland.

Fogassi, L., Gallese, V., Fadiga, L., Luppino, G., Matelli, M., and Rizzolatti, G. (1996). Coding of peripersonal space in inferior premotor cortex (area F4). *Journal of Neurophysiology 76*, 141–57.

Gallese, V., and Goldman, A. (1998). Mirror neurons and the simulation theory of mind-reading. *Trends in Cognitive Sciences 12*, 493–501.

Gallese, V., and Lakoff, G. (2005). The brain's concepts: The role of the sensory-motor system in reason and language. *Cognitive Neuropsychology 22*, 455–79.

Gallie, W. B. (1955). Essentially contested concepts. In *Proceedings of the Aristotelian Society* (pp. 167–98). Harrison and Sons.

Gallie, W. B. (1956). Essentially contested concepts. *Proceedings of the Aristotelian Society, New Series 56*, 167–98.

Gedigian, M., Bryant, J., Narayanan, S., and Ciric, B. (2006). *Catching Metaphors* (paper presentation). Scalable Natural Language Understanding Conference, Boston.

Gerstner, W., Werner, M. K., Naud, R., and Paninski, L. (2014). *Neuronal Dynamics: From Single Neurons to Networks and Models of Cognition.* Cambridge University Press. https://neuronaldynamics.epfl.ch/index.html.

Gibbs, R. W. (2006). *Embodiment in Cognitive Science.* Cambridge University Press.

Gibbs, R. W. (2006). Metaphor interpretation as embodied simulation. *Mind and Language, 21*(3), 434–58.

Gibbs, R. W. (1984). *The Poetics of Mind: Figurative Thought, Language, and Understanding.* Cambridge University Press.

Gibbs, R. W., and Matlock, T. (1999). Psycholinguistics and mental representations. *Cognitive Linguistics 10*(3), 263–70.

Glenberg, A. M., and Kaschak, M. P. (2002). Grounding language in action. *Psychonomic Bulletin and Review 9*, 558–65.

Glenberg, A. M., and Robertson, D. (2000). Symbol grounding and meaning: A comparison of high-dimensional and embodied theories of meaning. *Journal of Memory and Language, 43*, 379–401.

Glenberg, A. M., Sato, M., Cattaneo, L., and Riggio, L. (2008). Processing abstract language modulates motor system activity. *Quarterly Journal of Experimental Psychology 61*, 905–19.

Goldberg, A. E. (2006). *Constructions at Work: The nature of Generalization in Language.* Oxford University Press.

Graziano, M. S. A. (2014). Cortical action representations. In A. W. Toga and R. A. Poldrack (eds.), *Brain Mapping: An Encyclopedic Reference.* Elsevier.

Graziano, M. S. A. (2015). A new view of the motor cortex and its relation to social behavior. In S. S. Obhi and E. S. Cross (eds.), *Shared Representations: Sensorimotor Foundations of Social Life* (pp. 38–58). Cambridge University Press.

Graziano, M. S. A., and Aflalo, T. N. (2007). Mapping behavioral repertoire onto the cortex. *Neuron 56*, 239–51.

Graziano, M. S. A., and Aflalo, T. N. (2007). Rethinking cortical organization: Moving away from discrete areas arranged in hierarchies. *Neuroscientist 13*, 138–47.

Graziano, M. S. A., Yap, G. S., and Gross, C. G. (1994). Coding of visual space by premotor neurons. *Science 266*, 1054–57.

Gould, S. J., and Vrba, E. S. (1982). Exaptation: A missing term in the science of form. *Paleobiology 8*(1), 4–15.

Harmon-Jones, E., Gable, P. A., and Price, T. F. (2011). Leaning embodies desire: Evidence that leaning forward increases relative left frontal cortical activation to appetitive stimuli. *Biological Psychology 87*, 311–13.

Hubel, D. H. (1995). *Eye, Brain, and Vision*. W. H. Freeman.

Hubel, D. H., and Wiesel, T. N. (2005). *Brain and Visual Perception: The Story of a 25-Year Collaboration*. Oxford University Press.

Iacoboni, M. (2008). *Mirroring People*. Farrar, Strauss and Giroux.

Jameson, K. A. (2007). Tetrachromatic color vision. In P. Wilken, T. Bayne, and A. Cleeremans (eds.), *The Oxford Companion to Consciousness*. Oxford University Press.

Jeannerod, M. (1994). The representing brain: Neural correlates of motor intention and imagery. *Behavioral Brain Science 17*, 187–245.

Johnson, C. (1999). *Constructional Grounding* (doctoral dissertation). University of California, Berkeley.

Johnson, M. (1993). *Moral Imagination: Implications of Cognitive Science for Ethics*. University of Chicago Press.

Jostmann, N. B., Lakens, D., and Schubert, T. W. (2009). Weight as an embodiment of importance. *Psychological Science 1*, 1169–74.

Kahn, I., Knoblich, U., Desai, M., Bernstein, J., Graybiel, A. M., Boyden, E. S., Buckner, R. L., and Moore, C. I. (2013). Optogenetic drive of neocortical pyramidal neurons generates fMRI signals that are correlated with spiking activity. *Brain Research 1511*, 33–45.

Kandel, E. R. (2001). The molecular biology of memory storage: A dialog between genes and synapses. *Bioscience Reports 21*, 565–611.

Kandel, E. R. (2006). *In Search of Memory: The Emergence of a New Science of Mind*. W. W. Norton.

Kandel, E. R., Schwartz, J. H., Jessell, T. M., Siegelbaum, S. A., and Hudspeth, A. J. (2012). *Principles of Neuroscience* (5th ed.). McGraw-Hill.

Katz, P. S., and Frost, W. N. (1996). Intrinsic neuromodulation: Altering neuronal circuits from within. *Trends in Neurosciences 19*(2), 54–61.

Kay, P., and McDaniel, C. K. (1978). The linguistic significance of the meanings of basic color terms. *Language 54*(3), 610–46.

Keltner, D., and Ekman, P. (2015, July 3). The science of inside out. *New York Times.* http://www.nytimes.com/2015/07/05/opinion/sunday/the-science-of -inside-out.html.

Keltner, D., Marsh, J., and Smith, J. A. (2010). *The Compassion Instinct.* W. W. Norton.

Keysers, C., Wicker, B., Gazzola, V., Anton, J., Fogassi, L., and Gallese, V. (2004). A touching sight: SII/PV activation during the observation and experience of touch. *Neuron 42*, 335–46.

Kosslyn, S. M., Alpert, N. M., Thompson, W. L., Malijkovic, V., Weise, S., Chabris, C., Hamilton, S. E., Rauch, S. L., and Buonanno, F. S. (1993). Visual mental imagery activates topographically organized visual cortex: PET investigations. *Journal of Cognitive Neuoroscience 5*, 263–87.

Kosslyn, S. M., Ball, T. M., and Reiser, B. J. (1978). Visual images preserve metric spatial information: Evidence from studies of image scanning. *Journal of Experimental Psychology: Human Perception and Performance 4*, 47–60.

Kosslyn, S. M., and Thompson, W. L. (2000). Shared mechanisms in visual imagery and visual perception: Insights from cognitive science. In M. S. Gazzinga (ed.), *The Cognitive Neurosciences* (2nd ed.). MIT Press.

Kövecses, Z. (1990). *Emotion Concepts.* Springer.

Labov, W., and Waletzky, J. (1968). Narrative analysis: Oral versions of personal experience. *Journal of Narrative and Life History 7*, 3–38.

Labov, W. (1973). The boundaries of words and their meanings. In C.-J. N. Bailey and R. W. Shuy (eds.), *New Ways of Analyzing Variation in English.* Georgetown University Press.

Lakoff, G. (1964). Structural complexity in fairy tales (paper presentation). Summer Meeting of the Linguistic Society of America.

Lakoff, G. (1986). *Frame Semantic Control of the Coordinate Structure Constraint* (paper presentation). Proceedings of the 21st Annual Meeting of the Chicago Linguistic Society.

Lakoff, G. (1991). The invariance hypothesis. *Cognitive Linguistics 1*(1), 39–74.

Lakoff, G. (1993). The contemporary theory of metaphor. In A. Ortony (ed.), *Metaphor and Thought* (2nd ed., pp. 202–51). Cambridge University Press.

Lakoff, G. (1993). *Grounded Concepts without Symbols* (paper presentation). Proceedings of the Cognitive Science Society.

Lakoff, G. (1993). How metaphor structures dreams: The theory of conceptual metaphor applied to dream analysis. *Dreaming 3*(2), 77–98.

Lakoff, G. (1994). What is metaphor? In J. Barnden and K. Holyoak (eds.), *Analogical Connections*. Vol. 3, *Advances in Connectionist Theory*. Addison-Wesley.

Lakoff, G. (1997). How unconscious metaphorical thought shapes dreams. In D. J. Stein (ed.), *Cognitive Science and the Unconscious*. American Psychiatric Press.

Lakoff, G. (2005). The cognitive foundations of mathematics. In J. Campbell (ed.), *Handbook of Mathematical Cognition*. Psychology Press.

Lakoff, G. (2005). Why literal meaning? In S. Coulson and B. Lewandowska-Tomascyk (eds.), *The Literal-Nonliteral Distinction*. Peter Lang.

Lakoff, G. (2009). The neural theory of metaphor. In R. Gibbs (ed.), *The Cambridge Handbook of Metaphor and Thought*. Cambridge University Press.

Lakoff, G. (2012, October 30). Global warming systemically caused Hurricane Sandy. Huffington Post. http://www.huffingtonpost.com/george-lakoff/sandy -climate-change_b_2042871.html.

Lakoff, G., and Brugman, C. (1986). Argument forms in lexical semantics. *Proceedings of the Twelfth Annual Meeting of the Berkeley Linguistics Society*, 442–54.

Lakoff, G., and Dodge, E. (2005). The neural basis of image schemas. In B. Hampe and J. E. Grady (eds.), *From Perception to Meaning: Image Schemas in Cognitive Linguistics*. Benjamins.

Lakoff, G., and Johnson, M. (2002). Why cognitive linguistics requires embodied realism. *Cognitive Linguistics 13*(3), 245–64.

Lakoff, G., and Kövecses, Z. (1984). *The Cognitive Model of Anger Inherent in American English*. Cognitive Science Report No. 10. University of California, Berkeley. Reprinted in D. Holland and N. Quinn, *Cultural Models in Language and Thought*, Cambridge University Press, 1987.

Lakoff, G., and Narayanan, S. (2010). *Toward a Computational Model of Narrative* (paper presentation). Proceedings of the AAAI Fall Symposium.

Lakoff, G., and Núñez, R. (1997). The metaphorical structure of mathematics: Sketching out cognitive foundations for a mind-based mathematics. In L. English (ed.), *Mathematical Reasoning: Analogies, Metaphors, and Images*. Erlbaum.

Lakoff, G., and Núñez, R. (1998). What did Weierstrass really define? The cognitive structure of natural and ∈-δ continuity. *Mathematical Cognition 4(2)*, 85–101.

Langacker, R. W. (1987). *Foundations of Cognitive Grammar*. Vol.1, *Theoretical Prerequisites*. Stanford University Press.

Langacker, R. W. (1990). *Concept, Image, and Symbol: The Cognitive Basis of Grammar*. Cognitive Linguistics Research 1. Mouton de Gruyter.

Langacker, R. W. (1991). *Foundations of Cognitive Grammar*. Vol. 2, *Descriptive Application*. Stanford University Press.

Langacker, R. W. (2008). *Cognitive Grammar: A Basic Introduction*. Oxford University Press.

Lee, S. W. S., and Schwarz, N. (2011). Wiping the slate clean: Psychological consequences of physical cleansing. *Current Directions in Psychological Science 20(5)*, 307–11.

Lee, S. W. S., and Schwarz, N. (2012). Bidirectionality, mediation, and moderation of metaphorical effects: The embodiment of social suspicion and fishy smells. *Journal of Personality and Social Psychology 103*, 737–49.

Lindner, S. (1981). *A Lexico-semantic Analysis of Verb-Particle Constructions with* Up *and* Out (doctoral dissertation). University of California, San Diego.

Loenneker-Rodman, B., and Narayanan, S. (2012). Computational models of figurative language. In M. Spivey, M. Joannisse, and K. McRae (eds.), *Cambridge Encyclopedia of Psycholinguistics*. Cambridge University Press.

Maeda, F., Kanai, R., and Shimojo, S. (2004). Changing pitch induced visual motion illusion. *Current Biology 14(23)*, R990–91.

Marmor, G. S., and Zaback, L. A. (1976). Mental rotation by the blind: Does mental rotation depend on visual imagery? *Journal of Experimental Psychology: Human Perception and Performance 2*, 515–21.

Marslen-Wilson, W. D., and Tyler, L. (2007). Morphology, language and the brain: The decompositional substrate for language comprehension. *Philosophical Transactions of the Royal Society B 362*, 823–36.

Matlock, T. (2002). *How Real Is Fictive Motion?* (doctoral dissertation). University of California, Santa Cruz.

Matlock, T. (2004). Fictive motion as cognitive simulation. *Memory and Cognition 32*, 1389–1400.

McCawley, J. D. (1980). *Everything Linguists Have Always Wanted to Know about Logic . . . but Were Afraid to Ask*. University of Chicago Press.

Melzack, R., and Wall, P. D. (1965). Pain mechanisms: A new theory. *Science 150*, 971–79.

Mervis, C. (1986). Child-basic object categories and early lexical development.

In U. Neisser (ed.), *Concepts and Conceptual Development: Ecological and Intellectual Factors in Categorization* (pp. 201–33). Cambridge University Press.

Miller, K. D., Keller, J. B., and Stryker, C. D. (1989). Ocular dominance column development: Analysis and simulation. *Science 111*, 123–45.

Minsky, M. (1974). A framework for representing knowledge. MIT-AI Laboratory Memo 306. Reprinted in P. Winston (ed.), *The Psychology of Computer Vision*, McGraw-Hill 1975; J. Haugeland (ed.), *Mind Design*, MIT Press, 1981; and A. Collins and E. E. Smith (eds.), *Cognitive Science*, Morgan-Kaufmann, 1992.

Narayanan, S. (1999). *Moving Right Along: A Computational Model of Metaphoric Reasoning about Events* (paper presentation). Proceedings of the National Conference on Artificial Intelligence AAAI-99, Orlando, FL.

Narayanan, S. (2009). *A Computational Model of Narrative Interpretation* (paper presentation). Proceedings of the MIT Symposium on the Computational Modeling of Narrative. Edited proceedings also available as CSAIL Report No. 2009-063: MIT Computer Science and Artificial Intelligence Laboratory, Cambridge, MA.

Narayanan, S. (2010). Mind changes: A simulation semantics account of counterfactuals. ICSI, internal publication.

Narayanan, S. (2012). Mind changes: A simulation semantic model of counterfactuals. Extended version of *Cognitive Science* article.

Narayanan, S., and Hong, J. (2013). *MetaNet: A Multilingual Metaphor Repository* (paper presentation). Proceedings of the International Cognitive Linguistics Conference.

Nersessian, N. (2008). *Creating Scientific Concepts*. MIT Press.

O'Donnel, T. J. (2015). *Productivity and Reuse in Language: A Theory of Linguistic Computation and Storage*. MIT Press.

Palmer, S., Rosh, E., and Chase, P. (1981). Canonical perspective and the perception of objects. In J. Long and A. Baddeley (eds.), *Attention and Performance IX* (pp. 135–51). Erlbaum.

Petruck, M. (ed.). (2016). MetaNet [Special issue]. *Constructions and Frames 8* (2).

Propp, V. (1928). *Morphology of the Folktale*. Leningrad. Reprinted by Mouton, 1958; and University of Texas Press, 1968.

Pulvermuller, F. (2005). Brain mechanisms linking language and action. *Nature Reviews Neuroscience 6*, 576–82.

Pulvermuller, F., Harle, M., and Hummel, F. (2001). Walking or talking? Behavioral and neurophysiological correlates of action verb processing. *Brain and Language 78*, 143–68.

Pulvermuller, F., Huss, M., Kherif, F., Moscoso, F., Martin, P., Hauk, O., and Shtyrov, Y. (2006). Motor cortex maps articulatory features of speech sounds. *Proceedings of the National Academy of Sciences of the USA 103*(20), 7865–70.

Raposo, A., Moss, H. E., Stamatakis, E. A., and Tyler, L. K. (2009). Modulation of motor and premotor cortices by actions, action words and action sentences. *Neuropsychologia 47*, 388–96.

Regier, T. (1995). A model of the human capacity for categorizing spatial relations. *Cognitive Linguistics 6*(1), 63–88.

Rizzolatti, G., Fogassi, L., and Gallese, V. (2001). Neurophysiological mechanisms underlying the understanding and imitation of action. *Nature Reviews Neuroscience 2*, 661–70.

Rizzolatti, G., and Gallese, V. (2004). Do perception and action result from different brain circuits? The three visual systems hypothesis. In L. van Hemmen and T. Sejnowski (eds.), *Problems in Systems Neuroscience*. Oxford University Press.

Rizzolatti, G., Luppino, G., and Matelli, M. (1998). The organization of the cortical motor system: new concepts. *Electroencephalography and Clinical Neurophysiology 106*, 283–96.

Rizzolatti, G., and Sinigaglia, C. (2010). The functional role of the parieto-frontal mirror circuit: Interpretations and misinterpretations. *Nature Reviews Neuroscience 11*, 264–74.

Rosch, E. H. (1975). Cognitive representation of semantic categories. *Journal of Experimental Psychology 104*(3), 192–233.

Rosch, E. H., and Lloyd, B. (eds.). (1978). *Cognition and Categorization*. Erlbaum.

Rosch, E., and Mervis, C. B. (1975). Family resemblances: Studies in the internal structure of categories. *Cognitive Psychology 7*(4), 573–605.

Rosch, E. H., Varela, F., and Thompson, E. F. (1991). *The Embodied Mind*. MIT Press.

Rosenbloom, P. (1950). *The Elements of Mathematical Logic*. Dover.

Rumelhart, D. E. (1980). On evaluating story grammars. *Cognitive Science 4*, 313–16.

Sapolsky, R. (2010). This is your brain on metaphors. *New York Times*, November 4.

Schank, R., and Abelson, R. (1977). *Scripts, Plans, Goals and Understanding: An Inquiry into Human Knowledge Structures*. Erlbaum.

Searle, J., and Vanderveken, D. (1985). *Foundations of Illocutionary Logic*. Cambridge University.

Sinha, S. (2008). *Answering Questions about Complex Events* (doctoral dissertation). University of California, Berkeley.

Sweetser, E. (1990). *From Etymology to Pragmatics*. Cambridge University Press.

Talmy, L. (1983). How language structures space. In H. L. Pick Jr. and L. P. Acredolo (eds.), *Spatial orientation: Theory, Research, and Application* (pp. 225–82). Plenum.

Taub, S. F. (2001). *Language from the Body: Iconicity and Metaphor in American Sign Language*. Cambridge University Press.

Taylor, J. G., and Taylor N. R. (2000). Analysis of recurrent cortico-basal ganglia-thalamic loops for working memory. *Biological Cybernetics 82*, 415–32.

Thibodeau, P. H., and Boroditsky, L. (2013). Natural language metaphors covertly influence reasoning. *PLOS One 8*(1), e52961.

Tranel, D., Kemmerer, D., Damasio, H., Adolphs, R., and Damasio, A. R. (2003). Neural correlates of conceptual knowledge for actions. *Cognitive Neuropsychology 20*, 409–32.

Varela, F., Thompson, E., and Rosch, E. (1991). *The Embodied Mind: Cognitive Science and Human Experience*. MIT Press.

Williams, L. E., and Bargh, J. A. (2008). Experiencing physical warmth influences interpersonal warmth. *Science 322*, 606–7.

Wilson, N., and Gibbs, R. (2007). Real and imagined body movement primes metaphor comprehension. *Cognitive Science 31*, 721–31.

Woods, W. A., and Schmolze, J. G. (1992). The KL-ONE family. *Computers and Mathematics with Applications 23*(2–5), 133.

Yeshurun, Y., Swanson, S., Simony, E., Chen, J., Lazaridi, C., Honey, C., and Hasson, U. (2017). Same story, different story: The neural representation of interpretive frameworks. *Psychological Science 28*(3), 307–19.

Zeilhofer, H. U., Wildner, H., and Yévenes, G. E. (2012). Fast synaptic inhibition in spinal sensory processing and pain control. *Physiological Review 92*(1):193–235.

Zhong, C. B., and Leonardelli, G. J. (2008). Cold and lonely: Does social exclusion feel literally cold? *Psychological Science 19*, 838–42.

Zhong, C. B., and Liljenquist, K. (2006). Washing away your sins: Threatened morality and physical cleansing. *Science 313*, 1451–52.

Zimler, J., and Keenan, J. M. (1983). Imagery in the congenitally blind: How visual are visual images? *Journal of Experimental Psychology: Learning, Memory, and Cognition 9*, 269–82.

Zwaan, R. A. (2004). The immersed experiencer: Toward an embodied theory of language comprehension. In B. H. Ross (ed.), *The Psychology of Learning and Motivation*, vol. 43 (pp. 35–62). Academic Press.

Zwaan, R. A., and Madden, C. J. (2005). Embodied sentence comprehension. In D. Pecher and R. A. Zwaan (eds.), *Grounding Cognition: The Role of Perception and Action in Memory, Language, and Thinking* (pp. 224–45). Cambridge University Press.

Index

Page numbers followed by *f* or *t* refer to figures or tables, respectively.

anaphora, 305–11

Arbib, M. A., 201

Aristotle, 109, 112, 121–23, 148, 160

artificial intelligence (AI): backpropagation and, 189–90; ChatGPT, 149; cognition and, 341–43; computational modeling and, 190, 341; connections and, 343; constructions and, 342; deep learning and, 341–43; emotion and, 342; evolution and, 343; Go and, 189; GPT-X, 342; linguistics and, 343; memory and, 342; multimodality and, 342; neural firing and, 188–90; Open AI and, 149; Persian Gulf War and, 189; reinforcement learning and, 190; reward and, 342; scales and, 342; supervised learning and, 190; unsupervised learning and, 190

asterisk, 350n6

"Athematic Metaphor" (Rhodes and Lawler), 66

Austin, John L., 28, 58

axons: backpropagation and, 189–90, 341, 349n4; extensions of, 12, 16; Hodgkin-Huxley model and, 349n2; neural firing and, 175–80, 184–87; neural networks and, 12–13, 16–19

Aziz-Zadeh, L., 202

backpropagation, 189–90, 341, 349n4

Badre, David, 31, 33, 62, 304, 347n7, 348n5

Bailey, David, 87

Ballard, D. H., 221–22

Bargh, J. A., 228

basic circuits: activation and, 192–99; binding circuits and, 194; cascades and, 193, 195; cognition and, 191, 198–200; composition and, 191; computation models and, 191 99; connections and, 191–99; connectivity and, 198; constructions and, 191; disinhibition hypothesis and, 212–13; electrochemistry and, 191, 194; embedding and, 194; embodiment and, 191, 194, 199; emotion and, 191; evolution and, 191, 199; examples of, 193–95;

exaptation and, 191; executing networks (X-Nets) and, 192, 259–60; frames and, 191, 199; functional circuits and, 193, 196–97; gating and, 194, 196–99; generalization and, 191; genetics and, 190; gestalt circuits and, 194, 198–99; grammar and, 259–65; ideas and, 190–93, 259–60; image schemas and, 194; imagination and, 191; inhibition and, 194–99; integration and, 101; ions and, 196; language form and, 191; learning and, 191; linguistics and, 191, 198; mapping and, 194, 199; neural firing and, 195–99; neurons and, 192, 196–97; neurotransmitters and, 196; recruitment and, 191–92, 198; repurposing and, 191; reward and, 192–93; scales and, 197; sensorimotor control and, 191; spikes and, 195–96; synapses and, 190–91, 193, 196–98; topographic maps and, 194

basic-level categories, 53, 59, 163

basilar membrane, 20

BCM theory, 188

beeps, 69–71

Bergen, Benjamin, 67, 243, 297

Bernanke, Ben, 312

Bhandari, A., 347n7

Bienenstock, E. L., 188

Bill of Rights, vii–viii, 173

binding circuits, 174; agreement and, 263; basic circuits and, 194; case studies and, 320; combinatorial circuits and, 207–12; conscious self and, 208; coordination and, 226, 228; dynamic, 209; frames and, 211; grammar and, 255, 258t, 259, 263, 265, 268, 277, 303t, 304; metaphor and, 135, 141, 168; multiple, 211; multiple mapping circuits and, 215–16; neural theory applications and, 246; part-to-part, 102; part-whole, 110; schemas and, 102, 110; simple, 209–11; simulation and, 232–33; visual-to-acoustic, 69

binding constraints, 101–2

biochemical interactions, 10, 173, 180–81

bleeding, 339–40

blindness, 3, 102–3

Bobrow, Daniel, 116

Body in the Mind, The (Johnson), 104

Boroditsky, Lera, 5

boundary layers, 218–19

Brain Activity Map project, 11

brain-body connections, 31, 336

Brain from Inside Out, The (Buzsáki), 230

BRAIN Initiative, 11

Broca's area, 250

Brugman, Claudia, 261, 320

Bryant, J., 297

Buzsáki, György, 27, 230

calcium channels, 17, 181, 184

calcium ions, 13–15, 179

cascade circuits: basic circuits and, 193; case studies and, 313, 318–19; concept of, 43; convergence-divergence zones (CDZs) and, 78; coordination and, 206, 222; meaningful thought and, 335; metaphor and, 141, 168; neural firing and, 174; simulation and, 233–34, 245; verb meanings and, 337

cascades: activation and, 43–49; active, 43, 193, 229; basic circuits and, 193, 195; behavior control and, 200–207; bindings and, 50; case studies and, 311–13, 318–20, 325; cognition and, 44, 48; combinatorial circuits and, 207; computation models and, 43, 348n4; connections and, 44–49; consciousness and, 49; control and, 60–61; convergence-divergence zones (CDZs) and, 52–53, 57–58, 78; coordination and, 200–207, 222, 225–26, 229, 231; Dehaene on, 43–44; embodiment and, 48–49; emotion and, 48; executing networks (X-Nets) and, 336; eyes and, 45, 48; frames and, 113, 120; functional circuits and, 43; general features of, 48–49; gestalt circuits and, 47–48; grammar and, 255–60, 268, 277, 280, 285, 290–91, 303t; ideas and, 311–20; imagination and, 48–49; inhibition

and, 44–46, 49; integration and, 49, 58, 73, 103, 113, 120, 156–57, 168, 174, 231–35, 240, 245–47, 255–59, 268, 280, 290, 311, 320, 335–37; ions and, 45; language types and, 326; learning and, 49; lexical, 48, 337; localization and, 49; meaningful thought and, 336–37; metaphor and, 135, 141, 143, 149–50, 156–57, 168, 311–20, 351n17; modules and, 49; neural, 43–50, 61, 73, 222, 240, 244, 256, 260, 312–13, 336–37; neural firing and, 45–50, 174, 177, 185; neural theory applications and, 246–47; neurotransmitters and, 45; primary visual cortex (V1) and, 43–48; reading, 44–48; schemas and, 103, 348n4; sensorimotor control and, 48; simple, 43–44; simulation and, 231–35, 240, 243–45; simultaneity and, 49–50; synapses and, 49; topographic maps and, 44, 46, 348n4; unconsciousness and, 73, 149, 234, 277, 337

case studies: activation and, 309–11, 315; adjectives and, 321; anaphora, 305–11; binding circuits and, 320; cascade circuits and, 313, 318–19; cascades and, 311–13, 318–20, 325; categories and, 320–22; causation and, 314, 317; cognition and, 311; composition and, 311, 313; conceptual metaphor and, 146, 312–14, 317–20, 323; connections and, 316, 320; consciousness and, 311; convergence-divergence zones (CDZs) and, 55–56; by Dahl and Adachi, 139–40; embodiment and, 311; emotion and, 311, 320; executing networks (X-Nets) and, 311, 320; frames and, 307–10, 315–16, 325–26; generalization and, 310; gestalt circuits and, 320; Goldberg and, 309–10, 322–23; ideas and, 311, 318, 320; inhibition and, 309–11, 316; integration and, 311, 314, 317–20, 325; by Lakoff, 261, 320–21; learning and, 315; lexical items and, 320; linguistics and, 308–11, 318, 320; looking over, 320–26; mapping and, 314, 316, 319–20; memory case studies (*continued*)

and, 309; neural bindings and, 311, 316; neural firing and, 308; neural plasticity and, 315; primary metaphor and, 312, 315, 319–20; scales and, 313–19; semantics and, 307–9, 320–21, 325; spatial event structure system and, 313–15; special cases and, 310, 318; spikes and, 315; synapses and, 315; unconsciousness and, 311, 320

categories: basic-level, 163–64; best examples of, 161–62; case studies and, 320–22; cluster-based, 165; color and, 2–7; complexity of, 156–57; control and, 59; convergence-divergence zones (CDZs) and, 53–56; coordination and, 202; essence and, 121–22, 160–61; essential, 161; frames and, 164–66; general properties, 159; grammar and, 158–59, 260–67, 271–72, 300–305; hedges and, 163; incidental, 161; language types and, 327; linear scales in, 161; mapping-based, 166; meaningfulness of, 155–56, 167–71; members and, 157–58; metaphor and, 121–22, 142, 154–71; metonymic, 166; overlapping, 164; paragons and, 163; partitive construction and, 158–59, 261; part-whole relation and, 158–59; properties of, 159–60; radial, 165–67; reality and, 156; reference points and, 162–63; substitution-based, 165–66; *Women, Fire, and Dangerous Things* and, 156

category-as-container metaphor, 157

Catlin, J., 161

caudal area: generalization and, 32–34, 62–63, 77, 81; integration and, 33–34; prefrontal cortex and, 32–34, 62–63, 77, 81

causation: case studies and, 314, 317; event structure and, 336; frames and, 117; grammar and, 287, 298–302; metaphor and, 122, 128, 136, 168; motor control and, 91; neural firing and, 188; schemas and, 100

cause, 304

center-surround cells, 44–45, 47

central pattern generators, 226

cerebrosides, 14

cessation, 29–30, 222

Chang, Eddie, 250

Chang, N., 297

ChatGPT, 149

chemical channels, 15

"Chimpanzees Process Structural Isomorphisms across Sensory Modalities" (Ravignani and Sonnweber), 140

chlorine ions, 13–15, 349n7 (chap. 3)

Chomsky, Noam, 121, 332–33, 336, 338, 350n2

chromosomes, 3, 7, 81

chunking, 204–6

Churchland, Patricia, 82

circuit nodes, 41, 198, 214, 223

Citron, Francesca, 144–45

CNN, 115–16

cochlea, 20, 45, 348n19

cognition: artificial intelligence (AI) and, 341–43; basic circuits and, 191, 198–200; cascades and, 44, 48; case studies and, 311; combinatorial circuits and, 209, 213; control and, 38, 42–43, 59, 62; convergence-divergence zones (CDZs) and, 50–58; coordination and, 201, 203, 207, 222, 229; dual models and, 24–27, 31; embodiment and, 10–11, 31, 42, 58–59, 81, 102, 119, 182, 199, 246, 285, 292, 297, 341–42; frames and, 119–20; grammar and, 251, 283, 285, 292, 297–98; ideas and, 8–10; metaphor and, 26, 31, 42, 140, 142, 148, 153, 155, 162–63, 174, 183, 207, 245–46, 285, 297; multiple mapping circuits and, 221; neural firing and, 182–83; neural networks and, 11; neural phonology and, 332–34, 339; neural theory applications and, 246; part-whole relation and, 9, 25, 90, 105, 109–10, 113, 141, 158, 261, 347n5; reference points and, 162–63; schemas and,

93, 102, 111; simulation and, 232, 238–39, 242–45; sound symbolism and, 63, 65, 68–69

cognitive linguistics: basic circuits and, 198; case studies and, 311; coordination and, 229; dual models and, 27; grammar and, 251, 283, 398; ideas and, 8, 27, 69, 80, 111, 182, 198, 229, 246, 251, 283, 298, 311; neural firing and, 182; neural theory applications and, 246; perception and, 8, 27, 60, 80; schemas and, 111

cognitively special properties, 59

"Cognitive Phonology" (Lakoff), 333–34, 339

cognitive science programs, 102

color: activation and, 2; categories and, 2–7; combinatorial circuits and, 208; cones and, 2–5, 7, 45, 81–82; embodiment and, 4, 7; ganglion cells and, 2, 7, 44–46; light and, 1–7; metaphor and, 150, 159; neural bindings and, 70; neural firing and, 2; neural language and, 335; neural theory applications and, 246; neurophysiology of, 2–5; nonexistence of, 1–2; nonspectral, 4; perception of, 1–7, 70–71, 79–82; retina and, 2–4; simulation and, 232; tetrachromates and, 3–4; wavelength and, 1–7, 81; words for in different languages, 5

color blindness, 3

color wheel, 5–6

combinatorial circuits: binding circuits and, 207–12; cascades and, 207; cognition and, 209, 213; color and, 208; composition and, 210; computation models and, 207, 209, 211–13; connections and, 211; consciousness and, 208; embodiment and, 213; emotion and, 210; frames and, 207, 209–14; gating and, 211–12; generalization and, 209; gestalt circuits and, 210–13, 222; ideas and, 214; image schemas and, 207; inhibition and, 211–13; integration and, 207–8; learning and, 209, 213; mapping

and, 207, 213–14; mathematics and, 209; memory and, 212–13; metonymy and, 214; Narayanan and, 212; neural bindings and, 208–9, 212; neural firing and, 208, 211–12; neural plasticity and, 209; reward and, 210; scales and, 207, 213; semantics and, 210; special cases and, 214; synapses and, 209; topographic maps and, 207; unconsciousness and, 209, 214

communication metaphors, 127

composition: basic circuits and, 191; case studies and, 311, 313; combinatorial circuits and, 210; embedding and, 273–76; grammar and, 253–54, 273, 281–82; images and, 96–98; lexicon and, 281–83; mechanistic constraints on, 103–4; metaphor and, 143, 169; schemas and, 91, 96–99, 102–4, 108; semantics and, 281–83; simulation and, 233; sound symbolism and, 67

computational bridging theory, 81, 246–47

computation models, 174; artificial intelligence (AI) and, 190, 341; basic circuits and, 191–99; bridge and, 89–92; cascades and, 43, 348n4; combinatorial circuits and, 207, 209, 211–13; coordination and, 200–201, 226; dual models and, 25–27, 29–31, 79–80; Feldman and, 339; frames and, 116; functional model and, 181–82; gating and, 80; gestalt circuits and, 25–27; grammar and, 250, 260, 297, 303; metaphor and, 121, 144, 150, 168; motor control and, 83–91, 168; multiple mapping circuits and, 214, 221; neural firing and, 180–85, 188–90; neural phonology and, 333, 339; neural theory applications and, 246–47; perception and, 11–12, 80; reality and, 181–82; schemas and, 108, 111, 348n4; semantics and, 80; simulation and, 231, 239, 243, 245; structured neural computation (SNC) model, 89–92, 181–83, 189–90 conceptual metaphor: case studies and, 146, 312–14, 317–20, 323; categories and, 167; complex, 131; control and, 32, 42; Dahl and Adachi study on, 139–40; dual models and, 26; embodiment and, 134–43; essence of, 160; executing networks (X-Nets) and, 171, 335; frames and, 80, 113, 169, 174; Goldberg and, 146; grammar and, 257–59, 285, 296–97, 302, 304; hierarchy and, 168; ideas and, 8–10; idioms and, 129–30, 170; independent discoveries and, 123–25; integration and, 149–51; linear scale metaphor and, 132–34; linguistic metaphor and, 126; as logic-preserving, 134–35; love and, 128–29; mapping and, 8; metonymy and, 166; motor control and, 89, 92, 234, 240–42; multiple mapping circuits and, 216; neural networks and, 12–13; neuroscience and, 144–49; role of culture in, 348n20; schemas and, 97–100, 104, 108; system of, 126–28; systems and, 127

"Conceptual Metaphorical Mapping in Chimpanzees (*Pan troglodytes*)" (Dahl and Adachi), 139–40

conceptual organizations, 303t, 304

conceptual variations, 303t, 304

concurrent circuits, 29, 194, 222

conditional circuits, 29, 194, 223, 226–27, 246–47

conduit metaphor, 125–27, 134, 241, 318

cones, 2–5, 7, 45, 81–82

connections: artificial intelligence (AI) and, 343; basic circuits and, 191–99; cascades and, 44–49; case studies and, 316, 320; combinatorial circuits and, 211; control and, 31–32, 36–37, 41–42, 60–63; convergence-divergence zones (CDZs) and, 50–57; coordination and, 200, 204, 206, 222–23, 228; dual models and, 26; emotion and, 76, 145, 191, 320; function and, 187–88; grammar and, 261, 265, 272; hierarchy of, 51–52, 57, 62, 76, 347n7; ideas and, 329; learning

and, 15, 37, 75–76, 138, 151, 191, 214, 343; metaphor and, 138, 141, 145–51; motor control and, 85; multimodality and, 36, 69, 76; multiple mapping circuits and, 214–19, 222; neural firing and, 15–16, 37, 41, 46–49, 57, 61, 77, 80, 113, 175, 181, 184–89, 196, 228; neural networks and, 10–24; neural theory applications and, 246; prefrontal cortex and, 20, 24, 31, 62, 74, 79, 81, 347n7; schemas and, 95–96, 103, 105, 113; simulation and, 233, 239, 241; somatosensory cortex and, 23, 31, 74–75, 105; synapses and, 15–16, 32, 41, 78–81, 138, 184–85, 187, 191, 223, 329; synaptic weight of, 187; topographic maps of, 18, 20, 23, 44, 75, 103, 105, 147, 218

consciousness: cascades and, 49; case studies and, 311; combinatorial circuits and, 208; control and, 36, 38; convergence-divergence zones (CDZs) and, 57; integration and, 149–55; metaphor and, 149–55, 171; mind-body problem and, 183, 350n11; multiple mapping circuits and, 217; neural firing and, 183; neural networks and, 11; perception and, 11, 36, 38, 49, 57, 68–72; simulation and, 234, 242

Consciousness in the Brain (Dehaene), 68

conscious self, 208

constructions: adjective, 288, 297–98; amalgam, 290–91; artificial intelligence (AI) and, 342; basic circuits and, 191; bleeding effect and, 340; change-state, 337; coordination and, 206–7; dual models and, 28; emotion and, 252; equi, 274; executing networks (X-Nets) and, 251–52; fixed circuitry and, 253; grammar and, 251–57, 260, 263–73, 279–80, 283–88, 292–97, 300, 303–5, 336; integration and, 264–65; language types and, 327–28; lexical items and, 337; mapping circuits and, 254–55; metaphor and, 146, 150; neural phonology and, 331–34, 338–39; new meanings and, 255–56; organized thought and, 284–87; pragmatic network of, 251–59, 288, 290; semantic role modifiers and, 268; simulation and, 236; sound symbolism and, 256–57; why (not), 288–90

Constructions (Goldberg), 293, 300–305

containment: control and, 32, 35, 38; generalization and, 76; grammar and, 261; metaphor and, 124; multiple mapping circuits and, 218–20; schemas and, 9, 105, 110–11; simulation and, 235, 241; sound symbolism and, 63, 78

control: activation and, 33, 35; cascades and, 60–61; categories and, 59; central pattern generators and, 226; circuit structure limitations and, 228–29; cognition and, 38, 42–43, 59, 62; conceptual metaphor and, 32, 42; concurrent actions and, 224–26; conditional circuits and, 226–27; connections and, 31–32, 36–37, 41–42, 60–63; consciousness and, 36, 38; containment and, 32, 35, 38; coordination and, 222–31; electricity and, 36, 60; embodiment and, 34–35, 58–60; emotion and, 31, 40, 59, 61; eyes and, 31; frames and, 32, 34, 42; functional magnetic resonance imaging (fMRI) and, 31, 62; generalization and, 31–39; gestalt circuits and, 59–63; ideas and, 32, 38–43; imagination and, 59–61; inhibition and, 34; insula and, 31; integration and, 32–39, 42; iteration and, 222, 224; learning and, 34–38; localization and, 41; mathematics and, 34, 37; memory and, 40, 222, 228–30; mental imagery and, 60–63; modules and, 60; motor programs and, 59, 61–62; multimodality and, 36, 39, 59, 61, 69, 76; neural bindings and, 33; neural firing and, 37, 41, 61; neural optimization and, 36–37; neural plasticity and, 34, 37; packages and, 34; parietal cortex and, 32, 60–61; prefrontal cortex and, 31–32, 38, 62, 74, 79; primary visual cortex, 60; primitives and, 38; recruit-

ment and, 32–39; reinforcement and, 37; reward and, 37, 58–59, 87; semantics and, 38, 42; sensorimotor control and, 61; sequencing circuits and, 223; somatosensory cortex and, 31; spikes and, 34, 37; substitution circuits and, 227–28; synapses and, 32, 35, 38, 41; unconsciousness and, 34–39, 61

control by disinhibition, 104

control circuits: coordination and, 206, 222–27; dual models and, 29; emotion and, 59; location of, 31; motor control and, 87; prefrontal cortex and, 24; simulation and, 233

convergence-divergence zones (CDZs): activation and, 50–53, 57; cascades and, 52–53, 57–58, 78; case studies and, 55–56; categories and, 53–56; cognition and, 50–58; concept of, 50–51; connections and, 50–57; consciousness and, 57; Damasios and, 50, 52–57; dorsal stream and, 54; embodiment and, 57; eyes and, 51–52; face naming and, 51–54; frames and, 50, 57; functional magnetic resonance imaging (fMRI) and, 58, 72, 78; gestalt circuits and, 50–52; hierarchical linkage and, 50–52; ideas and, 50; integration and, 50–54, 57–58; learning and, 58; lesions and, 50, 53–58; lexical naming and, 54–55; memory and, 52, 57–58; motor programs and, 54–55; neural bindings and, 50, 52, 57; neural firing and, 52, 57; parietal cortex and, 54–55, 58; role of explanation, 58; semantics and, 50; somatosensory cortex and, 53, 55; unconsciousness and, 57; warehouse theory and, 54, 56–58, 77–78, 336

conveyed meaning, 252

Cooper, L. N., 188

coordination: activation and, 200–201, 205–8, 211–13, 222–23, 226–29; behavior control and, 200–207; binding circuits and, 226, 228; cascades and, 200–207, 222, 225–26, 229, 231;

evolution (*continued*)
and, 91; perception and, 3, 22, 37, 75; repurposing through, 75–76; schemas and, 101, 108

exaptation: basic circuits and, 191; coordination and, 207, 230–31; grammar and, 251; integration and, 230–31; metaphor and, 149, 168, 191; motor control and, 83, 91; as repurposing, 191; schemas and, 101, 108; simulation and, 245

executing networks (X-Nets): basic circuits and, 192, 259–60; cascades and, 336; case studies and, 311, 320; cause and, 304; central pattern generators and, 226; conceptual metaphor and, 171, 335; constructions and, 251–52; coordination and, 200, 223–27; dual models and, 89; embedding and, 90; embodiment and, 234–35; failure and, 91; fixed circuitry and, 253; frames and, 117–19; grammar and, 251–52, 258t, 260, 278, 280, 282, 297, 299, 303t, 304, 336–37; ideas and, 259–60; imperfective, 90; Lakoff and, 89; language types and, 84–86, 326, 337; lexical items and, 282; metaphor and, 108–9, 149, 168; motor control and, 87–92, 336; neural phonology and, 334; perfective, 90; purposes and, 91; schemas and, 95, 99–100, 107–12; sequences of, 90; simulation and, 234, 242–45

extensor muscles, 23

eyes: cascades and, 45, 48; color and, 1–7; control and, 31; convergence-divergence zones (CDZs) and, 51–52; coordination and, 200, 224, 230; language types and, 323; metaphor and, 127–28, 160; neural networks and, 21; perception and, 1–7, 21, 31, 45, 48, 51–52, 74; REM, 230; schemas and, 104

failure, 91

Farah, Martha, 60, 77, 91

Fauconnier, Gilles, 153–54, 243–44, 298, 302

Federenko, Eva, 250

Feldman, Jerome: combinatorial circuits and, 207–8, 350n12; computation models and, 339; dual models and, 24–25; *From Molecule to Metaphor*, 239, 243; grammar and, 297; Lakoff and, 83, 116, 297, 333, 339; mind-body problem and, 350n11; motor control and, 83–84, 89; multiple mapping circuits and, 221–22; neural phonology and, 333; Neural Theory of Language (NTL) and, 83–84; simulation and, 239–40, 243; structured neural computation (SNC) model, 190

Fillmore, Charles: deixis and, 335; frames and, 114–16, 336; grammar and, 292, 303, 305; metaphor and, 169; neural phonology and, 333; schemas and, 98

filters, ix–x, 336

firing rules, 11

flashes, 69–71

Foster, D. J., 238–39

frame circuits, 113–14, 121, 169, 213

FrameNet, 116

frames: activation and, 118, 120; basic circuits and, 191, 199; binding circuits and, 211; cascades and, 113, 120; case studies and, 307–10, 315–16, 325–26; categories and, 164–66; causation and, 117; cognition and, 119–20; combinatorial circuits and, 207, 209–14; complex, 32, 113, 169, 234; computation models and, 116; conceptual, 8–10, 26, 32, 34, 42, 80, 104, 113–16, 119–22, 149, 167–69, 174, 213, 240, 269, 271, 281, 283, 286, 303t; control and, 32, 34, 42; convergence-divergence zones (CDZs) and, 50, 57; coordination and, 228; dual models and, 24–25; embodiment and, 118–19; executing networks (X-Nets) and, 117–19; genetics and, 164; gestalt circuits and, 118–19; Goffman on, 349n7 (chap. 2); grammar and, 252, 258–65, 268–83, 286–88, 292, 296, 301–4, 350n7; ideas and, 8–10, 24–26; inhibition and, 120; integration and, 113, 120; Lakoff and, 115–16; language types and, 327, 329; linguistics and, 115, 336; mapping and, 8, 120, 124–26, 165, 169, 171, 199, 207, 213–16, 245, 258, 280–81, 316, 337; mathematics and, 116; metaphor and, 8–10, 26, 31–32, 42, 80, 113–17, 122–27, 135, 149–50, 154–57, 162–71, 174, 183, 207, 215–16, 234, 240, 242, 258, 268, 273, 296, 301–2, 315–16, 325–26, 329, 335, 337; metonymy and, 120–21; motor programs and, 114; multiple mapping circuits and, 215–16; Narayanan and, 116; neural firing and, 183; Neural Theory of Language (NTL) project and, 116; part-whole relation and, 9, 113; primitives and, 117, 119; reasoning algorithm and, 347n5; reward and, 117; scales and, 117; schemas and, 96, 104, 113–18, 174; semantics and, 115–19; simulation and, 234, 240–45; special cases and, 113–14, 119–21; unconsciousness and, 114, 119

From Molecule to Metaphor (Feldman), 239, 243

functional circuits: basic circuits and, 193, 196–97; cascades and, 43; concept of, 41; coordination and, 202; fixed, 41–42; motor control and, 86; neural firing and, 189; trauma and, 38, 41

functional generalizations, 39–43

functional magnetic resonance imaging (fMRI): activation and, 72, 78, 144; control and, 31, 62; convergence-divergence zones (CDZs) and, 58, 72, 78; globality and, 77; metaphor and, 144

functional muscle synergies, 85

functional structured neural computation (fSNC) model, 181–86, 189

ganglion cells, 2, 7, 44–46

gating: basics of, 194, 196–99; combinatorial circuits and, 211–12; computation models and, 80; coordination and, 223, 226, 227f; grammar and, 250, 258t, 265, 272, 276; metaphor and, 141, 152,

168–69; multiple mapping circuits and, 214–16; nodes for, 77–78, 337; schemas and, 103–4; simulation and, 233

gaze, 127–28, 230, 323–25

Gazzaniga, Michael, 68

generalization: anatomy of, 34–35; basic circuits and, 191; case studies and, 310; caudal, 32–34, 62–63, 77, 81; combinatorial circuits and, 209; containment and, 76; control and, 31–39; dual models and, 26, 31; embodied primitives and, 32–33; functional, 39–43; grammar and, 256–57, 276, 296, 298, 303–5; hierarchical, 347n7; ideas and, 42–43; idioms and, 123; inevitability of, 32; integration and, 33–36; linguistic, 76; metaphor and, 123–30; multimodality and, 36; multiple mapping circuits and, 220; neural bindings and, 33; neural networks and, 81; neural phonology and, 331; neural theory applications and, 246; physical logic and, 36–39; rostral, 32–34, 62–63, 77, 81, 348n5; schemas and, 94, 106; special cases and, 76; specifics and, 33, 348n5

generalized containment, 63

genetics: basic circuits and, 190; chromosomes, 3, 7, 81; coordination and, 205; frames and, 164; metaphor and, 164–65; neural language and, 337–38; neural networks and, 16; optogenetic techniques and, 349n7 (chap. 3); perception and, 16

Gerstner, W., 349n2

gestalt circuits: basic circuits and, 194, 198–99; cascades and, 47–48; case studies and, 320; combinatorial circuits and, 210–13, 222; computation model of, 25–27; conceptual organizations and, 303t, 304; control and, 59–63; convergence-divergence zones (CDZs) and, 50–52; doublets and, 261–66, 270, 332, 334, 337, 339; dual models and, 24–25; frames and, 118–19; grammar and, 258t, 259–70, 273, 280, 284, 288, 299, 303t, 304,

gestalt circuits (*continued*)
334, 336; hierarchical structure of, 336; metaphor and, 141, 163, 169; multiple mapping circuits and, 224; neural firing and, 174, 190; neural phonology and, 332, 334; neural theory applications and, 246–47; schemas and, 95–96, 104

gestalt doublets: grammar, 261–66, 270, 337; neural phonology and, 332, 334, 339

gestalt nodes: basic circuits and, 198–99; combinatorial circuits and, 211; dual models and, 25; grammar and, 260–61, 264–65, 273, 280, 337; metaphor and, 141; neural firing and, 190; neural phonology and, 334; schemas and, 95; threshold level and, 25, 30, 184f

gestalt perception, 76–77; cascades and, 47–48; categories and, 163; control and, 59, 61–63; dual models and, 24–27; metaphor and, 163

gestalt property, 24–25

Gilbert, Aubrey, 5

globality, 77–78

Go, 189

Goffman, Erving, 115, 349n7 (chap. 2)

Goldberg, Adele: anaphora and, 309–10; case studies by, 309–10, 322–23; *Constructions*, 293, 300–305; grammar and, 293–95, 300–305; integration and, 300–305; metaphor and, 144–46, 167, 170; statistical preemption and, 295

Goldsmith, John, 333, 339

Google AlphaGo, 342

Google Gemini, 342

GPT-X, 342

Grafton, S. T., 202

grammar: absolute exceptions and, 272–73; activation and, 258, 261, 264, 272, 303; adjectives, 159, 261, 265–69, 272–73, 280–81, 287–88, 297–98; agreement and, 263; anaphora and, 305–11; basic circuits and, 259–65; binding circuits and, 255, 258t, 259, 263, 265, 268, 277, 303t, 304; cascades and, 255–60, 268,

277, 280, 285, 290–91, 303t; categories and, 158–59, 260–67, 271–72, 300–305; causation and, 287, 298–302; cognition and, 251, 283, 285, 292, 297–98; composition and, 253–54, 273, 281–83; computation models and, 250, 260, 297, 303; conceptual metaphor and, 257–59, 285, 296–97, 302, 304; connections and, 261, 265, 272; constructions and, 251–57, 260, 263–69, 279–80, 283–88, 292–97, 300, 303–5, 336; containment and, 261; conveyed meaning and, 252; coordination and, 276–79; defaults and, 262–63; embedding and, 260, 273–76, 290; embodiment and, 249–51, 285, 287, 292, 297, 303t, 304; emotion and, 252, 266, 274, 276, 389; exaptation and, 251; executing networks (X-Nets) and, 251–52, 258t, 260, 278, 280, 282, 297, 299, 303t, 304, 336–37; expressing quality and, 266–68; form-meaning relations and, 261; frames and, 258–65, 268–83, 286–88, 292, 296, 301–4, 350n7; gating and, 250, 258t, 265, 272, 276; generalization and, 256–57, 276, 296, 298, 303–5; gestalt circuits and, 258t, 259–70, 273, 280, 284, 288, 299, 303t, 304, 334, 336; Goldberg and, 293–95, 300–305; ideas and, 249–53, 256–59, 265–73, 278–80, 285, 287, 292, 297, 299, 303; image schemas and, 257, 261, 268, 296; inflectional structures and, 260, 262; inhibition and, 258, 264–65, 272, 276, 285, 297; integration and, 255–59, 262–73, 277, 280–84, 290–91, 298–304, 335–37; Lakoff and, 261, 272–73, 286, 292, 297–98, 301–2, 350n4; learning and, 250, 253, 267, 275, 283, 296; lexical items and, 48, 78, 80, 191, 249, 261, 265–69, 279–84, 298, 320; linguistics and, 249–58, 263–66, 274, 279, 284–88, 292–98; localization and, 250; mapping and, 253–59, 280–81, 300, 303; meaning and form levels, 251, 255, 260, 264–65, 275, 279, 291, 296, 300; metaphor and,

257; metonymy and, 268, 280–81, 303t; minor rules and, 272, 337; morphology and, 249–50, 260, 266–67; names and, 265; Narayanan and, 297; neural bindings and, 255, 263, 265, 268, 271, 274–75, 280–81, 299, 303t, 304; neural circuitry and, 257–58; neural firing and, 255; neural optimization and, 255, 274–76, 279–80, 295–97, 303t; neural plasticity and, 259; new meanings and, 255–56; organized thought and, 284–87; packages and, 285; part-whole relation and, 261; phrasal structures and, 260; pragmatic network of, 251–59, 288, 290, 350n2; prefrontal cortex and, 304; primary schemas and, 303t, 311, 320; primitives and, 286–87, 298–99; quality description and, 272–73; recruitment and, 296; scales and, 268, 271, 283; semantics and, 265–66, 268–69, 274, 278–84, 287–90, 294–305; sentence adverbs and, 266; shared nodes and, 277–79; sound symbolism and, 256–57; special cases and, 262–63, 269, 274, 280, 304; spikes and, 259; statistical preemption and, 295; synapses and, 284, 296; syntax and, 273, 300, 335, 350; unconsciousness and, 252, 272, 277, 279, 283, 285

gratings, 64
Graybiel, A. M., 91, 203–4
Graziano, M. S. A., 202–3
Grice, Paul, 28
grid cells, 238
Grim Reaper, 151–53
Gross, C. G., 203
Gruber, Jeffrey, 299–300, 305, 336

hair cells, 20
Halle, M., 332–33
Harris, Zellig, 350n3
Hebb, Donald, 15–16, 188
Hebbian learning: backpropagation and, 190; combinatorial circuits and, 209; control and, 34, 37; ideas and, 247; met-

aphor and, 137; motor control and, 87; plasticity and, 228; synapses and, 315
hedges, 163
Hodgkin-Huxley model, 186, 349n2
"Homing in on Consciousness in the Nervous System" (Morsella et al.), 68–69
hormones, 14–15, 59
how pathway, 21
Hubel, D. H., 45–46
Human Brain Project, 11, 173, 250
Human Connectome Project, 11
Huron, David, 348n20
Hyman, Larry, 333, 339
hyperpolarization, 177, 179

ideas: basic circuits and, 190–93, 259–60; basics of, 8–10; before learning language, 253; cascades and, 311–20; case studies and, 311, 318, 320; circuits for, 265–73; cognitive linguistics and, 8, 27, 69, 80, 111, 182, 198, 229, 246, 251, 283, 298, 311; combinatorial circuits and, 214; connections and, 329; control and, 32, 38–43; convergence-divergence zones (CDZs) and, 50; dual models and, 24–28, 31–32; electrochemistry and, 8, 10; emotion and, 7; executing networks (X-Nets) and, 259–60; frames and, 8, 24–26, 31; generalizations and, 42–43; grammar and, 249–53, 256–59, 265, 273, 278–80, 285, 287, 292, 297, 299, 303; imagination and, 10; integration and, 9–10; lexicon and, 328; linguistics and, 8; mapping and, 8; metaphor and, 8–10, 13, 26, 31, 37, 42, 125–28, 135, 146, 150, 157, 161, 167–68, 171, 174, 246, 257, 273, 297, 311–20, 323, 329; motor control and, 83, 92; neural firing and, 183; neural language and, 249–53, 256–58, 265, 273, 278–80, 285, 287, 292, 297, 299, 303–4, 318, 320, 328–33, 338–39; neural mechanisms and, 173–74, 183, 190–93, 214, 234, 246–47, 349n7 (chap. 3); neural networks and, 11; neural phonology and, 333; neural theory applications ideas (*continued*)

and, 247; part-whole relation and, 9; perception and, 7–11, 24–28, 31–32, 38–43, 50, 73–88; power of, 265–73; primary, 8; recruitment and, 328–29; reflexivity and, 329–31; reinforcement and, 330; schemas and, 93; simulation and, 234, 240; spread of, 328–31; structured, 311–20; synapses and, 329, 331; thought mechanisms and, 83, 92–93, 125–38, 146, 150, 157, 161, 167–68, 171; unconsciousness and, 8–9, 329–30

idioms: conceptual metaphor and, 170; generalization and, 123; meaning of, 128–29; metaphor and, 123, 129–30, 146–48, 155, 170–71; warnings on, 146–48

image schemas: basic circuits and, 194; blindness and, 102; combinatorial circuits and, 207; composition and, 96–98; coordination and, 207; dual models and, 26; embodiment and, 59, 68, 78, 95, 104, 111, 117, 149, 171, 257, 326, 328, 337; executing networks (X-Nets) and, 107; frames and, 113, 117–18, 174; generalization and, 106; grammar and, 257, 261, 268, 296; Kant and, 92–95; language types and, 326, 328; metaphor and, 132, 149, 171; multiple mapping circuits and, 217–18, 220; simulation and, 234, 245; sound symbolism and, 67–68, 78; special cases and, 110; topographic maps and, 348n4

imagination: basic circuits and, 191; cascades and, 48–49; control and, 59–61; coordination and, 230; experience and, 60–61; grammar and, 278; ideas and, 10; metaphor and, 150, 153; metonymy and, 61; neural unity of, 61–63; schemas and, 68, 112–13; simulation and, 235–36, 239, 244

imperfective states, 90

inferior temporal cortex, 21, 53–56

inflection, 260, 262

influx, 178

inhibition: basic circuits and, 194–99; binding circuits and, 212–13; cascades and, 44–46, 49; case studies and, 309–11, 316; combinatorial circuits and, 211–13; control and, 34; coordination and, 205, 222–27; disinhibition hypothesis and, 212–13; dual models and, 26, 29, 80; frames and, 120; grammar and, 258, 264–65, 272, 276, 285, 297; ions and, 15, 185, 349n7 (chap. 3); lateral, 220–22; memory and, 349n7 (chap. 3); metaphor and, 141, 150–54, 157, 165, 168; multiple mapping circuits and, 217–21; neural firing and, 45–46, 79–80, 104, 185, 196–97, 205, 211–12; neural networks and, 15, 23; neural phonology and, 331; neural theory applications and, 246–47; part-whole relationship and, 347n5; schemas and, 96, 104; scotoma and, 70; simulation and, 233; touch experiments and, 70–71

input activation, 199

insula: control and, 31; location of, 31; neural language and, 335; neural networks and, 14, 20, 23; perception and, 14, 20, 23, 31, 74; schemas and, 105

integration: basic circuits and, 101; binding constraints and, 101–2; building complexity and, 234; cascades and, 49, 58, 73, 103, 113, 120, 156–57, 168, 174, 231–35, 240, 245–47, 255–59, 268, 280, 290, 311, 320, 335–37; case studies and, 311, 314, 317–20, 325; circuit patterns in, 153–54; combinatorial circuits and, 207–8; conceptual, 8–10, 32, 36, 42, 104, 149–57, 167–71, 174, 217, 231, 242, 258t, 283, 298, 303t, 304, 318; consciousness and, 149–55; constructions and, 264–65; control and, 32–39, 42; convergence-divergence zones (CDZs) and, 50–54, 57–58; coordination and, 207, 230–31; embodiment and, 234–35; exaptation and, 230–31; executing networks (X-Nets) and,

leaky integrate and fire model, 349n2
learning: artificial intelligence (AI) and, 341–43; basic circuits and, 191; cascades and, 49; case studies and, 315; combinatorial circuits and, 209, 213; connections and, 15, 37, 75, 76, 138, 151, 191, 214, 343; control and, 34–38; convergence-divergence zones (CDZs) and, 58; coordination and, 204–5, 224, 228; deep, 341–43; early childhood education and, 21–22, 38; grammar and, 250, 253, 267, 275, 283, 296; Hebbian, 15, 34, 37, 87, 137, 190, 209, 228, 247, 315; metaphor and, 137–43, 151, 170; motor control and, 84, 87; multiple mapping circuits and, 214, 222; neural firing and, 174, 181, 187–90; neural networks and, 15; neural plasticity and, 16, 34, 37, 78–79, 87, 181, 188, 209, 228, 259, 315; neural theory applications and, 247; simulation and, 231–32, 236; synapses and, 15, 35, 38, 75–76, 138, 174, 181, 187–91, 232, 296
Leibniz, Gottfried Wilhelm, 159
lesions: convergence-divergence zones (CDZs) and, 50, 53–58; perception and, 50, 53–58, 78
lexical items: cascades and, 48; case studies and, 320; composition and, 281–83; constructions and, 337; executing networks (X-Nets) and, 282; grammar and, 48, 78, 80, 191, 249, 261, 265–69, 279–84, 298, 320; ideas and, 328; neural firing and, 191; neural phonology and, 334
light: color and, 1–7; coordination and, 205; creating perception and, 1–7, 69, 71; metaphor and, 127, 152; neural theory applications and, 246; perception and, 1–7, 69, 71, 81; sound symbolism and, 66–67; ventromedial prefrontal cortex and, 205
Lindeman, L. M., 145–46
linguistics: artificial intelligence (AI) and, 343; asterisk and, 350n6; basic circuits and, 191, 198; case studies and, 308–11,

318, 320; cognitive, 8, 27, 69, 80, 111, 182, 198, 229, 246, 251, 283, 298, 311; coordination and, 207, 229; creating perception and, 69; dual models and, 27, 80; frames and, 115, 336; generalization and, 76; grammar and, 249–58, 263–66, 274, 279, 284–88, 292–98; ideas and, 8; language types and, 326–27; metaphor and, 122–28, 143–46, 170; motor control and, 88, 90; neo-Whorfian movement and, 285–86, 351n11; neural firing and, 182–83; neural mechanisms and, 174, 338; neural modeling and, 297; neural networks and, 22; neural theory applications and, 246–47; schemas and, 97, 102, 111; simulation and, 244; sound symbolism and, 63–65; syntax and, 273, 300, 335, 350; unconscious, 63–65, 69, 320, 335
localization: cascades and, 49; control and, 41; globality and, 77; grammar and, 250; simulation and, 233–34, 238, 245; thought and, 74
Loftus, Elisabeth, 229
logic schemas: embodiment and, 107; executing networks (X-Nets) and, 109, 111–12; part-whole structure and, 110; simulation and, 111–13
Louder Than Words (Bergen), 243
love, 123–24, 128–31

MacDonald, John, 71
Maeda, F., 64
manner-verb languages, 326–27
"Many Ways to Search for a Frog, The" (Slobin), 327
map-nets, 217–20
mapping: basic circuits and, 194, 199; case studies and, 314, 316, 319–20; combinatorial circuits and, 207, 213–14; coordination and, 206–7; frames and, 8, 120, 124–26, 165, 169, 171, 199, 207, 213–16, 245, 258, 280–81, 316, 337; grammar and, 253–54, 257–59, 280–81, 300, 303; ideas and, 8; metaphor and, 124–30,

133, 137–43, 153, 165–66, 169, 171; motor control and, 89; multiple mapping circuits and, 214–22; neural language and, 253–54, 257–59, 280–81, 300, 303, 314, 316, 319–20, 337; neural networks and, 11, 13; perception and, 8, 11, 13, 80; simulation and, 232–33, 245; topographic, 18 (*see also* topographic maps)

Marx, Chico, 1

mathematics: combinatorial circuits and, 209; control and, 34, 37; dual models and, 79; frames and, 116; Lakoff and, 82, 149; language types and, 328; metaphor and, 135, 143, 148–49, 159–60, 171; multiple mapping circuits and, 216–17; neural firing and, 187, 189; neural theory applications and, 246; *Where Mathematics Comes From*, 82, 149

maximal knowledge, 163

McGurk effect, 66, 71

meaning and form levels: coordination between, 206; grammar and, 251, 255, 260, 264–65, 275, 279, 291, 296, 300; language types and, 327; neural phonology and, 332, 334, 337–38

Meaning of the Body, The (Johnson), 104

memory: artificial intelligence (AI) and, 342; case studies and, 309; combinatorial circuits and, 212–13; control and, 40, 222, 228–30; convergence-divergence zones (CDZs) and, 52, 57–58; coordination and, 222, 228–30; inhibition and, 349n7 (chap. 3); metaphor and, 78, 170; neural networks and, 11, 21; schemas and, 104; simulation and, 229–30, 237–38, 243; warehouse theory and, 58, 78, 336

mental imagery: categories and, 163; control and, 60–63; metaphor and, 163; multimodality and, 76–77; schemas and, 92

Mervis, C. B., 161

metaphor: activation and, 137–41, 144–50, 153–54; adjectives and, 159; Aristotle and, 109, 112, 121–23, 148, 160; binding circuits and, 135, 141, 168; cascades and, 135, 141, 143, 149–50, 156–57, 168, 311–20, 351n17; case studies and, 312, 315, 319–20; categories and, 121–22, 142, 154–71; causation and, 122, 128, 136, 168; chimpanzees and, 139–40; cognition and, 26, 31, 42, 140, 142, 148, 153, 155, 162–63, 174, 183, 207, 245–46, 285, 297; color and, 150, 159; complexities and, 314–20; composition and, 143, 169; computation models and, 121, 144, 150, 168; conceptual, 8–10 (*see also* conceptual metaphor); conduit, 125–27, 134, 241, 318; connections and, 138, 141, 145–51; consciousness and, 149–55, 171; constructions and, 146, 150; containment and, 124; correlations in experience of, 136–37; embodiment and, 134–46, 168–70; emotion and, 135–36, 144–46, 150, 170; essence, 121–22, 160–61; evolution and, 168; exaptation and, 149, 168, 191; executing networks (X-Nets) and, 108–9, 149, 168; eyes and, 127–28, 160; frames and, 8–10, 26, 31–32, 42, 80, 113, 115, 117, 122–27, 135, 149–50, 154–57, 162–71, 174, 183, 207, 215–16, 234, 240, 242, 258, 268, 273, 296, 301–2, 315–16, 325–26, 329, 335, 337; functional magnetic resonance imaging (fMRI) and, 144; gating and, 141, 152, 168–69; gaze and, 127–28, 230, 323–25; generalization and, 123–30; genetics and, 164–65; gestalt circuits and, 141, 163, 169; Goldberg and, 144–46, 167, 170; ideas and, 8, 10, 13, 26, 31, 37, 42, 125–28, 135, 146, 150, 157, 161, 167–68, 171, 174, 246, 257, 273, 297, 311–20, 323, 329; idioms and, 123, 129–30, 146–48, 155, 170–71; image schemas and, 132, 149, 171; imagination and, 150, 153; independent discoveries of, 123–25; inhibition and, 141, 150–54, 157, 165, 168; integration and, 149–57, 167–71; Lakoff and, 123–25, 128–31, 135–36, 145–49, 151, 156, 163–65, 351n16; learning and, 137–43, 151, 170; light and, metaphor (*continued*)

127, 152; linguistics and, 122–28, 143–46, 170; logic of, 131–35; love relationships and, 123–24, 128–31; mapping and, 124–30, 133, 137–43, 153, 165–66, 169, 171; mathematics and, 135, 143, 148–49, 159–60, 171; memory and, 78, 170; mental imagery and, 163; metonymy and, 152, 161, 166, 268; motor programs and, 163; multimodality and, 142; Narayanan and, 89, 125; neural bindings and, 142–43, 157–58, 164; neural firing and, 137–38; neural theory of, 137–39; neuroscience and, 144–49; packages and, 123–24; partitive construction and, 158–59, 261; part-whole relation and, 141, 158–59; prefrontal cortex and, 145; primary, 125 (*see also* primary metaphor); primary motor cortex and, 147–48; primary schemas and, 142, 169; primitives and, 167, 171; reality and, 143; recruitment and, 151; Reddy on, 123–28; reinforcement and, 149; scales and, 131–34, 161–64; semantics and, 122–24, 127, 168; sensorimotor control and, 142, 149; somatosensory cortex and, 145; sound symbolism and, 66–68; source domains and, 124–26, 134–35, 139, 143, 157, 318; special cases and, 126–27, 130, 132, 147, 151–52, 161, 163, 166; spike-timing-dependent plasticity (STDP) and, 138–40, 143, 170; synapses and, 138; systems for, 126–28, 313–14; target domains and, 124–25, 135, 157, 259; topographic maps and, 147; unconsciousness and, 123, 129, 135, 137–38, 142, 148–58, 170–71; *Women, Fire, and Dangerous Things* and, 156

Metaphors We Live By (Lakoff and Johnson), 124

metonymy: categories and, 166; combinatorial circuits and, 214; frames and, 120–21; grammar and, 268, 280–81, 303t; imagination and, 61; metaphor

and, 152, 161, 166, 268; as symbolic action, 61

middle temporal (MT) region, 20–21

mind-body problem, 183, 350n11

minor rules, 272, 337

Minsky, Marvin, 115–16

mirror neurons, 60–61, 77

modulators, 185, 204

modules: cascades and, 49; control and, 60; human reason and, 76–77; multiple mapping circuits and, 221–22; neural language and, 335; neural networks and, 10, 18

molecular gates, 14

Morgan, Jerry, 299

morphology: adjectives and, 266; derivational, 266; grammar and, 249–50, 260, 266–67; language types and, 327; neural phonology and, 333, 338

Morsella, Ezequiel, 68–69

Moser, Edvard I., 238

motor control: activation and, 85–88, 90; aspect and, 88–92; bridge and, 89–92; causation and, 91; computation models and, 83–91, 168; conceptual metaphor and, 89, 92, 234, 240–42; connections and, 85; control circuits and, 87; convergence-divergence zones (CDZs) and, 54; embodiment and, 83; evolution and, 91; exaptation and, 83, 91; executing networks (X-Nets) and, 87–92, 336; functional circuits and, 86; functional muscle synergies and, 85; as hierarchical, 86; ideas and, 83, 92; Lakoff and, 89; language differences and, 84–86; learning and, 84, 87; linguistics and, 88, 90; mapping and, 89; motor programs and, 88; Narayanan and, 89; neural firing and, 86; neural plasticity and, 87; Neural Theory of Language (NTL) project and, 83–84, 87–89; parietal cortex and, 86; part-whole relation and, 90; primary motor cortex and, 85; reinforcement and, 87; schemas and, 68, 86,

neural firing (*continued*)

functional model and, 181–82; gestalt circuits and, 174, 190; grammar and, 255; Hodgkin-Huxley model and, 186, 349n2; ideas and, 183; influx and, 178; inhibition and, 45–46, 79–80, 104, 185, 196–97, 205, 211–12; integration and, 185; ions and, 175–79, 184–85; leaky integrate and fire model, 349n2; learning and, 174, 181, 187–90; lexicon and, 191; linguistics and, 182–83; mathematics and, 187, 189; metaphor and, 137–38; modulators and, 185; motor control and, 86; multiple mapping circuits and, 216, 219; negative charge and, 175–79, 184; neural networks and, 11–17; neural plasticity and, 181, 188; neurotransmitters and, 175–81, 184–87; polarization and, 177–80, 184; positive charge and, 176–79, 184, 186; postsynaptic neurons and, 177, 180; process of, 175–76; reaching peak, 178; reality and, 181–82; regularity of, 16, 79, 137–38; reinforcement and, 189–90; repolarization and, 177f, 179; resting potential and, 14, 177–79; reward and, 189–90; schemas and, 95, 104, 113; semantics and, 183; sequential openings and, 178; simulation and, 238; spikes and, 37, 176–81, 184–89, 195, 205, 223; structured neural computation (SNC) model, 181–86; synapses and, 12–17, 79, 174–81, 184–89; threshold and, 178; time gaps and, 176–80; trauma and, 38, 41; unconsciousness and, 183; vesicle opening and, 179

neural language: case studies and, 305–26; color and, 335; embodiment and, 249, 311, 326, 328, 335, 337; frames and, 335, 337 (*see also* frames); genetics and, 337–38; grammar and, 249–305 (*see also* grammar); ideas and, 249–53, 256–58, 265, 273, 278–80, 285, 287, 292, 297, 299, 303–4, 318, 320, 328–33, 338–39; insula and, 335; kernel sentences and, 350n2; language types and, 326–28; mapping and, 253–54, 257–59, 280–81, 300, 303, 314, 316, 319–20, 337; meaning and, 284–305; modules and, 335; neural phonology and, 331–34, 338–39; reality and, 285, 335–36; recruitment and, 297, 328–29, 335, 337–38; sensorimotor control and, 258t

neural mechanisms: basic circuits and, 191–200; combinatorial circuits and, 207–14; coordination and, 200–207, 222–31; Damasio and, 173; ideas and, 173–74, 183, 190–93, 214, 234, 246–47, 349n7 (chap. 3); linguistics and, 174, 338; multiple mapping circuits and, 214–22; neural firing and, 175–91; neural theory applications and, 246–47; simulation and, 231–46

neural networks: activation and, 12–13, 20–23; axons and, 12–13, 16–19; back-propagation and, 189–90, 341, 349n4; body as, 22–24, 74; cognition and, 11; computation models and, 11–13; connections and, 10–24; consciousness and, 11; dendrites and, 12–13, 16–17; dorsal stream and, 21; early childhood education and, 21–22; electricity and, 13, 15; embodiment and, 18–22; eyes and, 21; firing rules and, 11; generalization and, 81; genetics and, 16; hormones and, 14–15, 59; ideas and, 11; inhibition and, 15, 23; insula and, 14, 20, 23; ions and, 13–17, 20; learning and, 15; linguistics and, 22; mapping and, 11, 13; memory and, 11, 21; modules and, 10, 18; negative charge and, 13–14; neural firing and, 11–17; neural plasticity and, 16; neurons and, 10–20, 23; neurotransmitters and, 12–17; parietal cortex and, 20; positive charge and, 13–14; primary motor cortex and, 19, 31; primary visual cortex (V1) and, 18–21; primitives and, 13; recruitment and, 17–18, 22; somatosensory cortex and, 19, 23; synapses and, 12–17, 22; synergies and, 23; topographic maps and, 18–23;

trajector-landmark relation and, 21; what pathway and, 21

neural optimization: control and, 36–37; coordination and, 227; grammar and, 255, 274–76, 279–80, 295–97, 303t; integration and, 35–37, 227, 255, 274–76, 279–80, 295–97, 303t, 337–38; as least energy use, 35–36, 337; recruitment and, 337–38

neural phonology: activation and, 334; bleeding effect and, 339–40; Chomsky and, 332–33; cognition and, 332–34, 339; computation models and, 333, 339; constructions and, 331–34, 338–39; executing networks (X-Nets) and, 334; F-level, 331–32, 334, 339–40; generalization and, 331; gestalt circuits and, 332, 334; Halle and, 332–33; ideas and, 333; inhibition and, 331; integration and, 332; intuition and, 242–43; Lakoff and, 333–34, 339; lexical items and, 334; meaning and form levels and, 332, 334, 337–38; morphology and, 333, 338; Narayanan and, 333; P-level, 331–32, 334, 339; pragmatics and, 338; replacement by, 332–34; Yeats and, 339

neural plasticity: case studies and, 315; combinatorial circuits and, 209; control and, 34, 37; coordination and, 228; grammar and, 259; motor control and, 87; neural firing and, 181, 188; neural networks and, 16; perception and, 16, 34, 37, 78–79; spike-timing-dependent plasticity (STDP), 87, 138–40, 143, 170, 188, 222, 247

neural theory: application of, 246–47; binding circuits and, 246; cascades and, 246–47; case studies for, 305–26; cognition and, 246; color and, 246; computation models and, 246–47; connections and, 246; constructions and, 284; frames and, 252; generalization and, 246; gestalt circuits and, 246–47; grammar and, 284, 305, 338, 350n2; ideas and, 247; inhibition and, 246–47;

integration and, 246–47; learning and, 247; linguistics and, 246–47; mapping and, 232–33, 245; mathematics and, 246; metaphor and, 89, 131, 135, 137–39; phonology and, 331; sensorimotor control and, 245; spike-timing-dependent plasticity (STDP) and, 247

Neural Theory of Language (NTL) project: frames and, 116; motor control and, 83–84, 87–89; schemas and, 102; structured neural computation (SNC) and, 89–92

Neuronal Dynamics (Gerstner et al.), 349n2

neurons: axons, 12 (*see also* axons); back-propagation and, 189–90, 341, 349n4; basic circuits and, 192, 196–97; cascades and, 43–50; combinatorial circuits and, 209, 211–12; computation and, 11, 13, 80, 180–81, 185, 188, 203; connections and, 10–12 (*see also* connections); control and, 32, 36–41, 60–61; coordination and, 203, 205, 222–23, 229–30; dendrites, 12 (*see also* dendrites); flow of activation and, 13; grammar and, 259; gut, 23; ideas and, 8, 10, 329; metaphor and, 138; mirror, 60–61, 77; multimodality and, 36, 69, 76; multiple mapping circuits and, 218–19; neural firing and, 175–90; neural networks and, 10–20, 23; path of activated, 20; postsynaptic, 12–17, 177, 180, 186; presynaptic, 12–17, 177, 180, 185, 196; schemas and, 96, 103, 113; simulation and, 238, 241; size of, 347n4; synaptic terminals and, 12, 180

Neurophilosophy (Churchland), 82

neurotransmitters: basic circuits and, 196; binding and, 177–78; cascades and, 45; hormones and, 14–15, 59; neural firing and, 175–81, 184–87; neural networks and, 12–17; perception and, 12–17, 45, 79; release, 177; reuptake and, 179–80

Nixon, Richard, 115

Nobel Prize, 238

nonprototypical members, 158

nonspectral colors, 4

Norman, Donald, 116

norms, 58, 301, 325

Number Sense, The (Dehaene), 82

Núñez, R., 149

Och, May-Britt, 238

O'Connor, M. C., 292

O'Keefe, John, 238

old age, 22

Open AI, 149, 342

optimization of energy use, 76

optogenetic techniques, 205, 349n7 (chap. 3)

orientation-sensitive cells, 44, 47–48

packages: control and, 34; coordination and, 203–6; grammar and, 285; integration and, 233; language, 206; loops and, 205–6; metaphor and, 123–24; simulation and, 233, 236, 245; thought, 206

paragons, 163

parietal cortex: control and, 32, 60–61; convergence-divergence zones (CDZs) and, 54–55, 58; coordination and, 202; motor control and, 86; multiple mapping circuits and, 218; neural networks and, 20; simulation and, 236

partitive construction, 158–59, 261

part-whole relation: activation and, 347n5; cognition and, 9, 25, 90, 105, 109–10, 113, 141, 158, 261, 347n5; dual models and, 25; frames and, 113; grammar and, 261; ideas and, 9; inhibition and, 347n5; metaphor and, 141, 158–59; motor control and, 90; schemas and, 9, 95–96, 105, 109–10; semantics and, 347n3

path of activated neurons, 20

Paulus, W., 6

perception: action and, 61–62; cascades and, 43–50; color and, 1–7, 70–71, 79–82; computation models and, 11–12, 80; cones and, 2–5, 7, 45, 81–82; consciousness and, 11, 36, 38, 49, 57, 68–72; control and, 31–43, 58–63; convergence-divergence zones (CDZs) and, 50–58;

dual models and, 24–31; embodiment and, 4, 7, 18–22, 26, 31, 34–35, 48–49, 57–58, 73–74; evolution and, 3, 22, 37, 75; experiments in, 69–71; eyes and, 1–7, 21, 31, 45, 48, 51–52, 74; genetics and, 16; ideas and, 7–11, 24–28, 31–32, 38–43, 50, 73–88; insula and, 14, 20, 23, 31, 74; integration and, 69–73; ions and, 13–17, 20, 45, 79; lesions and, 50, 53–58, 78; light and, 1–7, 69, 71, 81; mapping and, 8, 11, 13, 80; neural networks and, 10–24; neural plasticity and, 16, 34, 37, 78–79; neurotransmitters and, 12–17, 45, 79; primary visual cortex, 18, 45, 60, 70, 75, 79; recruitment and, 17–18, 22, 32–39, 75–76, 79; retina and, 2–4, 18–19, 44–48, 60, 75, 81; rods and, 45; sound symbolism and, 63–68; unconsciousness and, 8–9, 27, 34–39, 57, 61–73

perfective states, 90

Persian Gulf War, 189

Pfeiffer, B. E., 238–39

Philosophy in the Flesh (Lakoff and Johnson), 148–49

phonaesthemes, 67–68

phrasal structures, 260–61

physical logic, 36, 39, 203

place cells, 238

Plato, 148

polarization, 17, 177–80, 184

Popper, Karl, 330

positive charge: ions, 13 (*see also* ions); neural firing and, 176–79, 184, 186; neural networks and, 13–14

positron emission tomography (PET) scans, 53

posterior temporal cortex, 54

postsynatpic neurons, 12–17, 177, 180, 186

potassium channels, 176

potassium ions, 13–15, 177f, 179

potential difference, 14

pragmatics: composition and, 253–54; conveyed meaning and, 252; fixed circuitry

94–95; recruitment and, 95; retina and, 103, 105; semantics and, 96; somatosensory cortex and, 105; sound symbolism and, 78; special cases and, 93–94, 107, 110; synergies and, 201; thinking with, 106–13; topographic maps and, 103, 105; trajector-landmark relation and, 100, 107; triangle, 93–94, 106, 113, 132; unconsciousness and, 100, 104, 106, 109–12; very basic, 100–101; viewpoint, 98–100

Schwartz, Alan, 126

scotoma, 70

Searle, John, 28, 58

secondary visual cortex (V2), 21, 43

Selfish Gene, The (Dawkins), 328–29

semantic role modifiers, 268

semantics: amalgam construction and, 290–91; basic, 287–88; case studies and, 307–9, 320–21, 325; combinatorial circuits and, 210; computational models and, 80; control and, 38, 42; convergence-divergence zones (CDZs) and, 50; frames and, 115–19; grammar and, 265–69, 274, 278–84, 287–90, 294–305; language types and, 326; metaphor and, 122–24, 127, 168; multiple mapping circuits and, 220; neural firing and, 183; part-whole relation and, 347n3; passive, 335; pragmatics and, 350n4; roles and, 335–37; schemas and, 96; simple clauses and, 287–88; simulation and, 239–44; sound symbolism and, 64, 67, 78; special cases and, 304; Talmy on, 96; unpassive adjective construction and, 288; why (not) construction and, 288–90

sensorimotor control: basic circuits and, 191; cascades and, 48; computational studies of, 83; control and, 61; coordination and, 201–3, 206, 222; metaphor and, 142, 149; neural language and, 258t; simulation and, 245

sentence adverbs, 266

sequencing circuits, 29, 194, 222–23

sequential openings, 178

shifter circuits, 29–30, 195

Shimojo, Shinsuke, 64, 69, 150

simulation: activation and, 27–28, 231–37; binding circuits and, 232–33; building complexity and, 234; cascade circuits and, 233–34, 245; cascades and, 231–35, 240, 243–45; cognition and, 232, 238–39, 244–45; color and, 232; composition and, 233; computation models and, 231, 239, 243, 245; connections and, 233, 239, 241; connectivity and, 239; consciousness and, 234, 242; constructions and, 236; containment and, 235, 241; control circuits and, 233; default networks (DNs) and, 236–38; dimensions of, 240; displacement and, 236–38; dorsal stream and, 236; embedding and, 233–34; embodiment and, 234–35, 238, 244; emotion and, 238; everyday, 239–40; exaptation and, 245; executing networks (X-Nets) and, 234–35, 242–45; frames and, 234, 240–45; gating and, 233; human v. computer, 243; ideas and, 234, 240; image schemas and, 234, 245; imagination and, 235–36, 239, 244; inhibition and, 233; integration and, 231–40, 242, 245; learning and, 231–32, 236; linguistics and, 244; localization and, 233–34, 238, 245; logic schemas and, 111–13; memory and, 229–30, 237–38, 243; mental life and, 238–39; neural bindings and, 232, 235; neural firing and, 238; neurons and, 238, 241; packages and, 233, 236, 245; prefrontal cortex and, 236; primitives and, 242; reality and, 240–41, 244–45; recruitment and, 236; semantics and, 239–44, 304; synapses and, 232; thought and, 239–44; unconsciousness and, 234–36, 242–43

simultaneity, 49–50, 211

sleep, 23, 32, 138, 230, 237, 297

Slobin, Dan, 256, 326–27, 351n11

Smith, K. S., 91, 204

sodium channels, 175–78
sodium ions, 13–15, 177f, 178
somatosensory cortex: connections and, 23, 31, 74–75, 105; control and, 31; convergence-divergence zones (CDZs) and, 53, 55; emotion and, 335; metaphor and, 145; neural networks and, 19, 23; schemas and, 105
Sonnweber, R., 140
Soros, George, 330
sound symbolism: attention shift and, 65; cognition and, 63, 65, 68–69; composition and, 67; constructions and, 256–57; containment and, 63, 78; embodiment and, 63–68; generalized containment and, 63; grammar and, 256–57; gratings and, 64; high tones and, 63–65; image schemas and, 67–68, 78; infants and, 65; integration and, 63–65; light and, 66–67; linguistics and, 63–65; McGurk effect and, 66; metaphor and, 66–68; motor control and, 65–66; mouth spatialization and, 66; phonaesthemes and, 67–68; semantics and, 64, 67, 78; unconsciousness and, 63–68
source domains: Lakoff and, 124; mapping to target domains, 124; metaphor and, 124–26, 134–35, 139, 143, 157, 318
spatial closeness, 19–20
spatial event structure system, 313–15
spatialization, 66
special cases: case studies and, 310, 318; caudal area and, 81; combinatorial circuits and, 214; defaults and, 262–63; frames and, 113–14, 119–21; generalization and, 76; grammar and, 262–63, 269, 274, 280, 304; metaphor and, 126–27, 130, 132, 147, 151–52, 161, 163, 166; schemas and, 93–94, 107, 110
spikes: basic circuits and, 195–96; case studies and, 315; control and, 34, 37; coordination and, 205, 223; grammar and, 259; metaphor and, 138; motor

control and, 87; neural firing and, 37, 176–81, 184–89, 195, 205, 223
spike-timing-dependent plasticity (STDP): metaphor and, 138–40, 143, 170; motor control and, 87; multiple mapping circuits and, 222; neural firing and, 188; neural theory applications and, 247
stapes, 20
statistical preemption, 295
stirrup bone, 20
structured neural computation (SNC) model: Feldman and, 190; functional, 181–86, 189; neural firing and, 181–86
"Studies in Lexical Relations" (Gruber), 299–300
substitution circuits, 30, 195, 223, 227–28
Sullivan, Karen, 146, 297
supervised learning, 190
suspension, 29, 222, 224
Sweetser, Eve, 126, 146
sweet talk, 144–45, 170
synapses: action potential and, 20, 176–81, 184–86; basic circuits and, 190–91, 193, 196–98; cascades and, 49; case studies and, 315; combinatorial circuits and, 209; connections and, 15–16, 32, 41, 80, 138, 184–85, 187, 191, 223, 329; connectivity and, 78, 81; control and, 32, 35, 38, 41; coordination and, 204, 223; depolarization and, 17, 177–78, 180, 184; grammar and, 284, 296; ideas and, 329, 331; inhibition of, 79–80, 185, 196–97, 349n7 (chap. 3); learning and, 15, 35, 38, 75–76, 138, 174, 181, 187–91, 232, 296; metaphor and, 138; modulators and, 185; neural firing and, 13–17, 79, 174–81, 184–89; neural networks and, 12–17, 22; neural plasticity and, 16, 34, 37, 78–79, 87, 181, 188, 209, 228, 259, 315; polarization and, 177–80, 184; postsynaptic neurons, 12–17, 177, 180, 186; presynaptic neurons, 12–17, 177, 180, 185, 196; simulation and, 232; strength of, 15–17, 32, 41, 49, 75–76, 79–80, 191, 204, 315, 329; transcription